The Philosopher's Toolkit: How to Be the Most Rational Person in Any Room

Patrick Grim, B.Phil., Ph.D.

THE
GREAT
COURSES

PUBLISHED BY:

THE GREAT COURSES
Corporate Headquarters
4840 Westfields Boulevard, Suite 500
Chantilly, Virginia 20151-2299
Phone: 1-800-832-2412
Fax: 703-378-3819
www.thegreatcourses.com

Patrick Grim, B.Phil., Ph.D.
Distinguished Teaching
Professor of Philosophy
State University of New York at Stony Brook

P rofessor Patrick Grim is Distinguished Teaching Professor of Philosophy at the State University of New York (SUNY) at Stony Brook, where he has taught since 1976. Having graduated with highest honors in both Anthropology and Philosophy from the University of California, Santa Cruz, Professor Grim was named a Fulbright Fellow to the University of St. Andrews, Scotland, from which he received his B.Phil. He received his Ph.D. from Boston University with a dissertation on ethical relativism and spent a year as a Mellon Faculty Fellow at Washington University. In addition to being named SUNY Distinguished Teaching Professor, he has received the President's and Chancellor's awards for excellence in teaching. Professor Grim was named Marshall Weinberg Distinguished Visiting Professor at the University of Michigan in 2006 and Visiting Fellow at the Center for Philosophy of Science at the University of Pittsburgh in 2007. He also has been a frequent Visiting Scholar at the Center for the Study of Complex Systems at the University of Michigan.

Professor Grim has published extensively in computational modeling and on such topics as theoretical biology, linguistics, decision theory, and artificial intelligence. His work spans ethics; philosophical logic; game theory; contemporary metaphysics; and philosophy of science, law, mind, language, and religion.

Professor Grim is the author of *The Incomplete Universe: Totality, Knowledge, and Truth* and the coauthor of *The Philosophical Computer: Exploratory Essays in Philosophical Computer Modeling* (with Gary Mar and Paul St. Denis); *Beyond Sets: A Venture in Collection-Theoretic Revisionism* (with Nicholas Rescher); and *Reflexivity: From Paradox to Consciousness* (also with Nicholas Rescher). He is the editor of *Mind and*

Consciousness: 5 Questions and *Philosophy of Science and the Occult* and a founding coeditor of more than 30 volumes of *The Philosopher's Annual*, an anthology of the best articles published in philosophy each year. He has taught two previous Great Courses: *Questions of Value* and *Philosophy of Mind: Brains, Consciousness, and Thinking Machines.* ■

Table of Contents

Table of Contents

Table of Contents

Answers for some sets of Questions to Consider and Exercises can be found in the "Answers" section at the end of this guidebook.

The Philosopher's Toolkit:
How to Be the Most Rational Person in Any Room

Scope:

Thinking is one of the things we do best. Wouldn't it be great if we could do it even better? This course, *The Philosopher's Toolkit*, gives you a set of thinking techniques designed with that goal in mind: tools for conceptual visualization, critical analysis, creative thinking, logical inference, rational decision, real-world testing, effective reasoning, and rational argument. The course uses interactive engagement to introduce a range of conceptual methods and perspectives: mind-stretching philosophical puzzles, mental exercises by example, thought experiments on which to test your powers, and deep questions to ponder.

You'll learn "hands on" the simple heuristics that make us smart; the basic strategies of decision theory and conceptual modeling; and how to handle everyday probabilities, track a train of thought, outwit advertisers, and detect a spin in statistics. The course also emphasizes rationality in the social context: how to use the wisdom of crowds; how to analyze the flow of thought in a debate; and how to defuse fallacious reasoning, deflate rhetoric, and break through opinion polarization. You can apply your new tools of analysis in evaluating arguments on both sides in a "great debate." Decision theory, game theory, probability, and experimental design are all introduced as part of the toolkit, but with a philosophical examination of their limitations, as well as their strengths.

In order to think better, we also have to understand why we make mistakes— the systematic ways in which our thinking goes wrong. You'll be surprised to find how poor we are at noticing continuity changes in both movies and in everyday life; in fact, we are "change blind." Our memories, however vivid, are created in large part by things that happened after the event: what questions were asked, how they were asked, and how the story was retold. Our estimates of risk and danger often have more to do with the vividness of a mental picture than with any clear-headed consideration of the probabilities. This course draws lessons from a wide range of literature in psychology and cognitive science in exploring the systematic conceptual biases that mislead us all—and how we can compensate for those biases. It also focuses on questions in philosophy of mind regarding the role of emotion in thought, individual and social rationality, and the comparative powers of gut instinct versus systematic analysis. Is emotion the enemy of rationality or a force we need to better understand as a conceptual resource in its own right?

All of the techniques in the course are outlined with the history of philosophy and science as a background. The power of visualization is introduced using insights in the history of thought from Pythagoras to von Neumann. The power of thought experiments is illustrated with examples from Galileo, from Einstein, and from contemporary philosophy of mind. Important perspectives and approaches are traced to Plato, Schopenhauer, Hobbes, Pascal, and Descartes. Concepts important for analyzing empirical data are drawn from the work of philosophers Karl Popper, John Stuart Mill, and Charles Sanders Peirce but are illustrated using Newton's experiments in a darkened room, R. A. Fisher's "Lady Tasting Tea" experiment, and Richard Feynman's experiment with a glass of ice water during televised hearings on the *Challenger* disaster. The history behind gambling and probability theory, game theory and the Cold War, and the strange tale of Aristotle's manuscripts all form part of the story.

The Philosopher's Toolkit is a perfect introduction to applied philosophy, designed for application in everyday life. The emphasis throughout is on practical conceptual strategies that are useful in any area of application, with interactive examples of logic in action taken from business, the media, and political debate. What are the advertising tricks that we all have to watch out for? What are the standard forms of bogus argument? How can we manage empirical data and statistical information without being misleading and without being misled? How can we have a rational discussion in a polarized environment? From these lectures, you'll gain a wide set of useful new thinking skills, methods, and techniques but also a deeper psychological, philosophical, and historical appreciation for the thinking skills you already have.

Why do we think the way we do? How can we think better? Those are the questions at the core of this course, examined using interdisciplinary input from cognitive science, psychology, logic, and philosophy of mind. The background is the full sweep of the history of thought. Through your involvement in these lectures, you'll have a deeper understanding of your own thinking, both good and bad. You'll also have a "philosopher's toolkit" of ways to make your thinking more effective, more flexible, more logical, more creative, and both more realistic and more rational. ∎

How We Think and How to Think Better
Lecture 1

Thinking is one of the things we do best, but wouldn't it be great if we could do it even better? *The Philosopher's Toolkit* is a set of tools designed with that goal in mind: tools for creative conceptualization, critical analysis, logical inference, rational decision making, effective reasoning, and rational argument. With those tools, we can be better thinkers—more creative, logical, inventive, and rational. This course is about thinking in two different but complementary ways. It's about how we think (a descriptive task) and how we can think better (a normative task). Our aim in the course will be to develop a set of conceptual skills that is useful in all kinds of thinking.

The Descriptive and the Normative

- Let's start with a simple example of the **descriptive** and **normative** tasks in thinking: Imagine we have a bat and a ball. The bat costs $1.00 more than the ball. Together, they cost $1.10. How much does the ball cost? Many people give 10 cents as the answer. That's the descriptive side of the story.

- But if the ball costs 10 cents, and the bat costs $1.00 more than the ball, then the bat will cost $1.10, and the bat and the ball together will cost $1.20, not $1.10. Even if you saw that immediately, it's likely that the first figure that popped into your head was 10 cents.

- There are two normative lessons to draw from this example. The first is that the intuitive and immediate answers we jump to are often wrong. The second is that it's easy to do better. In this case, to do better, we just need to check the first answer. If the total is $1.10, then the ball must cost 5 cents.

Thinking in Patterns

- Among the topics that will come up repeatedly in this course are patterns. It is in our nature to think in patterns. That's how we make sense of things—we look for the pattern that underlies them.

- A passage about kites illustrates this point. Without any context, the passage seems nonsensical, but once you know the subject of the text, everything falls into place. What you needed was a picture.

- Our aptitude for patterns is remarkable, but that propensity to look for patterns also has a downside. It's possible to draw connections that aren't really present.

- You may have experienced a situation similar to this: You take out the trash and suddenly can't find your car keys. You assume that you must have thrown your car keys away.
  - You have just jumped to a causal connection between two events because they happened close together. The Latin name for this kind of fallacious thinking is ***post hoc ergo propter hoc*** ("after it, therefore because of it").

  - There are many biases built into our thinking. We think in pictures and patterns, and we are extremely liable to be influenced by context.

- Here's another question to consider: According to the Old Testament, how many animals of each kind did Moses load into the ark? The answer is not two but zero, because it was Noah, not Moses, who loaded animals into the ark. Loading animals into the ark brings up a vivid picture, and we tend to focus on the picture, neglecting the fact that something's a bit off in the description.

- Much of the effort in advertising and marketing is designed to exploit our conceptual biases. Researchers have found, for example, that a pricing strategy of "four for $2.00," rather than "50 cents each" leads people to buy more than one item.

- Context can bias your thinking even when you're not aware of it, perhaps especially when you're not aware of it.
  - In experiments that involved "priming" undergraduates with words associated with the elderly, researchers found that mental context results in physical differences in the undergraduates' behavior—they walked slightly slower after hearing the words *Florida*, *gray*, *bingo*, and so on.

  - The reverse was also found to be true: physical context made a difference in mental behavior.

Working toward Social Rationality

- Much of this course is about individual **rationality**: developing conceptual tools and recognizing and avoiding conceptual traps. But much of it is also about social rationality. In argument and debate, rationality occurs in a social setting. In public thinking, there are different traps but also different possibilities.

- Argument and debate bring **rhetoric** along with them. The territory of this course includes examples of both good and bad rhetoric—debating techniques and how to detect them, rhetorical fallacies and how to defuse them. It also includes standard tricks used in advertising and how to recognize "spin" in statistics.

- *The Philosopher's Toolkit* includes normative tips and techniques for working toward enhanced social rationality. Sometimes, we have to dig out of polarization in order to give important topics the open and rational consideration they deserve.
  - Negotiations of a test-ban treaty between the United States and the Soviet Union during the Cold War serve as an example. In these negotiations, onsite inspections were a major verification issue, with each side focused on the number of inspections to be allowed each year.

  - Both sides went into the negotiations with a number in mind, but the negotiations broke down because neither side would compromise.

- With the benefit of hindsight, this looks like a case of crippling polarization—of social irrationality. The negotiations would likely have gone better if the conversation could have been shifted toward the real concerns: protecting sovereignty and national security on both sides while minimizing an overkill arms race.

Acquiring the Techniques of Philosophy

- The best philosophy borrows from and builds on diverse areas of thought, including psychology, statistics, mathematics, economics, the history of science, and even physics. The techniques of psychology, in particular, are wonderful for the descriptive task, providing insights into how we think. The question of how we *should* think typically demands a philosophical approach.

- Philosophy has been concerned with normative evaluation of thinking from its very beginning. In Plato's dialogues, Socrates encounters various people in the Athenian marketplace and attempts to find out what they think—about justice, or knowledge, or piety—and why.
  - Socrates then challenges that reasoning, probing it, testing it, attempting to show inconsistencies and irrationalities. The dialogues are exercises in the normative evaluation of conventional reasoning.

  - Interestingly, the context is always social. For both Socrates and Plato, rationality is something we work toward together.

- With Plato's student Aristotle, **logic** begins. In Aristotle, it is clear that the philosopher's task is not merely to figure out how people think but how to think better—more systematically, more validly, more logically. The enterprise of logic is a major thread in the history of philosophy that runs from the pre-Socratics down to the present day.

- In the beginning, with such thinkers as Pythagoras, Euclid, Ptolemy, and others, all intellectual disciplines counted as philosophy. Other disciplines later split off as well-defined topic areas and developed recognized techniques of their own.

- The remaining core is philosophy, which is, in large part, normative, concerned with how things ought to be.
  o That's why **ethics** is a major subfield of philosophy, concerned not merely with how people act but how they should act.

Pythagoras was considered a philosopher, although his thought is an amazing mixture of mysticism, musical theory, and mathematics.

  o That's why logic is a major thread, concerned not merely with what we think but how we should think.

  o Whenever we want to think better, in general and across all the disciplines, it's important to return to that normative core.

- The American philosopher Charles Sanders Peirce is thought of as the founder of philosophy of science, though he never held an academic position.
  o In 1877, Peirce wrote an article entitled "The Fixation of Belief," evaluating different techniques of thinking side by side.

  o Only the technique of science, he claimed, is able to recognize its own fallibility and use it to its own advantage. Only science offers a way of constantly testing and criticizing current beliefs in order to work toward better ones.

- The best philosophy has always been informed by the other disciplines. Indeed, some of the best of 20th-century philosophy has been done within other disciplines. Economists, physicists, and social scientists all have tools that should be part of the philosopher's toolkit.

Slow Down Your Thinking

- Thinking is something you can only understand by doing. Ours is not just a theoretical exploration of thinking; it is also intended as a practical exploration, with an emphasis on applications. That means that this course is a joint enterprise; we will work through exercises together.

- Skill with any conceptual tool, as with any physical tool, comes with practice. You should attempt to apply the take-home lessons of each lecture in this course to the rest of your thinking outside the course. And don't be afraid to pause the lectures; thinking often takes time.

- Let's return to the example we started with—the bat and ball. This example comes from the work of Shane Frederick, a professor of marketing at Yale. It is used as part of a theory of thinking by Daniel Kahneman, a psychologist who received the Nobel Memorial Prize in Economic Sciences in 2002.
  - Kahneman speaks of two systems in our thinking. System 1 is fast, intuitive, good for emergencies—and often wrong. When you jump to conclusions, you're using system 1.

  - System 2 is slower, more critical, and more systematic; here is where you check your answers.

  - It is very much in the spirit of this course that we need to use both systems in the appropriate sphere. In an emergency, a decision may have to be made fast, and our intuition may reflect a great deal of relevant past experience. But in many cases, what's required is careful reflection instead. Much of the philosopher's toolkit falls into that category.

- Kahneman suggests a technique for emphasizing whichever system is required: When what is called for is creative, exploratory, and intuitive thinking, make yourself smile. When what is called for is critical calculation and reflection, furrow your brow and pucker your lips in concentration. Give this approach a try and see if it works for you.

- In the next lecture, we'll look more closely at system 1 and system 2 thinking, exploring them in terms of rationality and emotion: cool rationality and hot thought.

Terms to Know

descriptive: Used to designate a claim that merely reports a factual state of affairs rather than evaluating or recommending a course of action. Opposed to normative.

ethics: The field of philosophy that focuses on moral issues: ethically good actions, ethically right actions, rights, and obligations.

logic: The study of patterns of rational inference and valid argument.

normative: Used to designate a claim that is evaluative in nature or recommends a course of action, as opposed to descriptive.

post hoc ergo propter hoc: "After it, therefore because of it"; a fallacy based on the claim that because something followed another thing, it must have been because of that other thing. This fallacy overlooks the possibility of coincidental occurrence. Abbreviated as *post hoc*.

rationality: Exercising reason, that is, analytical or logical thought, as opposed to emotionality.

rhetoric: The skills of effective speaking and presentation of ideas; also, the techniques of persuasion, fair or foul, for either good ends or bad.

Suggested Reading

Bargh, Chen, and Burrows, "Automaticity of Social Behavior."

Frederick, "Cognitive Reflection and Decision Making."

Kahneman, *Thinking Fast and Slow*.

Mussweiler, "Doing Is for Thinking!"

Questions to Consider

1. An important distinction in this lecture is between descriptive and normative approaches. To solidify this distinction, try these two exercises:
   (a) Give a purely descriptive account of when and where you grew up.

   (b) Give an account of when and where you grew up that includes normative elements, as well.

2. Charles Sanders Peirce prizes science because it recognizes its own fallibility. Why is it a "plus" to recognize fallibility? Does that apply to people, too?

3. This lecture included an example of change blindness. Can you offer a hypothesis for why we tend to be change blind?

Exercise

Below is an exercise in pattern recognition in literature, even for those who don't know anything about literature. Three lines in the following poem are lifted from poet A, three from poet B, three from poet C, and three from poet D. Try to identify the lines that come from the same poet. (See "Answers" section at the end of this guidebook.)

A Grim Amalgamation
by Libby Jacobs

Away! away! for I will fly to thee,
So deep in luve am I;
For somewhere I believe I heard my mate responding to me
When the evening is spread out against the sky
And the rocks melt wi' the sun.
There will be time, there will be time
To take into the air my quiet breath,
But not altogether still, for then she might not come
     immediately to me
Before the taking of a toast and tea.
And fare thee weel, my only luve!
The love in the heart long pent, now loose, now at last
     tumultuously bursting,
Now more than ever seems it rich to die.

How We Think and How to Think Better
Lecture 1—Transcript

Professor Grim: Thinking is one of the things we do best. Wouldn't it be great if we could do it even better? The philosopher's toolkit is a set of tools designed with that in mind: tools for a creative conceptualization, for critical analysis, for logical inference, for rational decision, for effective reasoning and rational argument. With those tools, we can be better thinkers—more creative, more logical, more inventive, more rational, sometimes faster, often better at hitting the target. Thinking is what philosophy is all about. In the beginning, all intellectual disciplines counted as philosophy. Philosophy just means "love of wisdom"—any kind of wisdom—and philosophers were just "wise guys." The intellectual history of the West is the history of other disciplines splitting off from that core: first mathematics, then astronomy, physics, eventually psychology, economics, sociology.

I'm Patrick Grim. I've been teaching critical reasoning for a long time, in addition to advanced logic, philosophy of science, and philosophy of mind. When it comes to thinking skills in general, philosophy is still the core. But the best philosophy has always kept its finger on the pulse of work in other disciplines. The best philosophy has always been intellectually omnivorous. The philosopher's toolkit includes tips and examples from statistics, mathematics, psychology, economics, history of science, and even a little physics. The course is about thinking in two different but complementary ways. It's about how we do think. But it's also about how we can think better. The first is what philosophers call a descriptive task—a description of how people actually think. The second is what philosophers call a normative task. It's about how we ought to think. In a normative approach to thinking we're out to evaluate thinking, both good and bad. That leads directly to an attempt to do better thinking ourselves, to develop techniques and strategies for thinking more clearly, thinking with greater focus, more effectively, more efficiently, building stronger and more rational patterns of argument. We don't want to stop with just a picture of how we do think. We want to think better. This is an applied course in better thinking, in all of the thinking we do. The aim is to develop a set of conceptual skills useful in all kinds of thinking. You don't need to be a philosopher to use the conceptual tools that we'll be working with.

So why do I call it the philosopher's toolkit? Part of that is probably just me. I'm a philosopher, and I'll be your guide; part of that is the fact that philosophy is omnivorous—it builds from and includes work from other disciplines. The philosopher's toolkit is intended as an inclusive toolkit. It's also the case that the history of western thought flowers from a central philosophical core. Many of the thinking skills we'll be developing are part of that central core. We'll be returning repeatedly to tips from the great thinkers. Many of those—including, not merely Plato and Aristotle, but Galileo and Newton—thought of themselves as philosophers. But it's the normative point that may be the most important. We're out not merely to explore the descriptive question—how people do think—but the normative task: how we can think better. That normative task, both theoretical and applied, the evaluation and development of good thinking, is characteristically philosophical—philosophical through and through. Let me give you a simple example of both sides of what we'll be doing: the descriptive and the normative. Most of what we do won't depend on numbers; this one calls a little addition. We have a bat and a ball. The bat costs $1.00 more than a ball. Together they cost $1.10. How much does the ball cost? If you said 10 cents, you've got good company. That's the answer given by about half the university students at Harvard, MIT, and Princeton. That's how people think. That's the descriptive side of the story. But take a minute and really think about it. If the ball costs 10 cents and the bat costs $1.00 more than the ball, then the bat will cost $1.10. But then the bat and the ball together will cost $1.20, not $1.10. You may be smarter than half of the students at Harvard, and MIT, and Yale. You may have seen that immediately. But even if you did, it's a good bet that the first figure that popped in your head was the figure that occurs to just about everybody, 10 cents.

There are two normative lessons to draw from that example. The first lesson is that the intuitive and immediate answers we jump to are often wrong. The second lesson is that it's easy to do better. In this case, all we needed to do better was to check on the first answer. 10 cents? No, that won't work. If the ball is 10 cents the bat has to be $1.10, giving us a total of $1.20. That must be wrong. Let's revise our answer. The correct answer must be five cents. Pretty simple, huh? Check your answer. Clearly half of the students at Harvard, MIT, and Yale didn't take time to do that. That's what we're after: not just a report on how we think but an evaluation of our thinking and a toolkit to make our thinking better.

One of the topics that comes up repeatedly in this course is patterns. It's in our nature to think in patterns. That's how we make sense of things—we look for the pattern that underlies them. Let me show you how good you are at seeing patterns in context. Here is an image of a thorn bramble. But there's a bunny in there. Can you see it? Now watch it move. Suddenly a few of those lines form a bunny as clear as day. but when I stop the image, the bunny disappears. Because of the motion, you see those few scratches as a pattern; they stick out from the context and you can see the bunny. Without the motion, you can't isolate them as a specific pattern any longer, and the bunny disappears.

Here's another example. Listen to the following passage: "A newspaper is better than a magazine. A seashore is a better place than the street. At first it's better to run than to walk. It takes some skill but it's easy to learn, even for young children. It's true that you need lots of room. Beware of rain; it ruins everything. A rock will serve as an anchor. If things break loose from it, however, you won't get a second chance." What could that possibly mean? Does it make any sense? What you're looking for is a picture or a pattern in which all the pieces fit. That's what we're always doing. We're looking for meaning in a series of experiences, looking for the patterns in the world around us, looking for a picture in which all the pieces fit. Did you get it in this case? What that passage is about is kites: A newspaper is better than a magazine. A seashore is a better place than the street. At first it is better to run than to walk. It takes some skill but it's easy to learn, even for young children. It's true that you need lots of room. Beware of rain; it ruins everything. A rock will serve as an anchor. If things break loose from it, however, you won't get a second chance. Once you know it's about kites, it all falls into place. What you needed was a picture. Your imagination was much more visual that time around.

Our aptitude for patterns is remarkable. We'll be putting that aptitude to conceptual work. But that propensity for patterns also has a downside. We're great at drawing connections. Unfortunately, we're also great at drawing connections that aren't really there. I'm sure something like this has happened to you. You take out the trash and suddenly can't find your car keys. Oh my gosh! You must have thrown your car keys away. You have just jumped to a causal connection between two events just because they happened close together. We do it all the time. And we often do it wrongly. At a later point we'll study patterns of fallacious thinking. That's one of them. The Latin name for it is *post hoc ergo propter hoc*. That means "after it, therefore because of it"—precisely the fallacy we all tend to make with the car keys. There are plenty of biases built into our thinking. We think in pictures and patterns, and we are extremely liable to be influenced by context.

Here's another question for you. You ought to know this one. Quick, now: According to the Old Testament, how many animals of each kind did Moses load into the Ark? Did you say two? The answer is actually zero, I guess, because it wasn't Moses that loaded animals into the Ark. It was Noah. If you caught that slip, good for you. If you didn't, it's not surprising. Loading animals into the Ark brings up a vivid picture. We tend to go with the picture, neglecting the fact that something's a bit off in the description. It's also true that "Noah" and "Moses" are about the same length, which makes the substitution easier to miss. If I had said "how many animals of each kind did Methuselah load into the Ark?" I'll bet more people would have caught it. And of course, the context is Old Testament, and Moses isn't out of place in that general context.

Much of advertising and marketing are made to exploit our conceptual biases. Consider, for example, the pricing in supermarkets. How often have you wondered why the pricing is "four for $2.00" instead of "50 cents each"? Researchers have found that the mere suggestion of buying four leads people to buy more than one. In one test, they used 86 different grocery stores and tracked actual consumer purchases comparing multiple-unit pricing—like "four for $2.00"—with unit pricing, like "50 cents each." Multiple-unit pricing resulted in 32% higher sales.

Context can bias your thinking even when you're not aware of it, perhaps especially when you're not aware of it. Here's one of my favorite experiments, designed by John Bargh, now at Yale: Undergraduates at NYU were assigned a task billed as a test of language comprehension. They were given five words and told to make a sentence out of four of them as quickly as possible. Half of the undergraduates were given a set that contained words associated with stereotypes of the elderly: Florida, knits, courteous, grey, sentimental, bingo. The other half were given words that didn't include those stereotypes. When the experiment was complete, the participants were thanked for their participation. They were told that the elevator was down the hall, and they were then surreptitiously timed. How long does it take them to reach the elevator? Half of them had been contextually primed with words associated with the elderly, half had not been. Those who had just been doing an exercise involving "elderly" words walked slower. It took them longer to reach the elevator. It wasn't a huge difference—a little over eight seconds for one group as opposed to a little over seven seconds for the other—but results were consistent enough to show the clear impact of context. The reverse effect has been shown as well. When students were asked to walk around a room at the rate of 30 steps per minute—about two thirds slower than normal—they were quicker to recognize words related to old age: forgetful, old, lonely. Mental context made a physical difference. Physical context made a mental difference.

Much of the course is about individual rationality, developing conceptual tools and recognizing and avoiding conceptual traps. But much of it is also about social rationality. In argument and debate rationality occurs in a social setting. Rationality goes public. In public thinking there are different traps, but also different possibilities. Argument and debate bring rhetoric along with them. The territory of the course includes rhetoric both good and bad, debating techniques and how to detect them, rhetorical fallacies and how to defuse them. It also includes standard tricks used in advertising and in putting a spin on statistics.

The philosopher's toolkit includes normative tips and techniques for working toward better social rationality. Sometimes we have to dig out of polarization in order to give important topics the open and rational consideration they deserve. Here's the kind of situation we want to avoid: Under Kennedy, in the midst of the Cold War, the United States and the Soviet Union were negotiating a test-ban treaty. On-site inspections were a major verification issue. Each side wanted to be able to check up on whether suspicious seismic activity was just a small earthquake or a detonation in violation of the treaty. How many inspections in the other's territory were to be allowed each year? Both sides went into the negotiations with a number in mind. The United States wanted at least 10 inspections per year. The Soviet Union wanted no more than three. For days they argued. 10, 3, 10. 3. Negotiations broke down before anyone had even discussed what inspections were to be like. Negotiations broke down because both sides had committed themselves to a magic number in advance. With the benefit of hindsight, that looks like a case of crippling polarization. How much better it would have been if the conversation could have been shifted away from commitment to magic numbers toward the real concerns: protecting sovereignty and national security on both sides while cutting down on an overkill arms race. There's no guarantee, but with better social rationality history might have been different.

I'm a philosopher, but I've been drawing on results from social psychology. The techniques of psychology are wonderful for the descriptive task, for insights into how we actually do think. The normative question of how we should think goes farther. It demands a typically philosophical approach. Philosophy has been concerned with normative evaluation of thinking from its very beginning. In Plato's dialogues, Socrates talks with all kinds of people in the Athenian marketplace. He talks with Meno the slave boy, and Parmenides the philosopher, Protagoras the Sophist, and Gorgias the professor of rhetoric. His typical procedure is to find out what they think— what they think about justice, for example, or knowledge, or piety—and why. What all the dialogues have in common is that Socrates then challenges that reasoning, probing it, testing it, attempting to show inconsistencies and irrationalities. The dialogues are exercises in the normative evaluation of conventional reasoning. Interestingly, the context is always social. For both Socrates and Plato, rationality is something we work toward together.

It is with Plato's student Aristotle that logic begins. In Aristotle it's clear that the philosopher's task is not merely to figure out how people do think; the philosopher's task is to figure out how to think better, more systematically, more validly, more logically. The enterprise of logic is a major thread in the history of philosophy that runs from the pre-Socratics down to the present day.

I've said that in the beginning all intellectual disciplines counted as philosophy. Other disciplines split off as well-defined topic areas developed recognized techniques of their own. Pythagoras was thought of as a philosopher, and he thought of himself that way. Indeed, Pythagorean thought is an amazing mixture of mysticism, musical theory, and what we think of as mathematics. It is only with formalization and axiomatics—in Euclid—that mathematics breaks off as a discipline of its own. Astronomy breaks off with Ptolemy's *Almagest*.

If you asked Sir Isaac Newton what his occupation was, he wouldn't have told you he was a physicist. He would have told you he was a natural philosopher. His master work is entitled *Philosophiae Naturalis Principia Mathematica*—the Mathematical Principles of Natural Philosophy. Physics becomes a recognized discipline in its own right only after Newton and because of his techniques. Psychology as a discipline traces back to William James, professor of Philosophy at Harvard. Anthropology developed from what was called philosophy of man. And so it goes.

As disciplines split off, philosophy kept track of those intellectual children and grandchildren. That's what philosophy of science is all about, another part of the philosopher's toolkit. The American philosopher Charles Sanders Peirce is thought of as the founder of philosophy of science, though he never held an academic position. In an 1877 article he talks about "The Fixation of Belief." He follows a path laid down centuries earlier by Francis Bacon, evaluating different approaches side by side. How shall we fix our beliefs? The first technique Peirce he considers is tenacity. If you have a belief, stick with it. That offers decisiveness, he says, but it cuts you off from new information. Another technique is the method of authority: believe what Aristotle, or Isaac Newton, or Charles Sanders Peirce believes. That can work for a while, but every thinker is eventually superseded. Only the technique of science, he says, is able to recognize its own fallibility and use that fallibility to its own advantage. Only science, he says, offers a way of constantly testing and criticizing current beliefs in order to work toward better ones. The best philosophy has always been informed by the other disciplines. Indeed, some of the best of 20th century philosophy has been done within other disciplines—within economics, game theory, decision theory, and computer science, for example. Economists and physicists and social scientists all have tools that should be part of the philosopher's toolkit.

In the course of the lectures, I will occasionally turn to lessons from the great thinkers. There is no chronological order to these; the idea is simply to tie thinking techniques to the history of thought. We can draw lessons from the way Galileo thought, how he designed real experiments, but also how he argumentatively employed mere thought experiments. Pythagoras shows up in lecture three, Aristotle plays a major role in several lectures, but we'll talk about Francis Bacon, Pascal, Hobbes, Rousseau, and lots of others as well. The great thinkers roam far beyond traditional philosophy—Einstein is a great thinker that I return to several times. So are the physicist Richard Feynman and the polymath John von Neumann.

Thinking is something you can only understand by doing it. Ours is not just a theoretical exploration of thinking, although there is plenty of that too. Even in the foundational lectures, my emphasis will be on applied philosophy—on putting thinking to work. Each lecture offers a technique of some kind with examples and often interactive challenges. I don't want the philosopher's toolkit to be on display behind glass. I want you to put it to use. We'll talk about graphing the flow of argument and detecting fallacies, for example, but I'll then have you put those tools to use in analyzing and evaluating a staged debate. That means that this is a joint enterprise. We have to do it together. In most lectures I have something that you have to do. Here's an example—pay close attention to what's happening in this dialogue:

[Video start.]

Speaker 1: Hi Susie. It seems like ages since I've seen you. Have you been away?

Speaker 2: It was a long vacation but I'm really glad to be back.

Speaker 1: Did you actually make it to the Piedmont Inn? Did it look the way it looked when you were there with your parents 15 years ago.

Speaker 2: That was really interesting. The building looked exactly the same as I remembered it as a child but it seemed so much smaller.

Speaker 1: Well isn't that always the case? Children have a very different perspective. Well, I'm glad it was a good trip.

Speaker 2: Yes, but It's good to be back to reality. The Smoky Mountains are beautiful, and if you haven't been to the Piedmont Inn, it's worth the trip.

Speaker 1: Oh well that's good to know. Bill and I have a little time off coming up, so we'll check it out.

Speaker 2: You should.

Speaker 1: Good.

[Video end.]

Professor Grim: Did you notice anything change? Depending on how you count, there were four major changes in what you just saw. Here are two of them: The color of the plates changed in the course of the scene, and those coffee cups became wine glasses halfway through. That was a little unfair. I asked you to pay attention to what was happening in the dialogue, and it was what was happening in the background that changed. It's still amazingly hard to see things that you would think should be obvious. This is a demonstration in how we all tend to be "change blind." Watch the dialogue again, and see if you can find two more changes.

[Video start.]

Speaker 1: Hi Susie. It seems like ages since I've seen you. Have you been away?

Speaker 2: It was a long vacation but I'm really glad to be back.

Speaker 1: Did you actually make it to the Piedmont Inn? Did it look the way it looked when you were there with your parents 15 years ago.

Speaker 2: That was really interesting. The building looked exactly the same as I remembered it as a child but it seemed so much smaller.

Speaker 1: Well isn't that always the case? Children have a very different perspective. Well, I'm glad it was a good trip.

Speaker 2: Yes, but It's good to be back to reality. The Smoky Mountains are beautiful, and if you haven't been to the Piedmont Inn, it's worth the trip.

Speaker 1: Oh well that's good to know. Bill and I have a little time off coming up, so we'll check it out.

Speaker 2: You should.

Speaker 1: Good.

[Video end.]

Professor Grim: I hope you got those. There's a scarf that mysteriously appears and disappears on one of the participants. And the things on the plates seem to change places. Then, of course, there's that little stuffed animal.

Sometimes what I'll ask you to do in the course of a lecture takes more time and thinking than that, something you may want to do with pencil on paper or the back of an envelope. I'll let you know at the beginning whether a lecture is one that is going to benefit from that kind of work. Skill, with any conceptual tool, like with any physical tool, comes with practice. I will want you to apply the take-home lessons of each lecture in the rest of your thinking, outside of the course.

These lectures are intended as just a first step in exercising and expanding the way you think. That's where I hope they will really pay off. There is one button that is particularly useful for this course. It's the pause button. Thinking often takes a little time. I will often suggest a question and propose that you press the pause button in order to think about it. Please do. It is peculiar that our education so often emphasizes timed tests. That has to be for the convenience of the instructor. The pause button is your route to something better—to the careful thinking that we're after and a response to questions that doesn't have to come immediately.

Let's return to the example we started with, the bat and ball. A bat costs $1.00 more than a ball. Together they cost $1.10. How much does the ball cost? That example comes from the work of Shane Frederick, a professor of Marketing at Yale. It is used as part of a theory of thinking by Daniel Kahneman, a psychologist who received the Nobel Prize in economics in 2002. Kahneman speaks of two "systems" in our thinking. system one is fast, intuitive, great for emergencies, and often wrong. When you jump to conclusions, you're using system one. When you first thought about the bat and ball example and "10 cents" leapt to mind, you were using system one. System two is slower, more critical, and more systematic. That's where checking your answer comes in. When you do, of course, it becomes obvious that the answer can't be 10 cents. That is system two thinking, as is the further thinking that takes you to the right answer: the ball costs five cents.

It is very much in the spirit of this course that we need to use both systems in their appropriate sphere. In an emergency, a decision may have to be made fast, and our intuition may reflect a great deal of relevant past experience. But in many cases what's required is careful reflection instead. Much of the philosopher's toolkit falls into that category. Kahneman suggests a technique for emphasizing whichever system is required. I'll let you try it out on your own to see if it works. The technique is this. When what is called for is creative, exploratory, and intuitive thinking, make yourself smile. When what is called for is critical calculation and reflection, make yourself imitate the furrowed brow and puckered lips of concentration. If you put a pencil side to side in your mouth, your face goes into something like a smile. That's the facial posture recommended for creative and intuitive thinking. If you put a pencil sticking straight out of your mouth, your facial posture will be like that recommended for critical calculation.

I know. It sounds crazy. But there is some psychological evidence that this link between the physical and the mental actually works. The background theory links it to Bargh's elevator test. You remember, those undergraduates who had just dealt with words stereotypically associated with the elderly moved different physically; they walked more slowly to the elevator. Students who were forced physically to move slowly were quicker to recognize words associated with the elderly. That's the idea here. Just as concentration does tend to make you frown, frowning can tend to make you concentrate. I'm more confident of almost all the techniques in these lectures than I am of this one. This one's a wild card, but give it a try. See if it works for you. See what you think.

How do people think? That's a descriptive question. How can we think better? That question is both normative and applied. Do you want to know both—both how we think and how we can think better? Welcome to the course.

Next time I'll start off close to where we ended here, with systems one and two. Next time, however, I'll talk about those in terms of rationality and emotion: cool rationality and hot thought.

Cool Rationality and Hot Thought
Lecture 2

R obert Louis Stevenson's *Dr. Jekyll and Mr. Hyde* illustrates a standard construction in which reason and the emotions are opposing parts of the mind. In this view, reason is at war with the emotions. The ideal of rationality, or Dr. Jekyll, is also exemplified by Mr. Spock of *Star Trek*, devoted to a life of logic and suppression of emotion. In the work of Sigmund Freud, there is also a Mr. Hyde: the id, which Freud calls the "dark, inaccessible" part of our personality. In this lecture, we will talk about what can be said for such a view and explore ways in which it turns out to be incomplete.

Reason versus Passion

- The ancient Greek philosophers are often cited as sources for the view of reason at war with the emotions. In *The Republic*, Plato develops a **tripartite theory of the soul**. Its three elements are *nous* ("reason"), *epithumia* ("passion"), and *thumos* ("spirit"). Plato makes it perfectly clear which element he thinks should be in control: reason.

- That classic picture of reason at war with the passions, and too often losing, echoes down the centuries. Erasmus laments that we have been given far more passion than reason; he estimates the ratio as 24 to 1. Shakespeare has Hamlet say, "Give me that man / That is not passion's slave, and I will wear him / In my heart's core."

- Is emotion the enemy of rationality? To at least some extent, the answer is yes. In emotional contexts, such as conditions of stress, rationality can indeed suffer. In a series of tests conducted in the 1960s, soldiers under stress experienced a decrease in cognitive abilities by 10 percent; memory failed by almost a third.

- What holds for stress holds for strong emotion generally: It can negatively impact cognitive capacity, attention, memory—all the things required for rational decision making.

Reason without Emotion Is Blind

- There are certainly cases in which emotion runs amuck, swamping rationality. But it doesn't look like we should simply try to eliminate emotion. Where would we be without it?

- The neuroscientist Antonio Damasio speaks of a patient he calls Elliot who suffered damage to a central portion of his frontal lobes, a region associated with judgment. Elliot aced all the rationality tests; his logical reasoning was intact; and his memory functioned normally, but his life had become chaos because he couldn't order priorities.

- The diagnosis wasn't a lack of reason; it was a lack of emotion. The impact of that lack of emotion on personal decision making and planning capacity was devastating. Because everything had the same emotional tone for him, Elliot had no reason to prioritize—no reason to do one thing rather than another.

- Reason without emotion seems to be blind, because emotion is tied to **value**—what psychologists call "**valence**." The things we love are the things we value positively—those that have a strong positive valence. The things we hate or fear are those that have a strong negative valence. Valence is crucial for action; we need it in order to decide what to do and to follow through and take action. Without the sense of value that is embodied in emotion, we become like Elliot.

- **Heuristics** are simple rules of action, immediate and effective in ways that don't require calculation or deliberation. It has been proposed that this is what emotions are: heuristics that give us the immediate value judgments required for action.

- One of the earliest theoretical studies of the emotions was Charles Darwin's *The Expression of Emotions in Man and Animals*. As the title makes clear, Darwin thought of emotion as something inherited from our evolutionary ancestors.
  - Such a view makes sense if emotions are a fast and frugal heuristic for value in action. In dangerous situations, we may not have time to think through all options; we just need to get out of the way of danger. Fear motivates that reaction. Anger gives us a fast and frugal reaction when threatened.

  - Without emotions as value-heuristics, we couldn't act as quickly and decisively as we often need to. What we learn from Elliot is that without emotions we would lose decision making and appropriateness of action entirely.

- The 18th-century philosopher David Hume offers an extreme form of the claim that emotions should rule.
  - Value, Hume says, comes not from reason but from something else. It is not contrary to reason, Hume says, "to prefer the destruction of the whole world to the scratching of my finger." That is an issue of value, not an issue of reason.

  - Hume turns Plato's claim completely on its head. It is not reason that should rule, he says, but the passions. "Reason is, and ought only to be the slave of the passions, and can never pretend to any other office than to serve and obey them."

Reason or Emotion? Both.

- What's the answer? Are emotions the enemy of reason, to be subjugated and suppressed, or should emotions be in control? This question represents a logical **fallacy**, called a "false dilemma." A false dilemma relies on a bogus problem set-up with only two options, when in fact, those are not the only options. In this case, we need both reason and emotion—cool rationality and hot thought.

- For many people, at least up to a point, emotional arousal actually improves performance. This is a familiar fact among athletes. That is what pep talks in the locker room are all about: getting players emotionally pumped and raring to win. Athletes in competition often want performance stress. Beyond a certain point, however, stress becomes overstress. With too much stress, performance declines quickly.

- In the case of both emotion and stress, that pattern of performance is known as the "inverted U." Performance rises on one side of the upside-down U, stays fairly high for a stretch, but then plummets dramatically. A mild level of emotion and even stress seem necessary for alertness and full involvement. But in conditions of excessive emotion and stress, performance declines.

- Without stress, people tend to rely on analysis of the problem, both wide and deep. They consider many options and think each of them through. We can think of that as the purely analytical approach. Under stress, in contrast, people attend to the general outline of the problem, rather than a wide range of specifics. They focus on significantly fewer options. We can think of that as the **intuitive approach**.

- We can portray the intuitive approach with either a negative or a positive spin. In a negative light, there is a reduction in the careful consideration of alternative options, in the tracking of options at length, and in the use of working memory and attention to detail. On the positive side, there is an increase in the focus of attention to just the critical issues and elements, avoiding inessentials.

- Recent brain studies at the University of California seem to confirm the conclusion that we need both reason and emotion. Researchers there noted both the analytical and the intuitive in the context of gambling. Brain scans of gamblers revealed that the best players used both the **amygdala** (associated with emotion) and the **prefrontal cortex** (associated with control). Rational decision making seems to require both.

- Some decisions demand instantaneous action in an immediate context. For those, we can rely on the heuristics of gut instinct and emotional response. But some decisions—human decisions, at least—aren't about what to do immediately. Some decisions are about what to do in the next few months, or over the next five years, or in the course of a lifetime.

  © Brand X Pictures/Thinkstock.

  Brain scans of those who gamble intuitively and irrationally show primary activation in the amygdala (associated with emotion) and relatively little activation in the prefrontal cortex (associated with control).

  o For those kinds of decisions, we don't need to rely on emotion alone; we have every reason to recruit the analytical and reflective abilities of the prefrontal cortex. In those cases, it would be wrong not to consider numerous options or to think through potential consequences in detail and depth.

  o But for extremely short-range decisions, it's a good thing that we have immediate emotional responses.

How Can We Make Better Decisions?
- The fact that decision making can decay in conditions of extreme emotional stress leads to a few simple lessons we can apply in our lives.

- Avoid making decisions—particularly long-range decisions—in extreme emotional situations. Decisions about living wills and funeral arrangements, for example, should be made before the need arises so that the entire brain can be engaged.

- Write up a list of pros and cons, and then balance the pros against the cons.
  - o Clearly lay out goals and alternatives, and be careful to explore all alternatives. Expectation of consequences should be based on the best evidence available.

  - o Make sure value is clearly included. What is the importance of different goals? Use emotional response as an indicator, but be aware of ways in which short-term emotion might be a bad indicator for long-term goals.

- Monitor your own emotional reactions in the decision-making process. If the "rational" considerations seem to lean one way, are you disappointed? If they lean another way, does that make you anxious? Emotional response in the midst of the attempt at rational decision making is telling you something about what you value. You may discover that you value things in different ways than you thought you did.

Deferred Gratification

- A fascinating experiment regarding short-term emotional response and long-term control is the famous "marshmallow test" conducted at Stanford in the late 1960s and early 1970s.

- Individual nursery school children were brought into a room and seated at a table that had a marshmallow on it. The children were offered a choice: Eat the marshmallow right away, or wait for a few minutes while the researcher left the room. If they waited, they would get another marshmallow when the researcher came back and could eat them both. If they wanted to eat the first marshmallow before the researcher got back, they were told to ring a bell. The researcher would come back and they could eat the first marshmallow, but they wouldn't get a second one.

- Some children ate the single marshmallow immediately. Some restrained themselves by deliberately looking away or by toying with the marshmallow. That kind of restraint indicates a capacity for **deferred gratification**—for rational control of emotion in decision making.

- The children were tracked through later life, and researchers found that those who couldn't wait for the second marshmallow seemed to have more behavioral problems later—with maintaining attention, sustaining friendships, and handling stress. In some cases, those who were able to wait scored significantly higher on SATs.

- The lesson here is: In order to increase rationality in our decision making, we should perhaps all learn to wait for that second marshmallow.

Terms to Know

amygdala: A small region deep within the brain that is associated with emotion.

deferred gratification: The ability to restrain oneself from taking an immediate payoff in order to obtain a larger payoff later.

epithumia: Appetite or passion; according to Plato, the second element of the soul.

fallacy: A form of argument in which the premises appear to support a conclusion but, in fact, do not; the term is often used to refer to familiar types of logical mistakes that may be used to trick or mislead.

heuristics: Simple guides to action or rules of thumb that allow us to act or make a decision without calculation or deliberation.

intuitive approach: An approach to problem solving that focuses on the general outline of a problem, attending to a few critical issues and generating a quick decision after consideration of relatively few alternative solutions. Opposed to an analytic approach, which examines a wide range of specifics and generates an exhaustive list of alternative solutions.

nous: Reason; according to Plato, the first element of the soul and that which should rule.

prefrontal cortex: The anterior portion of the brain, which lies just behind the forehead, heavily involved in complex planning and decision making.

thumos: Spirit or fortitude shown in battle; according to Plato, the third element of the soul.

tripartite theory of the soul: Plato's theory that the soul includes three parts: *nous*, *epithumia*, and *thumos*.

valence: A psychological term describing the positive or negative value (often emotional) that we ascribe to things.

value: Desirability; psychological research indicates that relative valuations tend to be inconsistent over time and sensitive to context.

Suggested Reading

Damasio, *Descartes' Error*.

Hume, "Of the Passions."

Plato, *The Republic*.

Thagard, *Hot Thought*.

Questions to Consider

1. A major theme of this lecture is that emotion is a helpful "fast and frugal heuristic" for some kinds of decision making but detrimental for others.
   (a) From your own experience, give an example in which you think an immediate emotional decision was the right kind of decision to make.

(b) From your own experience, give an example in which you think an immediate emotional decision turned out badly.

2. The Stoics thought rationality should rule and the emotions should be suppressed. Hume claimed that reason should be the slave of the passions—that the emotions should rule and reason should be suppressed. If you had to choose one of those extreme positions, which would you choose?

3. What would life be like if you had no emotion regarding anything at all? What would life be like if all your reactions were purely emotional?

Exercises

The Spanish artist Francisco de Goya (1746–1828) has a famous etching called *The Sleep of Reason Brings Forth Monsters* (view it at http://www.rijksmuseum.nl/images/aria/rp/z/rp-p-1921-2064.z.). What does the title tell you about Goya's stance on reason and the emotions?

Watch one of the many replications of the Stanford marshmallow experiment available online (http://www.youtube.com/watch?v=6EjJsPylEOY or http://vimeo.com/7494173). What do you think the children in your family would do in that situation? What do you think you would have done as a child?

Cool Rationality and Hot Thought
Lecture 2—Transcript

Professor Grim: In Bournemouth, England, in 1885, Robert Louis Stevenson was recovering from an illness. Trapped in a nightmare, he cried out in his sleep. His wife, Fanny, woke him up, and he immediately began to write out the nightmare. We know it as *The Strange Case of Dr. Jekyll and Mr. Hyde*. There are some great movie versions of the tale. My favorites are the classics: the 1931 Frederick March and the 1941 Spencer Tracy versions. But if you've only seen the movies, you may not know that in the original story, Mr. Hyde is significantly smaller than Dr. Jekyll. Hyde lives inside Jekyll, a metaphor for an undisciplined monster of unfettered passion that lives within us all. Jekyll doesn't so much become Hyde, as let him out.

In the work of Freud there's a Mr. Hyde too. It's the Id, what he calls "the dark, inaccessible part of our personality, a chaos, a cauldron full of seething of excitations." Is that what emotion is? That part of us that we have to suppress? There is a standard picture in which reason and the emotions are opposing parts of the mind, or the soul. On that view reason is at war with the emotions. Pursuit of rationality demands that reason win and emotion lose. The ideal of rationality is Mr. Spock of Star Trek, devoted to a Vulcan life of logic and suppression of emotion. In this lecture I'll talk about what can be said for such a view and also ways in which it turns out to be incomplete.

The ancient Greek philosophers are often cited as sources for the view of reason as at war with the emotions. In *The Republic*, Plato develops a tripartite theory of the soul. Its three elements are *nous*, *epithumia*, and *thumos*. None of these is easily translated, though *nous* is roughly reason, *epithumia* is appetite or passion, and *thumos* is the spirit or fortitude shown in battle. Plato's tripartite soul is paralleled by his tripartite theory of society, which has corresponding social classes. But Plato makes it perfectly clear who he thinks should be in control. He pictures the triumvirate as a charioteer with two horses, reason reigning in the good horse of *thumos*—fortitude—and the bad horse of the passions.

This theme is strongest in the Stoics, both Greek and Roman. Epictetus talks of the stronger emotions—sorrows, griefs, and envies—as "passions that make it impossible for us to even listen to reason." Emotions corrupt judgment. The wise man would outgrow emotion, maintaining a stoically calm indifference in the face of both the greatest good fortune and the greatest disaster. The stoic sage was supposed to be equally indifferent to wealth and poverty, honor and disgrace, even pleasure and pain.

The passions are classically thought of as things that happen to us—things we suffer—rather than things we do. That classic picture of reason at war with the passions, and too often losing, echoes down the centuries. Erasmus laments that we have been given far more passion than reason. He estimates the ratio as 24:1. Shakespeare has Hamlet say "Give me that man that is not passion's slave, and I will wear him In my heart's core."

Is emotion the enemy of rationality? To at least some extent, the answer is yes. In emotional contexts rationality can indeed suffer. Consider stress. There were a series of army tests done in the 1960s to measure performance under stress. You could say performance under the emotion of fear. These are the Berkun studies. In one scenario, new military recruits were taken up in a plane and told they were participating in a study on the effects of high altitude. Once in the air, the plane began to yaw wildly. They saw one of engines sputter and then stop. Through the intercom they heard reports of repeated malfunctions, and in typical army style, they were then told to fill out some forms. One was labeled "Emergency Data Form." It looked like a last will and testament with complicated questions about what to do with their personal possessions. The other was a test measuring retention of the ditching procedures they had read before take-off. How well did they follow the procedures on the first form, compared to people who weren't under stress? How well did they remember the ditching procedures? Not well. The cognitive abilities required in the first task decreased by 10%. Memory failed by almost a $\frac{1}{3}$.

In another test, Berkun had individual soldiers led down winding roads to isolated wilderness areas as part of a simulated tactical exercise—a "war game." But they then heard radio reports that live shells were landing off target. A series of explosions went off around them, progressively coming closer. Their only way to communicate with headquarters was with a radio transmitter. But they discovered it was broken. They had to fix it. How long did it take to go through the steps required to fix the transmitter? Under conditions of no stress, the phone was fixed in a couple of minutes. When you think you're being shelled it takes significantly longer, five minutes, seven minutes..

Studies on eyewitness testimony have shown that stress reduces memory efficiency, memory accuracy. In a study by Johnson and Scott, people show up for an experiment and are left waiting in the receptionist's office. In one case, the receptionist goes into the next room. The subjects overhear a conversation about an equipment failure. A man comes out with a pen in his hand and his hands covered with grease. He says something, "Oh, brother", maybe, and leaves. In the other case, with a different group of subjects, they hear an angry interaction that ends with chairs crashing and bottles breaking. A man comes out with a letter opener in his hands and his hands covered with blood. He says "Oh, brother", and leaves. Subjects were then asked to pick the man they saw from a line-up. In the case with the pen and the grease, 49% were able to identify him correctly. In the case with the letter-opener and the blood, only 33% were able to do so.

What holds for stress holds for strong emotion generally: it can negatively impact cognitive capacity, attention, memory, all the things required for rational decision making. You can probably cite cases from your own experience when you've made the wrong decision because you were under stress at the time, or swayed by fear, or grief, or sexual passion, or incredibly angry. I know that I can.

Now I want to give you the other side of the picture. There are certainly cases in which emotion runs amuck, swamping rationality. But it doesn't look like we should simply try to get rid of emotion. Where would we be without it? The Neuroscientist Antonio Damasio speaks of a patient he calls Elliot who suffered damage to a central portion of his frontal lobes, just above the eye sockets. That's a region associated with judgment, but Elliot aced all the rationality tests. Logical reasoning was intact. His memory functioned normally. His language abilities were perfectly normal. He passed tests on means-end reasoning, and on expected consequences of particular actions. He passed tests on standard ethical principle ... all with flying colors. But Elliot's life had become a total mess. It was hard for him to get out of bed in the morning. He couldn't keep a job because he couldn't order priorities. He would spend incredible amounts of time on something trivial, while ignoring the things that were really important. He couldn't make effective plans for the next few hours, let alone the next few days or months. He didn't seem to learn from his mistakes. The diagnosis wasn't a lack of reason. It was a lack of emotion. Damasio says that he never saw a tinge of emotion from Elliot throughout their interaction, "no sadness, no impatience, no frustration." Everything had the same emotional tone to him. Personal triumph meant no more than personal disaster. Neither meant much of anything. Elliot was in a sense the perfect stoic: everything was met with the same passionless equanimity.

The impact of that lack of emotion on personal decision making and planning capacities was devastating. Because everything had the same emotional tone, Elliot had no reason to prioritize, no reason to do one thing rather than another. Even after tests that probed his ability to conceive alternatives and to trace predictable consequences he would say, "But I still wouldn't know what to do." Reason without emotion seems to be blind. Why? Because emotion is tied to value. The psychologists call it valence. The things we love are the things we value positively—those that have a strong positive valence. The things we hate or fear or despise are those that have a strong negative valence. Valence is crucial for action; we need it in order to decide what to do, and we need it in order to follow through and do it. Without the sense of value that's embodied in emotion we become like Elliot.

Heuristics are simple rules of action or rules of thumb, immediate and effective in ways that don't require calculation or deliberation. It has been proposed that that's what emotions are: heuristics that give us the immediate value judgments required for action. One of the earliest theoretical studies of the emotions came from Charles Darwin. Thirteen years after he wrote *The Origin of Species*, thirteen years before Stevenson's Jekyll and Hyde, Darwin wrote *The Expression of Emotions in Man and Animals*. As the title makes clear, Darwin thought of emotion as something inherited from our evolutionary ancestors. Such a view makes sense if emotions are a fast and frugal heuristic for value in action. In dangerous situations we may not have time to think through all options, weighing their comparative consequences, figuring out the best long-term action. In dangerous situations, as the train is roaring toward us, for example, we just need to get out of the way fast. Fear motivates precisely that. Anger gives us a fast and frugal reaction when threatened. Without emotions as value heuristics, we couldn't act as quickly and decisively as we often need to. What we learn from Elliot is that without emotions we would lose decision making and appropriateness of action entirely.

So perhaps emotions should rule. The 18th century philosopher David Hume offers an extreme form of that claim. Value, Hume says, comes not from reason but from something else. It is not contrary to reason, Hume says, "to prefer the destruction of the whole world to the scratching of my finger." That is an issue of value, not an issue of reason. So Hume turns Plato's claim completely on its head. It is not reason that should rule, he says, but the passions. "Reason is, and ought only to be the slave of the passions and can never pretend to any other office than to serve and obey them."

So which is it? Are emotions the enemy of reason, to be subjugated and suppressed? Or is Hume right? Is it emotions that should be in control? When it comes to fallacies, we'll call that a false dilemma. A false dilemma relies on a bogus problem setup. Which is it to be? A or B? A or B? When in fact those are not your only options. Are emotions the enemy of reason, to be subjugated and suppressed? Or should emotions be in control? The truth seems to be neither. In full rational decision and full rational action, both emotion and what we think of as pure logic have important parts to play. We need both cool rationality and hot thought.

Lab studies have shown that cognitive performance can decay in situations of high emotion and high stress. But looked at as a whole, the evidence on rationality and emotion is significantly more nuanced. For many people, at least up to a point, emotional arousal actually improves performance. This is a familiar fact among athletes. That is what pep talks in the locker room are all about: getting players involved, arousing aggression toward their competitors, getting them emotionally pumped and raring to win. Athletes in competition often want performance stress. It brings out their best performance. Beyond a certain point, however, stress becomes over-stress. With too much stress, performance declines quickly.

In the case of both emotion and stress, that pattern of performance is known as the "inverted U." Performance rises on one side of the upside-down U, stays fairly high for a stretch, but then

plummets dramatically. A mild level of emotion and even stress seems necessary for alertness and full involvement. But in conditions of excessive emotion and excessive stress, performance declines. The middle stretch of highest performance is surprisingly wide. Earlier in the lecture, I mentioned the Berkun tests on battlefield stress. As the explosions come nearer and nearer to your position, you get worse and worse at being able to fix the radio transmitter. Those studies were done in the early 1960s. Our ethical standards for human experimentation are stricter now. It's extremely unlikely that any human experimentation review board would allow those experiments today, even in the military. When researchers tried more recently to construct a study regarding cognitive performance under battlefield stress, they found their hands were tied. They weren't allowed to use the kind of deception that Berkun did and weren't allowed to put subjects under that kind of stress. They were restricted to fairly limited stress situations, and with limited stress, they saw no decrease in performance. In the course of testing, however, it just so happened that one of the soldiers thought he was having a serious medical problem. There the stress was high. There the stress was real. And there the researchers did see a drastic decline in cognitive performance.

Across the middle span of the inverted U, performance can stay high despite increase in emotional stress. But even if the level of performance doesn't decrease along that plateau, a number of investigators have noted that the kind of performance changes. The kinds of problem-solving strategies we employ tend to change as stress increases. Without stress, people tend to rely on analysis of the problem, both wide and deep. They consider lots of options and think each of them through. We can think of that as the purely analytical approach. Under stress, in contrast, people attend to the general outline of the problem, rather than a wide range of specifics. They focus on significantly fewer options. We can think of that as the intuitive approach.

You can portray that intuitive approach with either a negative or a positive spin. The negative spin: there is a reduction in the careful consideration of alternative options. There is a reduction in tracking options at length and a corresponding reduction in use of working memory and attention to detail. The positive spin: there is an increase in the focus of attention to just the critical issues and elements, avoiding the inessential parts.

Different people, of course, may react differently in emotionally loaded situations. In very similar circumstances, some people use an intuitive approach. Others use an analytic approach. Interestingly enough, both those strategies may prove successful. Here, for example, are two cases. The USS *Samuel B. Roberts* was operating in the Persian Gulf when it struck a mine, caught fire, and began to sink. The captain, Paul Rinn, analyzed the situation and pursued a course of action directly opposed to Navy protocol. He estimated how much water the ship could take on and still stay afloat and realized the ship would sink before the crew could put out the fire. He therefore directed the crew to make keeping the ship afloat the first priority. Putting out the fire became secondary. He is on record as saying that he arrived at that decision coolly and analytically. But there is also the case of a United DC-10 travelling from Denver to Chicago. The hydraulic system and maneuverability suddenly failed. Captain Al Hayes and his crew had to come up with a new way to fly the plane using the throttles and a limited number of environmental cues. That wasn't in their training. They landed safely, but attributed their success entirely to intuitive decision making under stress. They said "it wasn't analysis. It was intuition." One of the interesting things about those two cases is that an emotional decision in the first case— fire!—might have led to disaster. Time spent on analysis in the second case might have led to disaster as well.

So, should emotion be the slave of reason, or reason the slave of emotion? The answer seems to be neither. Both are involved in our decision making, though perhaps in different ways in different situations. Both should be involved in our decision making. The rational thing to do is to jettison neither cool analytical rationality, nor the fast and frugal heuristics of intuitive emotional reaction. The rational thing is to take advantage of both. There are two extremes that we should clearly avoid. One is Elliot's: decision making and living in general so devoid of emotion that is devoid of value. The other extreme is decision making under such extremes of emotion or stress that rationality can no longer get a grip. Human decision making seems best when it's between those extremes. We're at our best when we can use all the tools at our disposal.

Some recent research at the Institute of Neurology at the University of California seems to confirm that conclusion in terms of brain studies. Researchers there noted something like the two approaches—the analytical and the intuitive—though in the very different context of rational and irrational gambling. It turns out that some people gamble almost entirely with their gut. They handle risk inconsistently, and they make some fairly major mistakes. They do poorly. What the researchers found was that in those cases the primary activation was in the amygdala. The amygdala lies at the core of what is known as the "old brain," a feature we share with animals all the way down to reptiles. It has long been known to be a primary activator of emotion. Brain scans of those who gambled intuitively and irrationally showed fairly little action in the prefrontal cortex. The prefrontal cortex is a part of the "new" brain that distinguishes us and higher primates. That part of the brain is not developed in lower animals. It has long been associated with higher-order executive processes and emotional control.

So what does brain activity look like in more rational decisions? It's not that the prefrontal cortex is turned on and the amygdala is turned off. What the researchers found was that the amygdala was active in both groups. What the rational players had that the irrational "gut" players did not was the use of both parts of the brain, emotionally intuitive and rationally controlling. Players who used both parts of their brain did much better. Of course I'm not advocating gambling with either part or the whole of your brain. We'll return to the entire topic of probability in a later lecture. ß

The amygdala is associated with emotion, the prefrontal cortex with control. Rational decision making seems to require both. In Elliot's case the amygdala wasn't impacted. But in his case it appears that the ability to integrate emotion into decision making was. The fact that the "old brain" amygdala is activated in emotion certainly fits the Darwinian view that emotions are "fast and frugal" heuristics for quick action that we have inherited from our evolutionary ancestors. It also says something about how we, who have both amygdala and prefrontal cortex, should go about making different kinds of decisions. Some decisions demand instantaneous action in an immediate context. For those we can rely on the "fast and frugal" heuristics of gut instinct and emotional response. But some decisions—human decisions, at least—aren't about what to do immediately. Some decisions are about what to do in the next few months, or over the next five years, or in the course of a lifetime. For those kinds of decisions, we don't need to rely on emotion alone. For those kinds of decisions, we have every reason to recruit the analytical and reflective abilities of the prefrontal cortex. In those kinds of cases it would be wrong not to consider lots of options. It would be wrong not to think through potential consequences in detail and depth. For extremely short-range decisions it's a good thing that we have immediate emotional responses. For long-range decisions it's a good thing we have those other analytical abilities as well. At an earlier point, I appealed to your experience, to the fact that all of us have made poor decisions in the heat of passion, in extreme anger or grief, for example. If you think back on some of your bad decisions, I think you'll find that the problem may have been making long-range decisions with only a short-range tool.

So how should we make decisions in emotionally-stressed situations? We don't have all the answers, and there's a lot about emotion we don't know. But perhaps we can at least assemble a few pointers. One of the things we know is that decision making can decay in conditions of extreme emotional stress. So here's a simple lesson: Avoid decision making in those kinds of situations. Plan things so that you won't have to make decisions—particularly long-range decisions—in predictable conditions of extreme emotional stress. That's what pre-planning for funeral arrangements is all about. If you and those you love can sit down and make decisions now, you can make them more rationally than when, inevitably, you will be impacted by devastating loss and deep grief. That's also what living wills and declarations regarding your future health care are all about. If you make those decisions now, you can use your entire brain. If you are forced to make those decisions in a crisis, you won't have that luxury.

A standard picture of "rational" decision making has you write up a list of pros and cons. You are then supposed to balance the pros against the cons. Darwin actually did this when deciding whether he should get married. He wrote up the pros of staying a bachelor: freedom to go where he liked, "choice of society and little of it," he said, "conversation of clever men at clubs." He wrote up the cons: no children, no one to care for you in old age, what is the use of working? The contemporary philosopher Paul Thagard has paid a great deal of attention to "hot thought"—to a picture of decision making that integrates both emotion and analytic rationality.

Thagard thinks that list-making like Darwin isn't a bad idea, but he emphasizes several crucial elements. First, goals and alternatives have to be clearly laid out. You should be careful to explore all alternatives. Expectation of consequences should be based on the best available evidence. Second, value has to be clearly in there. What is the importance of different goals? Use emotional response as an indicator but be aware of ways in which short-term emotion might be a bad indicator for long-term goals. Here's the interesting part: As you consider things in this light, Thagard recommends monitoring your own emotional reactions in the process. If the "rational" considerations seem to lean one way, are you disappointed? If they lean another way, does that make you anxious? Emotional response in the midst of the attempt at rational decision making is telling you something; it is telling you something about what you value. You may discover that you value things in different ways than you thought you did. Pay attention to what those reactions are telling you about what you really value. Include that in the mix.

What I've been emphasizing as the take-home message is that optimal decision making seems to involve both reason and the emotions. Despite our initial portrayal of Plato in terms of a conflict between reason and emotion, I think that integration of the two is his message as well. The Stoics seem to want to flatten emotion entirely. But Plato's image is of a charioteer and two horses. He wants the charioteer to remain in control, but that chariot is going to go nowhere without the horses.

Let me leave you with a fascinating piece of experimentation regarding short-term emotional response and long-term control. This is the "marshmallow test" developed at Stanford in the late 1960s and early 1970s. Individual nursery school children were brought into a room. A marshmallow was put in front of them. They were offered a choice: they could eat that marshmallow right away, or they could wait for a few minutes while the researcher left the room. If they waited, they would get another marshmallow when the researcher came back and could eat them both. If they wanted to eat the first marshmallow before the researcher got back, they were told to ring a bell. The researcher would come back and they could eat that first marshmallow. But they wouldn't get a second one. Some kids ate the single marshmallow immediately. Some didn't even bother to ring the bell. Some kids restrain themselves by deliberately looking away, or by merely smelling the marshmallow, or poking it and licking their fingers. That kind of restraint indicates a capacity for "deferred gratification"—for rational control of emotion in decision making.

The interesting thing in the Stanford study is that those kids were tracked through later life. Researchers asked just about any question they could think of: How well did they do in school years later? How well could they cope with life's inevitable problems? What did they get on the SATs? What they found was that the kids who couldn't wait for the second marshmallow seemed to have more behavioral problems later. They had trouble maintaining attention. They had trouble maintaining friends. They didn't handle stress as well. The kid who could wait fifteen minutes for a second marshmallow had SAT scores two hundred points higher than the kid who could wait only thirty seconds. It may be that capacity for deferred gratification is set early in life, but it may be that we can cultivate it even now. In order to increase rationality in our decision making, we should perhaps all learn to wait for that second marshmallow.

In the next lecture I want to reach into the philosopher's toolkit in order to hand you what I think is the most powerful thinking tool of all: conceptual visualization.

The Strategy of Visualization
Lecture 3

The purpose of this course is to give you some conceptual tools for effective problem solving, logical thinking, and rationality in general. This lecture will help you to understand the most important tool of all for logical thinking and effective problem solving: visualization. As we'll see in this lecture, visualization is a tool you already have; our goal is to increase your respect for that tool and encourage you to put it to work.

What Is Visualization?

- Let's start with an example for you to try in your head. Imagine you have a long rectangular board. You cut it into two lengths, one of which is three times the length of the other. You now cut the smaller of those lengths again into two pieces, one of which is twice as long as the other. The smallest piece is now 8 inches square. How long was the board you started out with?

- Next, draw the board, following the instructions above. How long was the board? Once you have a picture, the answer may seem pretty obvious. The piece you divided into three must be three times the length of the 8-inch square, so it's 24 inches long. The original board must be four times that—96 inches, or 8 feet long.

- When you try to do a problem like this in your head, it's complicated, but it's not when you can see the problem—even in your mind's eye. That's the power of **visualization**.

Pattern Recognition

- People are much better than computers at **pattern recognition**. Think of our ability to recognize human faces. Newborns don't have any trouble recognizing a face as a face. We don't have any trouble recognizing the face of someone we know. We can sometimes recognize the face of a person we haven't seen for 20

years, even with a fleeting glance. Computer scientists have been trying to replicate that ability for decades.

- Computers are faster and more accurate at all kinds of symbolic calculation, but when it comes to information processing and pattern recognition, nothing beats the human visual system.

How Philosophy Uses Visualization

- There is a long history of visualization in the philosopher's toolkit. In *The Republic*, Plato outlines a theory of the relation between the visible and the intelligible world, between mere opinion and genuine knowledge. He maps out the theory visually, with an image known as Plato's "divided line"—a division between the visible and the intelligible, further divided on each side between superficial understanding and a deeper grasp. One can almost see Plato scratching that line in the Athenian sand.

- That tradition of visualization continues through the history of thought. Another philosopher-mathematician, René Descartes, saw that a link between such equations as $y = 2x^2 + 9x + 3$ and the beautiful arc of a parabola. By treating x and y as coordinates in a plane, algebra became geometrically visual. We still call those **Cartesian coordinates**. And the conceptual tool used by Descartes was visualization.

The Limits of Cognitive Capacity

- Why does it help to work with pencil and paper rather than entirely in your head? One reason is cognitive capacity. There are just so many things we can hold in conscious attention at one time.

- One of the most cited papers in psychology is a classic from the 1950s by George Miller: "The Magical Number Seven, Plus or Minus Two: Some Limits on Our Capacity for Processing Information." The thesis is that we can only manage about seven pieces of information at one time, plus or minus two. That's the limit of a phone number, without the area code.

- Not surprisingly, this thesis depends on context, on what the seven items are, and how the information is "chunked." The truth is that there is no single magic number that measures human cognitive capacity, but Miller makes this point that our cognitive grasp is limited.

- You run up against that capacity when you try to multiply two 3-digit numbers in your head. But if you write the two numbers down and multiply them on paper, you can do such problems easily. One benefit of writing problems out—using visualization—is that it frees up cognitive load.

Improving Cognitive Capacity

- Sometimes visualization can practically solve a problem by itself, as we see in the word problem concerning the roads in Brown County. If you've been paying attention to the theme of this lecture, you worked that problem out by making a sketch. Sketches can help in two ways.
  - First, once you codify each piece of information from a problem in a sketch, you no longer have to try to keep it in short-term memory.

  - Second, if the sketch captures all the information in the problem, it practically solves the problem itself.

- You can also use visualization for more abstract problems, such as the one concerning students and their quiz grades. This is really a problem in information management. Here again, visualization is the key, even though the tie between people and grades is more abstract than that between roads between towns.
  - This is a common type of problem on standardized tests. It's known as a "matrix problem" because it is most easily solved with a particular visualization: a rectangular array called a **matrix**.

  - The question here is really about how two things align: students and their grades. Visualization turns that abstract alignment into a spatial one: a checkerboard matrix with the grades lined

up along the top and the student names down the side. Spatial relationships are easier to handle than abstract ones—you can see them.

	A	B	C	D
Jim				
Billy				
Tom				
Lisa				
Molly				

Visualizing $E = mc^2$

- Einstein was one of the great visualizers. He said he never thought in symbols or equations but in images, feelings, and even musical architectures. The images, he said, came first. The words came later.

- In thinking through **relativity theory**, Einstein envisaged lightning strikes hitting the ends of a moving train. He asked himself what the timing of the strikes would look like to someone inside the train and outside the train.

- Einstein also employed much more abstract visualizations. He talks of gravity bending light, for example, and speaks of the curvature of space time. It has become common to visualize those ideas in terms of a bowling ball on a rubber sheet.
  - If you think of space as a square grid on a rubber sheet, what happens with light around massive bodies is similar to what the grid would look like if a bowling ball was dropped in the middle.

  - The gravitational force makes straight lines curved—just as light will curve around a massive body.

- That image ties in with one of the first and most dramatic tests of some of Einstein's work. Suppose gravity does bend light; then, the observed position of some distant source of light—such as the stars—should be different when there is and isn't a massive body, such as the sun, between us and them. In 1919, the British astronomer Arthur Eddington organized expeditions to Africa and Brazil in order to photograph the same portion of the sky before and during an eclipse. He was able to display the variance in the plates as a vindication of Einstein's vision.

- As Einstein himself emphasized: "Discovery is not a work of logical thought, even if the final product is bound in logical form."

Visualizing the Pythagorean Theorem

- The **Pythagorean theorem** has been called the single most important result in all of mathematics. And yet Pythagoras's proof was almost certainly visual, using no math at all. Pythagoras probably conceived of the theorem as follows: "For any right triangle, the square on the hypotenuse is equal to the sum of the squares on the other two sides." For "square" here, just think of a square.

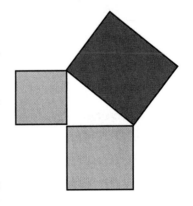

- The triangle in the middle is a right triangle—any right triangle. The large square on the upper right is the "square on the hypotenuse." What the theorem says is simply that the area of the square on the upper right is always the same as the sum of the area of the other two squares put together.

- Pythagoras himself saw the importance of this result. It is said that he sacrificed 100 oxen in celebration.

Mental Rotation

- It's important to remember that your thinking isn't limited to the space between your ears. There is no reason not to think of thinking as something that can happen using your eyes and with a pencil and paper in front of you. Thinking can indeed be more powerful in that format.

- Mental visualization is very much like using a pencil and paper. There has been a lot of work in psychology done on what is called "mental rotation." In it, you are given shape silhouettes in two-dimensional form and asked whether it's possible to convert one into the other by rotating them. For example, which of the shapes on the right are rotations of the one on the left?

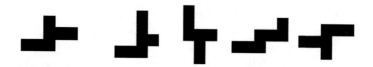

- In the three-dimensional version, you're given three-dimensional shapes and asked whether the shapes are the same but somehow rotated.

- Researchers have found that the time it takes you to discover whether two rotated shapes are the same is directly related to how far you'd have to turn one to get the other. Just as it is faster to rotate something 90 degrees than a full 180 degrees, people are faster at recognizing something as rotated 90 degrees than a full 180 degrees.

- Another interesting result is that some of the same portions of the brain that are used in actual rotation are used in imagined, or mental, rotation. Seeing and imagining recruit some of the same crucial brain circuits.

- The next lecture puts the tools of visualization to work in examining the fundamental elements of all reasoning: the atoms of thought.

Lecture 3: The Strategy of Visualization

Cartesian coordinates: The position of a point on a plane as indicated in units of distance from two fixed perpendicular lines, the x (horizontal) axis and the y (vertical) axis.

matrix: A rectangular array; a visualization technique using a checkerboard to represent how two variables align. One set of values is represented by rows in the matrix; a second set of values is represented by columns.

pattern recognition: The ability to recognize a set of stimuli arranged in specific configurations or arrays, for example, to recognize faces as faces rather than patches of color or melodies as melodies rather than merely sequences of notes.

Pythagorean theorem: In Euclidean geometry, the relation of the three sides of a right triangle, represented by the formula $a^2 + b^2 = c^2$. The area of the square whose side is the hypotenuse of the right triangle (the side opposite the right angle) is equal to the sum of the area of the squares of the other two sides.

relativity theory: Proposed as an alternative to Newtonian physics, Einstein's theory of relativity (general and special) holds that matter and energy are interchangeable, that time moves at a rate relative to one's rate of speed, and that space itself can be curved by gravity. In large part, relativity theory is a series of deductions from the assumption that some things are not relative. For example, assuming that the speed of light is a constant, Einstein showed that observable simultaneity will be relative to the comparative motion of two observers.

visualization: The process of using diagrams or imagery as an aid to thought in representing a problem, calculation, or set of possibilities.

Suggested Reading

Bronowski, *The Ascent of Man.*

Levine, *Effective Problem Solving.*

Polya, *How to Solve It.*

1. The central claim of this lecture is that problems are often easier if we can recruit our visual abilities.

   (a) Give an example of an everyday problem that you already tend to think about visually.

   (b) Are there things that you think about visually and other people do not?

   (c) Take an everyday conceptual problem you are dealing with now. Is there any way you might approach that problem visually?

2. Can you suggest a hypothesis to explain why we are so good at pattern recognition and not particularly good at either large-scale memory or symbolic computation?

3. Do you think it's true that visualizing something good happening can help make it happen? Can you suggest a hypothesis to explain why that might hold?

Exercise

John has a pretty good weekly salary. In fact, if the salary of his older brother Bill were doubled, John would make only $100 less than Bill. Bill's salary is $50 more than that of their youngest brother, Phil. John makes $900 a week. How much does Phil make?

   (a) First, try to do this problem in your head without a piece of paper in front of you.

   (b) Now try it with pencil and paper; draw a horizontal line to represent salaries, with dots for where people are on that line.

(See "Answers" section at the end of this guidebook.)

The Strategy of Visualization
Lecture 3—Transcript

Professor Grim: The purpose of this course is to give you some tools, conceptual tools from the philosopher's toolkit. Tools for effective problem solving, for logical thinking, tools for rationality in general. In this lecture I want to give you what I think is the most important tool of all. If I could hand you just one tool for logical thinking and effective problem solving, this would be it: the tool of visualization. Some of the lectures in this course work best if you can participate with pencil and paper along the way. This is one of those. If you can't work with pencil and paper right now, go on to another lecture and come back to this one.

Visualization. I'll start with an example that I want you to do entirely in your head. You have a long, rectangular board. You cut it into two lengths, one of which is three times the length of the other. You now cut the smaller of those, again, into two pieces, one of which is twice as long as the other. The smallest piece is now eight inches square. How long was the board you started out with? If you're like me, that's a medium-grade problem; not impossible, but it's easy to get lost.

Now I want you to draw the board. Draw a long, rectangular board. Divide it into two lengths, one of which is three times the length of the other. You'll draw a board with four equal parts. The smaller of those parts is cut into two pieces, one of which is twice as long as the other. If you draw it, your picture will probably show that smaller piece divided into three. The smallest piece of all is now 8 inches square. How long was the board? Looking at that picture, the answer is pretty obvious. The bit you divided into three has to be three times the length of your eight-inch square. So it's 24-inches long. The original board has to be four times that—96 inches, or eight feet, long.

When you try to do it in your head, it's complicated. But not when you can see it. Not even when you can see it "in your mind's eye." When you try to do it in your head, it's mathy. You end up with all those symbols. But it's a lot less mathy, a lot less intimidating, when you can see it. Then it's just a picture. If you're like me, pictures are a lot easier to think with than symbols are. I'm not out to teach you visualization. It's a tool you already have. What

I want to do is to increase your respect for that tool. To use it more. To put it to work.

There is something that people are extremely good at but computers aren't. What we're particularly good at is pattern recognition. Think of your ability to recognize human faces. Newborns don't have any trouble recognizing a face as a face. We don't have any trouble recognizing the face of someone we know. We can sometimes do that even if we haven't seen the person for twenty years, even if they're older, they've grown a moustache, and we see them fleetingly through a train window. Computer scientists have been trying to replicate that ability for decades. When it comes to information processing and pattern recognition, we have an amazingly fast and accurate resource at our disposal. Computers are faster and more accurate at all kinds of symbolic calculation. But when it comes to information processing and pattern recognition, nothing beats our visual system. Herbert Simon was a major figure in both computer science and cognitive science. In a paper called "Why a Diagram is (Sometimes) Worth Ten Thousand Words," he compares side by side a purely symbolic approach to a problem and an approach that uses a diagram. The symbolic approach runs to several detailed pages. The diagram, it turns out, is both more efficient and more convincing. Why is the diagram better? Simon thinks it's because we can see it, and seeing it allows us to exploit the visual thinking that's built in. Seeing it allows us to draw the "immediate visual inferences" that people are so good at.

There is a long history of visualization in the philosopher's toolkit. At the core of the *Republic*, Plato outlines a theory of the relation between the visible and the intelligible world, between mere opinion and genuine knowledge. How does he try to get that theory across? Visually. With a diagram. It's called Plato's "divided line," divided between the visible and the intelligible, further divided on each side between superficial understanding and a deeper grasp. One can almost see Plato scratching that line in the Athenian sand. And I'm sure he did. Of course Plato wasn't the first. Before Plato came Pythagoras, philosopher, mystic, and mathematician. Everyone knows the Pythagorean theorem; for any right triangle, the square of the hypotenuse is equal to the sum of the squares on the other two sides. Well, almost everybody knows it. At the end of the Wizard of Oz, the Scarecrow shows off his new brain by saying "The sum of the square roots of any two sides of an isosceles triangle is equal to the square root of the remaining side." That's just so wrong in so many ways. No, for any right triangle, the square of the hypotenuse is equal to the sum of the squares on the other two sides.

The Pythagorean theorem is the shining gem in Euclid's geometry. In geometry, you may have proven the Pythagorean theorem from axioms, as Euclid did. But it's pretty clear that Pythagoras didn't think of it in terms of axioms. The source of his insight was almost certainly visual. In the last part of the lecture I'll show you what Pythagoras's proof might have looked like. He almost certainly proved it by manipulating shapes with simply the power of visualization. Pythagoras's insight was literally something that he saw. That tradition of visualization continues through the history of thought. Another philosopher-mathematician, René Descartes, saw that there was a link between equations like $y = 2x^2 + 9x + 3$ and the beautiful arc of a parabola. By treating x and y as coordinates in a plane, algebra becomes geometrically visual. We still call those Cartesian coordinates. The conceptual tool? Visualization.

Why does it help to work with pencil and paper rather than entirely in your head? One reason is cognitive capacity. There are just so many things we can hold in conscious attention at one time. I'll give you two numbers. I want you to multiply them in your head. The numbers are 9 and 7. That was easy, right? 63. Here are two more numbers. Now multiply these in your head. 27 and 54. Ouch! If you got that, I'm sure it took longer. The number is 1,458. Let me give you a third pair of numbers, 259 and 376. Unless you're a savant, that would take longer. It would hurt even more, and your error rate would rise.

One of the most cited papers in psychology is a classic from by the 1950s by George Miller: "The Magical Number Seven, Plus or Minus Two: Some Limits on Our Capacity for Processing Information." The thesis is that we can only manage about seven pieces of information at one time, plus or minus two. That's the limit of a phone number, without the area code. Not surprisingly, the seven-plus-or-minus thesis depends on context, on what the seven items are, and how you "chunk" information. The truth is, there is no single magic number that measures human cognitive capacity. But Miller makes his point: our cognitive grasp is limited. It's probably much more limited than we'd like to think. You run up against that capacity when you try to multiply two three-digit numbers in your head. But now do something different. Write the two numbers down. 259 and 376. Multiply them on paper. That's going to be no problem at all. It'll take a minute, particularly since we're so used to grabbing a calculator, but with pencil and paper it's easy. One benefit of writing it out is that it frees up cognitive load. If you assign the memory function to a piece of paper, you can let your brain concentrate on the manipulation function. Like doing the actual multiplication.

But visualization has benefits far beyond memory. Sometimes a visualization can practically solve the problem by itself. Here's a problem like that. There are five roads in Brown County. One runs from Abbieville to Brownsville by way of Clinton. One runs from Clinton to Derbyshire by way of Fremont. One runs from Fremont to Brownsville by way of Abbieville. That's all the roads in Brown County, and all the roads in and out of those towns. Now let me give you three questions. Which towns have roads connecting them directly to three other towns? Which towns have roads connecting them directly to only two other towns? How many towns do you have to pass through to get from Brownsville to Derbyshire? If you want to try those, press pause for a minute and start again when you're ready.

If you've been paying attention to the theme of this lecture, you worked those problems out by making a sketch. Indeed I don't know why anyone would try to handle that kind of a complexity without a sketch. Two things happen with a sketch. The first is the simple matter of memory. Once you codify each piece of information in the sketch, you no longer have to keep it in short-term memory. Was that Abbieville to Clinton by way of Brownsville, or Abbieville to Brownsville by way of Clinton? The other thing that happens is that the sketch practically solves the problem by itself, because the information is all in the sketch. You can supply the information asked for just by looking at it. Which towns have roads leading directly to three other towns? Abbieville, Clinton, and whatever F stood for. Which towns have roads connecting directly to just two others? Just Brownsville. How many towns between Brownsville and Derbyshire? Two, by two separate routes, though you could also go through three towns if you wanted to. The diagram itself embodies the information you need. The sketch itself solves the problem.

You can use visualization for more abstract problems, as well. Here's one. Each of four students (boys Jacob and Eli and girls Lauren and Molly) got a different passing grade on a quiz (A, B, C, D). Both boys scored lower than Lauren. Eli got a B. The problem gives us that information about kids and grades, and asks us to answer some further questions. What grade did Lauren get? If the girls got both the best grade and the worst grade, what grade did Jacob get? If both boys had scored less than both girls, which of the initial statements I just gave you would have to be false? This is really a problem in information management. Use the information you have in order to squeeze out some more. And here, again, visualization is the key, even though the tie between people and grades is more abstract than roads between towns.

What I just gave you is a common type of problem on standardized tests. It's known as a "matrix problem" because it's most easily solved with a particular visualization, a rectangular array called a matrix. The question is really about how two things align. Students Jacob, Eli, Lauren, and Molly on the one hand, and the grades A, B, C, and D on the other. Visualization turns that abstract alignment into a spatial one, a checkerboard matrix with the grades lined up along the top and the student names down the side. Spatial arrangements, spatial relationships are easier to handle than abstract ones— you can see them. To solve the problem, we enter the information we have and read off the information we need. The three questions draw on the basic information we've been given. We get that down first. You should be aware that problems like this don't guarantee complete information. We want to tack down the information we have, see what possibilities that leaves open, and then turn to the particular questions asked. So, draw the grid, solve the problem, and come back. I'll be here working on it too.

So let's think this through on the matrix. Start with the pieces of information that are most specific and seem to offer an immediate payoff. It was the second piece of information that we were given, but "Eli got a B" is the most promising. Let's use a dot to indicate positive information—that Eli got a B. We drop that in the matrix. We'll use an X to indicate negative information. Since Eli got a B, we know he didn't get some other grade. So, X those out. Since we know that each student got a different grade, we know that no one else got a B. X those out. Our matrix just got smaller—we've filled in a lot of information with that one bit. Now enter the rest of our information. "Both boys scored lower than Lauren." Well, that doesn't tell us what both boys scored. But it does tell us what Lauren scored. Eli is a boy, Lauren did better than both boys, so Lauren had to get an A. Fill in the positive information. Now fill in the negative information. That's all the information we have, but now it's visual—we can literally see what we know and what we don't. So what about those two other questions? If the girls got both the best grade and the worst grade, what grade did Jacob get? Since Lauren got an A, that would mean that Molly got a D. That would X out everything on the graph except for one slot left for Jacob. He would have to get the C. And how about that last question? If both boys had scored less than both girls, which of the initial statements would have to be false? Well, it wouldn't have to be false that both boys scored less than Lauren—in fact that would have to be true. But it couldn't be true that Eli got a B, unless more than one student got the same grade. So in that case one of these initial pieces of information would have to be false, either that Eli got a B, or that each student got a different grade. As I say, that type of problem is a favorite in standardized testing. But the key is simply a particular kind of visualization. With the right representation, the abstract becomes spatial. When you approach it visually, the problem is easy.

People can train themselves in certain types of visualization and even in certain types of seeing. I'm sure this is true of chess masters, who can see the significance of board patterns that are invisible to the rest of us. An easy example is what's called an open file, a sequence of squares that are vacant, open to a rook or queen, for example, running from the player's side of the board toward the opponent's. What mastery demands is being able to visualize the board, not in terms of where pieces are, but where they can go, or even in terms of fields of force.

That leads me to another sense in which people talk about visualization. People sometimes talk about visualization as picturing what you want and claim that that will help you get what you want. I'm not sure that will work as a general technique. You can visualize being handed a million dollars tomorrow all you like. I don't think that will make it happen. But the story is told that the comedian Jim Carrey, still a struggling Canadian comic in the mid-1980s, wrote himself a check for ten million dollars "for acting services rendered," dated years into the future. He carried that check for years as a focus, an inspiration, and a promise to himself. Mentally rehearsing a musical passage can indeed make you better at it. Mentally envisaging how your tennis serve should look can help you make it that way. But, of course, mental rehearsal isn't the whole story. You have to pick up that tennis racket to put your visualization to work.

Einstein was one of the great visualizers. He said he never thought in symbols or equations, that he thought in images, feelings, and even musical architectures instead. The images, he said, came first. The words came later. In thinking through relativity theory, he envisaged lightning strikes hitting the ends of a moving train. He asked himself what the timing of the strikes would look like to someone inside the train, and how they would look to someone watching the train from the outside. That visualization is part of a thought experiment I'll explore in a later lecture.

Einstein also employed much more abstract visualizations. He talks of gravity bending light, for example, and speaks of the curvature of space time. It has become common to visualize those things in terms of a bowling ball on a rubber sheet. If you thing of space as a square grid on a rubber sheet, what happens with light around massive bodies is something like what that grid looks like when you drop a bowling ball in the middle. The gravitational force makes straight lines curved—just as light will curve around a massive body. That image ties in with one of the first and most dramatic tests of some of Einstein's work. Suppose gravity does bend light. Then the observed position of some distant source of light—like the stars—should be different when there is and isn't a massive body between us and them. Like the sun. If we could photograph the stars in an area of the sky when (a) the sun is next to them and (b) when it isn't, and compare those photographs, we should be able to see if the theory is true. Unfortunately, you can't see the stars next to the sun—the sun is too bright. You can see them, however, during a total eclipse. In 1919 the British astronomer Arthur Eddington organized expeditions to Africa and Brazil in order to photograph the same portion of the sky before and during an eclipse. He was able to display the variance in the plates as a vindication of Einstein's vision. Philosophers of science still talk about that expedition, in part because Eddington may have glossed over some ambiguities in the data. What no one denies is how productive that visualization of curved space time has turned out to be. The phenomenon is now called "gravitational lensing."

Sometimes Einstein's visualizations were intended to explore even the limits of possibility. This is how Einstein himself portrayed one of these earliest visualizations. At the age of 16, he tells us, he tried to visualize what he would see if he pursued a beam of light with the speed of light. In that "childlike thought experiment," he says, "one can already see the germ of relativity theory." In that context he again emphasizes visualization: "Discovery is not a work of logical thought, even if the final product is bound in logical form."

There's an interesting footnote here. How did Einstein acquire his mastery of visualization? At 16, he was a student at the Aarau school. The teaching at Aarau was based on the work of an educational reformer named Pestalozzi. What Pestalozzi wrote was that the learning of numbers and language must be definitely subordinated. "Visual understanding is the essential and only true means of teaching how to judge things correctly." An emphasis on visualization was an explicit part of Einstein's education.

Let me return to Pythagoras. The Pythagorean theorem has been called the single most important result in all of mathematics. And yet Pythagoras's proof was almost certainly visual, using no math at all. First, the theorem the way Pythagoras probably saw it. "For any right triangle, the square on the hypotenuse is equal to the sum of the squares on the other two sides." For "square," don't think of mathy things like numbers raised to the power of two. For "square," just think "square." The triangle in the middle is a right triangle. Any right triangle. That large square on the upper right is the "square on the hypotenuse," and what the theorem says is just that the area of that square is the same as the sum of the area of the other two squares put together. Always.

Pythagoras's visual proof probably went something like this. Take that square on the hypotenuse. Because it's a square, and because they're right triangles, we can fold four of these into that square. Like this. [silence during demonstration.] Those cover the square on the hypotenuse, except for a little square in the middle, which you'll notice has sides that are the difference between the long side and the short side of our original triangle. When we put that in the square, we've covered the area of the square on the hypotenuse. What we want to show is that that area is the same as these other two put together. Now because that is the area of the square on the hypotenuse, that area will remain the same even if we rearrange the pieces. [silence for demonstration.] Same area, rearranged. What we're out to show is that that area is the same as the area of these two squares put together. Take the little one. Its edge is the same as this edge over here. It will cover that area first. Now take this one over here. This is the same as the long edge of our triangle. With this difference between the short and the long edge added in. That covers this area perfectly. There you have it. The square on the hypotenuse equals the sum of the squares on the other two sides. Nothing in that visual proof depended on what kind of right triangle it was. So it will work for them all. Pythagoras himself saw the importance of the result. It is said that he sacrificed 100 oxen in celebration.

What I hope you'll take from this lecture isn't so much a thesis, as a tool, or increased respect for a tool you already have. Your thinking isn't limited to the space between your ears. There is no reason not to think of thinking as something that can happen using your eyes and with pencil and a piece of paper in front of you. Thinking can indeed be more powerful in that format. Mental visualization is very much like that. There has been a lot of work done in psychology on what is called mental rotation. In two-dimensional form, you are given two shape silhouettes and are asked whether it's possible to convert one into the other by rotating them.

Which of the shapes on the right are rotations of the one on the left? In three-dimensional form, you're given two three-dimensional shapes and asked whether the shapes are the same but somehow rotated.

Which of the shapes on the right, here, are rotations of the ones on the left? It has been found that the time it takes someone to discover whether two rotated shapes are the same is directly related to how far you'd have to turn one in order to get the other. Just as it is faster to rotate something 90° than a full 180°, people are faster at recognizing something as rotated 90° than a full 180°. Another interesting result is that some of the same portions of the brain that are used in actual rotation are used in imagined or mental rotation. Seeing and imagining recruit some of the same crucial brain circuits. Did you get those, by the way? For the first one, rotation will give you the second and the fourth images. For the second one, it's the first and the third.

The last interesting result is perhaps the most important for this lecture. Like all forms of visualization, mental rotation is something that gets better with practice. In the next lecture we'll put these tools of visualization to work in examining the fundamental elements of all reasoning: the atoms of thought.

Visualizing Concepts and Propositions
Lecture 4

T wo fundamental ideas are at work in this lecture. First is the idea that concepts, the atoms of thought, lie behind our words. Those concepts bring with them their own inherent structural relationships. A simple way of visualizing those relationships is through Venn diagrams: subcategories within larger categories, subsets within larger sets. An expanded visualization is in terms of concept trees: hierarchical organizations of categories and subcategories, sets and subsets. Second is the idea of propositions behind our sentences. These are the molecules of thought—the stuff of evidence and deliberation, the working material of all rational argument. Categorical propositions can also be visualized in Venn diagrams. We'll use that visualization to illustrate the logic of Aristotle.

What Are Concepts?

- **Concepts** are the most basic elements in all thinking. Concepts are ideas, and words represent those ideas. You can think of concepts as either things in your brain—your concepts—or as something we share as part of our culture—our concepts. But many people think of them as just flotsam and jetsam, floating randomly in a mental or cultural sea.

- The fundamental fact is that concepts do not float around independently. Physical atoms are organized by physical structure— the physical structure represented in the periodic table. As the atoms of thought, concepts come with their own inherent organizational structure, too. Crucial to all effective thinking is an ability to visualize that inherent conceptual structure and put it to use.

Words Are Not Concepts

- Is there a difference between **words** and concepts? Absolutely. Words are different from language to language, but concepts are not. It's *frog* in English but *Frosch* in German and *sapo* in Portuguese and *grenouille* in French. These are different words but just one concept. The recognition that concepts are distinct from words has been part of the philosophical toolkit at least since Aristotle, who said that, although speech is not the same for everyone, "the mental affections themselves, of which these words are the signs … are the same for all mankind."

- We use words to express concepts, but words are different from the concepts they express. Simple as it is, that point turns out to be crucial in building good reasoning, in detecting bad reasoning, and in reasoning effectively together. Concepts and words are different: a single concept may be expressed by many different words. Even more important, very different concepts may lie behind the same word.

The Extension of a Concept

- Concepts don't just float around in a random sea; they come with their own relationships—an inherent structure between them. For example, frogs are a kind of vertebrate; but vertebrates are not a kind of frog. Those concepts have a kind of logical order, a relational structure.

- What we're talking about, then, is really a three-part relationship: the relationship among words, concepts, and the things those concepts apply to. Words express concepts. Concepts apply to things. We might say that words apply to things, too, secondhand. Words express the concepts that apply to things. Philosophers speak of the totality of things that a concept applies to as its **extension**.

- We can express precisely the same thing in terms of **set theory**. What are all the things that the concept "frogs" applies to? The set of all frogs. Here's where visualization comes in. Try to imagine all the frogs, past, present, and future, gathered into one corral. What that corral contains is the extension of the concept "frog": the set of all frogs. What is the extension of the concept of "idea"? It's the set of all ideas—past, present, and future. Can you form a real physical corral of those? No, but you can form a conceptual corral; you can gather them in thought.

Connotation and Denotation

- There is a philosophical distinction that is important at this point: a distinction between the **connotation** and the **denotation** of a concept. That distinction has its roots in the medieval **Scholastics**, but it appears with particular prominence in the work of John Stuart Mill.

- Denotation concerns the things that a concept applies to—its extension. That's different from the connotation of words or concepts—their particular emotional associations.

- Here's a concept for you: your grandmother's cookies. The denotation of that concept is its extension—the cookies themselves, whether you liked them or not. But such phrases as "grandmother's fresh-baked cookies" or "grandmother's delicious fresh-baked cookies" clearly come loaded with a positive spin. That emotional take, whether positive or negative, is the connotation.

- Denotation is crucial to solid **rational argument**, reasoning in terms of facts about the things themselves. Connotation is the stuff of emotional persuasion, which we know can be far from rational. For that reason alone it's important to be able to separate the two.

Concept Trees: A Visualization Tool

- So far we've been talking about the organization of concepts and **categories** two at a time. But the relationships we're after form much larger structures, as well. These can be visualized in **concept trees**.

- There is an analogy between concept trees and family trees. If you drew a family tree for the descendants of Davy Crockett, you might have Davy as the trunk, his children as the first level of branches, their children as branches off those, all the way down to the twigs: the living descendants of Davy Crockett. There are hundreds of twigs.

- A concept tree looks exactly like that, except the trunk will be the most general of a set of categories. The first branch will be a first division of subcategories. The next set of branches will be subdivisions of those—all the way down to the twigs.

Propositions: The Idea behind the Words
- Words express concepts. But you can put words together to make sentences. And what sentences express are propositions.

- A **proposition** is a claim, a statement, or an assertion. It's what you mean when you say something—not the sentence you utter, not the sounds you make, but the message you're trying to get across. Propositions are the information packets that we use **sentences** to express.

- Remember that concepts aren't just words—they're something "behind" the words: what the words are used to convey. Propositions aren't just sentences. They're something "behind" the sentences. They're something extra-linguistic that we use language to convey.

- Seeing through sentences to the propositions behind them is hard for some people. What we want to know is whether a sentence asserts something. Consider this sentence: "The Surgeon General says that smoking is bad for your health." Does that sentence assert that smoking is bad for your health? No, it merely reports what the Surgeon General says.

- Here's why this is important: In everything we do, it is what is behind the words that matters, not the words themselves. It is the ideas that matter, not the words we use for those ideas. It is the informational content of sentences that matters, not the sentences themselves. When we talk about a solid argument, or a rational hypothesis, or a valid line of thought, it is the propositions involved that are the focus of attention.

Categorical Propositions and Venn Diagrams

- Concepts go together to make propositions in multiple ways, but we can look at some simple cases, as Aristotle did when he invented logic. These simple cases are called **categorical propositions** because they combine two categories. Here are some examples:
  - No frogs are philosophers.

  - All neurologists are MDs.

  - Some laws are unjust.

  - Some inventions are not patentable.

- The first thing to notice is that all four of these propositions link just two categories. Categories can be visualized as the circles of **Venn diagrams**. That visualization technique scales up to diagrams not only of concepts but also of their combinations. It is a diagram in logical space.

- All four of these propositions can be represented with overlapping circles for the two categories involved. Draw two overlapping circles. For the first proposition, mark the left circle "frogs" and the right one "philosophers." That will give you three internal spaces: an area of overlap, the frogs outside that overlap, and the philosophers outside that overlap. Logicians speak of "logical space." This is a diagram in logical space.

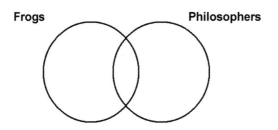

- A philosopher frog would be in the middle overlap area, a member of both circles. But the proposition tells us that there aren't any frog philosophers. It tells us that the central area is empty. Black out the overlap, and you have a visualization of how those two concepts go together in that proposition.

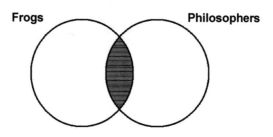

- Consider the second proposition: "All neurologists are MDs." Here again, we start with overlapping circles, labeled "Neurologists" and "MDs." The proposition is that all neurologists are MDs, which tells us that one of the three spaces is empty: There are no neurologists outside the MD corral.

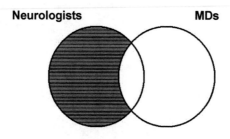

- The other two propositions give us positive information rather than negative.

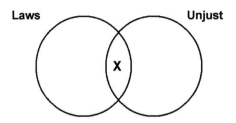

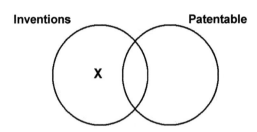

- With Venn diagrams, we can go from concepts to categorical propositions. Those will come in handy a few lectures down the line. Aristotle was the real inventor of logic, a logic that dominated Western philosophy for 2,000 years. It turns out that we can outline all the logic of Aristotle using the visualization you have at your fingertips.

Thinking Better: The Pitfalls of Categorization
- Categorization is an essential tool of thought, but—like other tools—we have to be careful how we use it.

- We group real things in our categories, but our categories are of our own making. We group things together in terms of what we take to be important similarities. But we could be dead wrong about what is really important.

- Categorization can also fail for ethical reasons. Sometimes, it's just plain wrong to think of people in terms of the groups they belong to rather than as individuals in their own right.

categorical proposition: In Aristotelian logic, a simple proposition that combines two categories using "all are," "none are," "some are," or "some are not." Categorical propositions can be visualized using two circles in a Venn diagram.

category: Any group of related things; the grouping is based on what are perceived as important similarities between those things.

concepts: Ideas, the basic elements or "atoms" of thought, as distinct from the words that represent those ideas.

concept tree: The hierarchical structure that visualizes the relationships within a set of related concepts.

connotation: The emotional tone or "flavor" associated with the ideas or things that words label.

denotation: The things that a concept or word applies to.

extension: In philosophy, the set or totality of things that a concept applies to.

proposition: A claim, statement, or assertion; the message or meaning behind the words in a written or spoken sentence; the information a sentence expresses or conveys.

rational argument: A way of presenting and supporting claims that relies on logical transition from premises to conclusion.

Scholastics: Philosophers practicing the dominant Western Christian philosophy of the European Middle Ages.

set theory: The branch of mathematics that studies the properties of sets and their interrelations—abstract properties of collections, regardless of what they contain.

sentence: A sentence is a series of words, spoken or written, that expresses a claim, statement, or assertion.

Venn diagram: A way of visualizing the relations between concepts and their extensions through the use of overlapping circles.

word: A linguistic representation of an idea; as distinct from concept.

Suggested Reading

Aristotle, *The Organon, or Logical Treatises, of Aristotle*; *On Interpretation*; *Categories*.

Carroll (Bartley, ed.), *Lewis Carroll's Symbolic Logic*.

Kelley, *The Art of Reasoning*.

Questions to Consider

1. In one of the last exercises in the lecture, you were asked to sort these words into three groups:

breaded interior facades instinct brains fondue

How did you sort them? Why do you think you sorted them that way?

2. Give an example of conceptual categories that you find useful in your everyday thinking. Give an example of conceptual categories that you think are socially harmful.

3. The contemporary philosopher David Lewis may have been following Plato in the *Phaedrus* in talking of concepts "cutting nature at the joints." How can we tell whether our concepts are really cutting nature at the joints or whether they're just perpetrating our own misconceptions of nature's joints?

Paradoxes are often tied to particular categories. Here is one:

> Some words apply to themselves. They "are what they say." "Short" is a short word, for example, "old" is an old word, and "English" is an English word. We'll call all the words like that "autological."

> Some words do not apply to themselves. They "are not what they say." "Long" is not a long word, for example, "new" is not a new word, and "German" is not a German word. We'll call all the words like that "heterological."

> Every word must either apply to itself or not do so. Thus, it looks like all the words divide between the autological and the heterological: the words that apply to themselves and those that don't.

The paradox arises with this question about the concepts just introduced: Is "heterological" heterological?

> If it is, then it looks like "heterological" "is what it says." But then it's autological, not heterological.

> Thus, it looks like "heterological" must not be heterological. But then it "isn't what it says." That's just what "heterological" means. Thus, if "heterological" is not heterological, it is heterological.

Do you think that this paradox has a solution? If so, what? If not, why not?

Visualizing Concepts and Propositions
Lecture 4—Transcript

Professor Grim: I have talked about the power of visualization. In this lecture I want to put visualization to work in examining some of the fundamental elements of all reasoning. You certainly know about atoms, and you can envisage them binding into molecules. The atoms of reasoning are concepts, the topic of this lecture. By the end we'll be talking about how those atom-concepts bind together into the molecules of reasoning, propositions. Concepts are the most basic elements in all thinking. Concepts are what we think with.

In *The Philosophical Investigations* the 20th century philosopher Ludwig Wittgenstein says, "The aspects of things that are most important for us are hidden because of their simplicity and familiarity. (One is unable to notice something — because it is always before one's eyes.)" Something like that applies here. In this lecture I want to emphasize some familiar facts and simple visualization techniques regarding the basic elements of thought, simple, but fundamental. We will rely on these facts and reach for these techniques repeatedly in further lectures.

Let's begin with concepts. Everyone knows that concepts are ideas and that words represent those ideas. But people tend to think that concepts just float around in a random sea. They think of the universe of concepts is something like this: Justice, presidential power, integrity, democracy, self determination, convenience, economics, mercy, beauty, efficiency. "Justice" is a concept, "presidential power" is a concept. "Self-determination" is a concept. Each of these is a word, each word represents a concept, and the concepts are just concepts. All about the same. You can think of concepts as either things in your brain—your concepts—or as something we share as part of our culture—our concepts. But it's still true that many people think of them as just flotsam and jetsam, floating randomly in a mental or cultural sea. The fundamental fact I want to emphasize is that concepts do not float around independently. Physical atoms are organized by physical structure. The physical structure represented in the periodic table. As the atoms of thought, concepts come with their own inherent organizational structure too. Crucial to all effective thinking is an ability to visualize that inherent conceptual structure and to put it to use.

What you actually heard a few minutes ago were words: Justice, presidential power, self determination. The words themselves are just combinations of sounds, or if written, certain combinations of marks. The concepts lie behind those words, not the words we use, but the ideas we express with those words. Is there a difference between words and concepts? Absolutely. The words are different from language to language; the concepts are not. It's "frog" in English but "frosch" in German and "sapo" in Portuguese and "grenouille" in French. Different words, but just one concept.

The recognition that concepts are distinct from words has been part of the Philosophical Toolkit at least since Aristotle, in 350 BC. The Greeks supposedly called other people barbarians including Egyptians, Etruscans, and Persians because they sounded like they were babbling, "bar bar bar." That makes the beginning passages of Aristotle's *On Interpretation* particularly remarkable. What Aristotle says is that words are symbols or signs for "affections or impressions of the soul." Speech is not the same for all races of men. But "the mental affections themselves, of which these words are the signs," Aristotle says, "are the same for all mankind." We use words to express concepts, but words are different from the concepts they express. Simple as it is, that point turns out to be crucial in building good reasoning, in detecting bad reasoning, and in reasoning effectively together.

Here's why. Because concepts and words are different, a single concept may be expressed by many different words. Even more importantly, very different concepts may lie behind the same word. Consider the word "bank," for example, and listen to the following story:

> The plane's engine suddenly cut out. The pilot put his plane into a bank in order not to crash into the town below. He guided his plane safely to the river bank so as not to hit the bank on Main street.

The word "bank" appeared three times in that story. You had no trouble in grasping the different concept behind the same word each time. This is a harmless example. But there are a range of debates in which different concepts may lie behind a different word. Consider the word "fairness." In a particular case we may have to ask whether "fair" is being used to mean to all alike, or according to prior agreement, or to each according to his need, or to each according to his investment, to each according to his effort, or to each according to some standard of merit. Because different concepts may lie behind the same word, we may be talking past each other unless we identify the particular concept we have in mind by a particular word. "Equality" and "liberty" and "terrorism" are important words but only because of the ideas behind them. Those are important ideas. All the more reason to be sure we can identify the specific concept at issue.

Concepts don't just float around in a random sea. They come with their own relationships, an inherent structure between them. Here are some questions, a simple example. Are frogs a kind of vertebrate, or are vertebrates a kind of frog? Are mammals a kind of squirrel, or are squirrels a kind of mammal? Easy. Frogs are a kind of vertebrate. It's not that vertebrates are a kind of frog. Mammals aren't a kind of squirrel; squirrels are a kind of mammal. Those concepts have a kind of logical order, a relational structure.

Words apply to concepts. Concepts apply to what? If we use the word "things" in full generality, recognizing not only physical things but things like ideas, what concepts apply to are things. The concept of "idea" applies to ideas. The concept of "frog" applies to frogs. What we're talking about, then, is really a three-part relationship: The relationship between words, concepts, and the things those concepts apply to. Words express concepts. Concepts apply to things. We might say that words apply to things too, second hand. Words express the concepts that apply to things. Philosophers speak of the totality of things that a concept applies to as its extension. What's the extension of a concept? Just all the things the word applies to. We can express precisely the same thing in terms of set theory.

What are all the things that the concept "frogs" applies to? The set of all frogs. Here's where visualization comes in. Try to imagine all the frogs, past, present, and future, gathered into one corral. What that corral contains is the extension of the concept "frog": the set of all frogs. In the same way, the extension of the concept of "vertebrates" is a corral with all vertebrates in it. The extension of the concept of "idea"? Well, that's an interesting one. It's the set of all ideas, past, present, and future. Can you form a real physical corral of any of these things? No. But you can form a conceptual corral. You can gather them in thought.

Venn diagrams are a way of visualizing the relations between concepts in terms of their extensions. Think of a corral for frogs—a corral with all the things the concept applies to. Now, without disturbing that first corral, think of another corral for vertebrates, inside of which are all the vertebrates—not only frogs, but sharks, and squirrels, and you. The only way to conceptualize that, to build a corral for vertebrates without disturbing the one you've already built for frogs, is to have one corral inside the other. Your original frog corral will be inside the larger vertebrates corral. One circle inside another. All squirrels are mammals? There will be a squirrel circle inside a larger mammal circle. All amphibians are animals? One circle inside another. We can think of what we're talking about as sets and subsets. The set of frogs is a subset of the set of vertebrates, or as the extension of one concept including the extension of another. or as the scope of one concept including the scope of another. Those all come down to the same relationship, and that relationship is captured by internal circles in a Venn diagram.

Aristotle approached the relations of concepts in terms of categories and sub-categories, genus and species. Today we think of genus and species as terms in biology. But in Aristotle the idea is much more general. Some categories include smaller and more specific categories within them. A larger inclusive category is the genus. The smaller categories inside—specified by further differentia—are the species. That idea, too, amounts to circles inside circles, just like the circles in Venn diagrams. The diagrams are named after John Venn, a 19th century logician. There was another logician about the same time who came up with a similar idea, though he used squares. That other logician was Charles Dodgson. Charles Dodgson may not be a familiar name to you, but you already know quite a bit about him. In addition to his work as a mathematical logician, Dodgson wrote peculiar fantasy stories for a little girl named Alice. He used the pseudonym Lewis Carroll.

There is a philosophical distinction that is important at this point: a distinction between the connotation and the denotation of a concept. That distinction has its roots in the medieval Scholastics, but it appears with particular prominence in the work of John Stuart Mill. Denotation concerns the things that a concept applies to, its extension, just the things themselves. That's different from the connotation of words or concepts, the fact that concepts often label things with a particular flavor or emotional association. So here's a concept for you: your grandmother's cookies. The denotation of that concept is its extension—the cookies themselves, whether you liked them or not. But phrases like "grandmother's cookies," let alone "grandmother's fresh-baked cookies," let alone "grandmother's delicious fresh-baked cookies" clearly come loaded with a positive spin. That emotional take, whether positive or negative, is the connotation. Denotation is crucial to solid, rational argument, reasoning in terms of facts about the things themselves. Connotation is the stuff of emotional persuasion, which we know can be far from rational. For that reason alone it's important to be able to separate the two.

So far we've been talking about the organization of concepts and categories two at a time. But the relationships we're after form much larger structures as well. A second form of visualization takes us farther: visualization in terms of concept trees. There is an analogy between concept trees and family trees. If you drew a family tree for the descendants of Davy Crockett, you might have Davy as the trunk, his children as the first level of branches, their children as branches off of those, all the way down to the twigs: the living descendants of Davy Crockett. There are hundreds of twigs. A concept tree looks exactly like that, except the trunk will be the most general of a set of categories. The first branch will be a first division of sub-categories. The next set of branches will be a sub-divisions of those, all the way down to the twigs.

If you're in a situation right now in which you can work with a pencil in hand, that's the best. If you're in a situation in which you can't do that, you'll have to do it in your mind. If you can imagine drawing the concept tree, you'll be doing exactly what I'm after. Try drawing a concept tree for this set of concepts: amphibians, vertebrates, squirrels, frogs, tree frogs, mammals, reptiles, turtles, snakes. Which of these concepts is the trunk of the tree? That means, which is the most general? Which is everything else a kind of? That's right. The trunk is going to be vertebrates—animals with backbones. All the rest are kinds of vertebrates. What is the next most general level? At that point it will branch into three: mammals, reptiles, and amphibians. Each is a type of vertebrate, none is a type of the other. Then on to the next level.

You can test any concept tree for accuracy by starting at any twig and going back to the trunk. It should work like this: tree frogs are a kind of frog, which is a kind of amphibian, which is a kind of vertebrate. If that works for every branch, you've captured the structure of that set of concepts as a tree diagram. Any set of related concepts can be visualized that way. Here is another one to work out on your own. What is the relationship between these concepts? Western religions, Christianity, Buddhism, religions, Eastern religions, Judaism, Protestantism, Presbyterianism, Confucianism, Zen Buddhism, Orthodox Judaism. Can you see that concept tree form as I say the words? If so, you're exercising precisely the kind of conceptual visualization I'm after. Western religions, Christianity, Buddhism, religions, Eastern religions, Judaism, Protestantism, Presbyterianism, Confucianism, Zen Buddhism, Orthodox Judaism. Our concepts aren't just flotsam in a mental or cultural sea. They come with inherent relationships that can be visualized in terms of Venn diagrams and concept trees.

Concepts or categories are just the atoms of thought. Rationality really operates on the next level up, the molecules of thought. Those are called propositions. Let's start again with words. Words express concepts. But you can put words together to make sentences. And what sentences express are— There's no ordinary word for what it is that sentences express. We'll use a not-so-ordinary word instead. The philosopher's word is propositions. The basic idea is easy to get, but it can be hard to explain. A proposition is a claim, or a statement, or an assertion. It's what you say when you say something—not the sentence you utter, not the sounds you make, but the message you're trying to get across. Propositions are the information packets that we use sentences to express. Remember, concepts aren't just words; they're something behind the words, what the words are used to convey. Propositions aren't just sentences. They're something behind the sentences. They're something extra-linguistic that we use language to convey.

Words and concepts don't map onto each other one to one. That was the lesson of the story about banks. Sentences and propositions don't map onto each other one to one either. Here's a sentence you might hear on the news. I want you to think beyond the sentence to the proposition being expressed: "The CIA has succeeded in killing the leader of a major terrorist organization." Here's a different sentence. Different words, different order, different sentence: "The head of one of the major terrorist organizations has been killed by the CIA." Do those two sentences express different propositions? Not unless you're extremely nitpicky. One sentence is in the active voice. One sentence is in the passive. One uses "leader" instead of "head of," "a major terrorist organization" instead of "one of the major." But the message is the same. It's the same packet of information—the same proposition—conveyed by two different sentences.

How about these two: "The deed was done in the billiard room with the rope by Colonel Mustard." "Colonel Mustard did it in the billiard room with the rope." In the game Clue, those two sentences give precisely the same information. If you'd win with one, you'd win with the other. So different sentences can express the same proposition. It's also true that the same sentence can express different propositions. Suppose, for example, that we have a tape recording of you yesterday. When we play it back, we hear this sentence: "I have a horrible headache." We also have a tape recording of your best friend three weeks ago. When we play it back, we hear this sentence: "I have a horrible headache." Those are precisely the same sentence, word for word. But it's obvious that they're saying different things. One of them is about you. One of them is about your friend. One of them is about you yesterday. One is about your friend three weeks ago. Different people. Different times. Different headaches. The same sentence, used to express different propositions in different contexts.

Seeing through sentences to the propositions behind them is hard for some people. What we want to know is whether a sentence asserts something. So consider this one: "The Surgeon General says that smoking is bad for your health." Does that sentence assert that smoking is bad for your health? It may say other things too. We want to know whether part of what it asserts is that smoking is bad for your health. No. That sentence just reports what the Surgeon General says. The phrase "smoking is bad your health" appears there, but the sentence doesn't say that smoking is bad for your health. That's not part of what the sentence itself asserts. It just says that's what the

Surgeon General says. You can't tell whether a sentence asserts that smoking is bad for your health by just trawling for that set of words.

You have to understand what the sentence really says in order to know what proposition is being expressed. So listen to the following sentences and tell me which of these assert that smoking causes cancer? (1) If smoking causes cancer, it is bad for your health. (2) Since smoking causes cancer, it's bad for your health. (3) Although smoking causes cancer, many people continue to smoke. If you said (2) and (3), you're right. With "since" and "although", (2) and (3) both commit themselves to the claim that smoking causes cancer. (1) says smoking is bad for your health if it causes cancer, but it doesn't commit itself to that "if." That's not part of what it asserts. How about these? "Either smoking causes cancer or car exhaust causes cancer." And this one, "Smoking and car exhaust both cause cancer." The first one is an either/or. It doesn't commit to one side in particular. The other commits to both sides. Among other things, then, it does assert that smoking causes cancer. Here's why this important: In everything we do, it's going to be what is behind the words that matters, not the words themselves. It is the ideas that will matter, not the words we use for those ideas. It will be the informational content of sentences that matter, not the sentences themselves. So when we talk about a solid argument, or a rational hypothesis, or a valid line of thought, it is going to be the propositions involved that are the focus of attention.

Concepts are the atoms of all logic. Propositions are the molecules. How do the atoms go together to make those molecules? How do concepts go together to make propositions? Let me count the ways. I give up, too many. But we can start with some simple cases. These are the simple cases that Aristotle started with when he invented logic. They're called categorical propositions because they combine two categories. Consider propositions that look like these: "No frogs are philosophers," "All neurologists are MDs," "Some laws are unjust," "Some inventions are not patentable." The first thing to notice is that all four of these propositions link just two categories. We know that categories can be visualized as the circles of Venn diagrams. That visualization technique scales up to diagrams not merely of concepts but of their combination. All four of these propositions can be represented with overlapping circles for the two categories involved. Draw two overlapping circles. For the first proposition, mark the left circle "frogs" and the right one "philosophers." That will give you three internal spaces, an area of overlap, the frogs outside that overlap, and the philosophers outside that overlap. Logicians speak of logical space. This is a diagram in logical space.

What that first proposition is telling us is that no frogs are philosophers. That means that one of those spaces is empty. Which one? No frogs are philosophers. A philosopher frog would be in that middle overlap area, a member of both circles. The proposition is telling us that there aren't any frog philosophers. It's telling us that that central area is empty. Black out the overlap and you have a visualization of how those two concepts go together in that proposition.

Take the second proposition: All neurologists are MDs. Here, again, we start with overlapping circles labeled "neurologists" and "MDs." Here, in this case, the proposition is that all neurologists are MDs. Here, again, it's telling us that one of those three spaces is empty. Which one? That's right. It's telling us that the leftmost space is empty, so scratch it out. All neurologists are MDs—there are no neurologists outside the MD corral.

The other two propositions are different. They're giving us positive information rather than negative. They're telling us that something is in a space. So draw overlapping circles for "laws" and "unjust." The third proposition is that some laws are unjust. That means there's something in the overlap. So put an x there. That just means that there is something in that middle area; there's some unjust law. Finally, "some inventions are not patentable." You figure that one out.

With Venn diagrams, we can go from conceptual atoms to categorical propositions. Those will come in handy a few lectures down the line. Aristotle was the real inventor of logic, a logic that dominated Western philosophy for 2,000 years. It turns out that we can outline all the logic of Aristotle using the visualization you have at your fingertips.

Throughout this lecture we've been talking about categories. Thinking probably couldn't occur at all without that basic logical tool. These days most tools come with warnings, and that applies to logical tools too. So here are some warnings regarding categories: It's real things that we group in our categories, but our categories are of our own making. We group things together in terms of what we take to be important similarities. But we could be dead wrong about what's really important. There is no guarantee that our current concepts "cut nature at its joints," in the wonderful phrase of the philosopher David Lewis.

Consider the Linnaeus system of biological taxonomy first set down by Carl Linnaeus in his *Systema Naturae* of 1735. That system divides all living things into phylums and classes, from there into orders, families, genus and species. It's built on eyeball judgments of anatomical similarity. If what we're interested in is the pattern of evolution, anatomical similarity may not always be a very good indicator. Today we can use DNA to track the point at which two species shared a common ancestor. In those terms, some species are more closely related than Linnaeus thought. Some species aren't so closely related. It turns out that the Linnaean categories didn't always cut nature at its joints.

I started with a random handful of words. Here is another random handful. I want you to take a minute and sort these into three groups. The words are breaded, interior, facades, instinct, brains, fondue. Press pause, sort those into three groups, and come back. So how did you categorize those words? Maybe like this, the b words: breaded, brains. the f words: facades, fondue. the i words: instinct, interior. But then again, maybe you sorted them like this, the food words: breaded, fondue, the architectural words: facades, interior, the mind words: brain, instinct. You might have sorted them like this, the six-letter words: brains, fondue. The seven-letter words: breaded, facades. The eight-letter words: interior, instinct. Which is the right way? There is no right way. Or better, there are lots of right ways. Which is the right way depends on the purpose of the categorization. If you are compiling a dictionary, you want the first way, the b words, the f words. If you're compiling a thesaurus, you want the second way, the food words, the brain words. If you're worried about message length, you want something more like the third.

Here is another warning. Categorization can fail for ethical reasons. We are all familiar with the ethnic and racial stereotypes that go hand in hand with discrimination and injustice. Those stereotypes generally come with a heavily negative connotation—with broad negative emotional implications. But there are also times when categorization itself may be unethical. Sometimes it's just plain wrong to think of people in terms of the groups they belong to rather than as individuals in their own right. Categorization is an essential tool of thought, but like other tools we have to be careful how we use it.

What I want you to take away from this lecture are two fundamental ideas. We we'll be using those ideas throughout the lectures that follow. First, the idea of concepts behind our words—the atoms of thought. Those concepts bring with them their own inherent structural relationships. A simple way of visualizing those relationships is Venn diagrams—sub-categories within larger categories, sub-sets within larger sets. An expanded visualization is in terms of concept trees, hierarchical organizations of categories and sub-categories, sets and sub-sets. Second, the idea of propositions behind our sentences. These are the molecules of thought. The stuff of evidence and deliberation, the working material of all rational argument. Categorical propositions can also be visualized in Venn diagrams. We'll use that visualization later as an easy in to all the logic of Aristotle. Before that, however, we'll explore the power of thought experiments.

The Power of Thought Experiments
Lecture 5

W e often think of the world of imagination as something different from the world of reality: an escape into a purely mental realm of our own creation. But really understanding what is happening around us—rather than just watching it—often requires thought that extends beyond the real into contemplation of what could be, or ought to be, or even what cannot be. This lecture presents some famous thought experiments as illustrations for more general strategies of effective problem solving. What these strategies have in common is the attempt to solve real problems with something distinctly unreal: the imaginative world of thought experiments.

Going to Extremes

- Here's a problem-solving example that emphasizes an extension of the visualization strategy: "going to extremes."

- There are two flagpoles, each 100 feet high. Tied between the tops of those poles is a 180-foot rope that dangles 10 feet off the ground at its lowest point. How far apart are the two flagpoles?

- Consider the extremes. What is the maximum distance the two flagpoles could be apart? The answer is 180 feet, because there is a 180-foot rope between them. At that extreme, the rope would be stretched taught, 100 feet from the ground at every point.

- How about the other extreme: the minimum distance apart? The two flagpoles could be right next to each other. The 180-foot rope would just dangle down, 90 feet on each side, 10 feet off the ground. We've found the answer by going to extremes: The two flagpoles have to be right next to each other.

Generalizations and the Counter-Case

- What we want in our theories—in both science and philosophy— are **universal generalizations**: natural laws that apply in all circumstances.

- If true, generalizations are tremendously powerful. Their Achilles' heel, however, is that it takes only one counter-case to refute them. If we say that something holds true for all of a class, it will take just one exception to prove us wrong.

- In a physical experiment, Galileo dropped a single cannonball and small musket ball from the top of the Leaning Tower of Pisa. The physics of the time was Aristotle's physics, and what Aristotle had said was that heavy objects will fall faster.

- But when Galileo dropped the cannonball and musket ball, they hit at the same time. All it took was a single experiment, one simple case, and Aristotle's physics was soundly refuted.

- Although Galileo's was a physical experiment, he also offered a pure thought experiment for the same conclusion without actually dropping cannonballs: If we tie an 8-pound cannonball to a 4-pound cannonball and drop them, what will happen?

- Aristotle says that the 4-pounder will travel slower. It seems that it will hold back the 8-pound cannonball. On this reasoning, the two cannonballs tied together should move slower than the 8-pound cannonball alone. But tied together, the two cannonballs make a 12-pound weight; thus, the two together should move faster than the 8-pound cannonball alone.

- Aristotle can't have it both ways. Without dropping anything—with just a thought experiment—Galileo put all of Aristotelian physics into doubt.

$E = mc^2$ Thought Experiment

- Next to philosophers, those who love thought experiments most are physicists. Maxwell's Demon was a thought experiment that played a major role in the history of **thermodynamics**. Schrödinger's Cat was a thought experiment central to the debate over **quantum mechanics**.

- Here's a thought experiment that was crucial in the development of Einstein's theory of relativity. Just as he was a master of visualization, Einstein was a master of the thought experiment.

- Suppose a train car is moving along a track, with Mike in the middle of the car. He passes Fred, who is standing on the platform. We assume that the speed of light is a constant. Bolts of lightning hit both ends of the train car. The light from the two bolts reaches Fred at the same time. He sees them hit simultaneously.

- What does Mike see, in the middle of the car? The car is moving toward one lightning bolt, running up to and meeting the light coming from it. Thus, Mike, in the middle of the car, will intersect light from that bolt first. The car is running away from the other bolt, adding distance for the light from it to travel. That light won't reach Mike until a little later. From Mike's perspective in the moving train, the bolt at the front of the train hits before the one at the back.

- Einstein's theory of relativity is, in large part, a series of deductions from the assumption that some things are not relative. What Einstein's thought experiment shows is that if the speed of light is not relative, observable simultaneity—whether two things are seen as happening at the same time—will be.

- Einstein also followed a firm methodological principle: that theories must be written in terms of what is observable. When you rewrite this thought experiment purely in terms of what is observable, you can see how time and simultaneity will be relative to the comparative motion of two observers. That is the intuition at the core of Einstein's theory of special relativity.

Simplify, Simplify, Simplify

- If the problem you face seems resistant to your best attempts, think of a problem that's like it but simpler. That's the strategy of analogy.

- Some problems are difficult simply because they have many parts. For those, a strategy of divide and conquer can prove effective. Solve the parts independently, and then put them back together again. In this way, you turn a complex solution for a complex problem into simple solutions for simple ones.

- Both strategies can be illustrated with the classic Tower of Hanoi problem.
  - Imagine that you have three posts in a row. The right and middle posts are empty. On the left post is stacked a series of six disks, ordered from largest at the bottom to smallest at the top. Your task is to end up with all six disks, in the same order as they are now, on either of the two other poles.

  - But there are restrictions. You can move only one disk at a time. You can move a disk only from one pole to another—you can't just put it to the side, for example, or hold it in your hand. And you can never put a larger disk on top of a smaller one. How would you proceed? How many moves do you think it would take you to get all six disks from one pole to another?

  - First, try visualization; you want to be able to see the problem. Draw a sketch of the three poles and the six disks. It's even better if you can manipulate a three-dimensional version. All you really need are six disks, each smaller than the one before it. Your three posts can be three spots on a table in front of you.

- If a complex problem has you blocked, try an analogous problem in a simpler version. What if the problem was about just two stacked disks or four stacked disks?

- Do you see the pattern? In your solution, the problem with two disks took 3 moves. The three-disk problem took that many moves, plus moving a bigger disk, plus that many moves again: 7 moves. The problem with four disks was really the three-disk solution, plus moving the largest, plus the three-disk solution: 15 moves.

- Scale that up and you can see the pattern for six disks. The minimum number of steps for five disks is the number of steps to move four, plus a move for the fifth one, plus the number of steps to move the four on top of it again: 31. Six disks will be that number of moves, plus one, plus the number of moves again: 63.

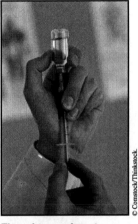

© Comstock/Thinkstock.

A Thought Experiment in the Ethical Realm

- Not all thought experiments are out to prove something or to find the right answer. Sometimes, thought experiments are explorations in a very different sense. Here is a thought experiment in the ethical realm: the Runaway Trolley.

Thought experiments can be imaginative explorations in any number of fields; in epidemiology, for example: If you didn't have enough vaccine for the whole population, where would it do the most good?

- A trolley is running out of control down a track. The bad news is that it is hurtling toward five people tied to the track by a mad philosopher. The good news is that you are standing next to a rail switch. All you have to do is pull the lever and the trolley will go down a second track instead. The bad news is that the mad philosopher has also tied one person to the track before it gets to the divider.

- What do you do? Do you do nothing, allowing the five to die? Or do you pull the switch to save them, thereby sacrificing the one?

- This thought experiment was originally presented by the philosopher Philippa Foot. It was designed merely to point up differences in ethical theory. It turns out that it may also reveal differences in the way different people approach ethical problems. In ethics, utilitarians hold that the right thing to do is that which results in "the greatest good for the greatest number." They would clearly pull the switch, sacrificing one in order to save five. What would you do?

Terms to Know

quantum mechanics: Developed early in the 20th century, quantum mechanics is a sophisticated theory of physics at the subatomic scale; a mathematically elegant, empirically established but philosophically puzzling theory of the very small.

thermodynamics: A branch of physics that is concerned with the relationships of heat, pressure, and energy.

universal generalization: A statement about "all" of a particular class: "All X's are Y's."

Suggested Reading

Baggini, *The Pig That Wants to Be Eaten.*

Einstein (Harris, trans.), *The World As I See It.*

Sorensen, *Thought Experiments.*

1. What were your answers in the Runaway Trolley case?

   (a) The trolley is rushing toward five victims on one track. By pulling the switch, you can divert it toward one victim on another track. Would you do so? Do you think you should?

   (b) The trolley is rushing toward five victims on one track. By pulling the switch, you can divert it toward one victim on the other track, but that victim is your daughter. Would you pull the switch? Do you think you should?

   (c) The trolley is rushing toward five victims on the track. You can stop it, but only by pushing the fat man next to you onto the tracks. Would you? Should you?

2. Frank Jackson's black-and-white Mary thought experiment goes like this:

   (a) Mary lives in a black-and-white world, but through her black-and-white reading has learned all that science has to teach regarding color vision.

   (b) One day, she walks into the world of color and suddenly learns something new: what red *looks* like.

Jackson concludes that there are facts of phenomenal consciousness—such as what red looks like—that are beyond the reach of science.

What do you think of Jackson's thought experiment? Does it prove his conclusion?

3. How can we explain the power of thought experiments? How can it be that thinking about something wildly hypothetical tells us something about the reality around us?

Here is a famous thought experiment regarding perception. It was sent in a letter from William Molyneux to John Locke on July 7, 1688, and is considered in Locke's *An Essay Concerning Human Understanding*. The question is made more poignant by the fact that Molyneux's wife had gone blind during the first year of their marriage:

Suppose a person born blind learns to feel the differences between shapes, such as spheres and cubes. If his sight were suddenly restored, would he be able to tell without touching them which were the spheres and which were the cubes?

What do you think?

The Power of Thought Experiments
Lecture 5—Transcript

Professor Grim: We often think of the world of imagination as something different from the world of reality, an escape into a purely mental realm of our own creation. But in fact, imagination can offer a wonderful route into reality. Really understanding what's happening around us, rather than just watching it, often requires thought that extends beyond the real into contemplation of what could be, or ought to be, or even into what can't be.

In this lecture I'll use some famous thought experiments as illustrations for more general strategies of effective problem solving. What these strategies have in common is the attempt to solve real problems with something distinctly unreal—the imaginative world of thought experiments. In a previous lecture I talked about visualization and gave you a proof of the Pythagorean theorem with no math at all. Later in this lecture I'll outline an Einsteinian thought experiment that lies at the core of relativistic physics, again with no math.

I'll start with an example that emphasizes an extension of the visualization strategy. It's the strategy of going to extremes. How do you solve a jigsaw puzzle? Most people start at the edges. They go to the extremes because they're easier. Once they know the layout there, all the rest becomes easier. So try going to extremes in this example: There are two flagpoles, each 100-feet high. Tied between the tops of those poles is a 180-foot rope, which dangles 10 feet off the ground at its lowest point. How far apart are the two flagpoles? If you draw a picture, it will be a picture of two flagpoles with a curved rope between them. At the bottom of the curved rope you'll mark that the bottom is 10 feet off the ground. But now how far apart are the two flagpoles? If you want to try the problem on your own, press pause and come back.

Some people might try to solve that problem by getting out a textbook and looking up parabolas. But there's a much easier way. Consider the extremes. What is the maximum distance the two flagpoles could be apart? One hundred and eighty feet, since there is a 180-foot rope between them. At that extreme the rope would be stretched taught, 100 feet from the ground at every point. How about the other extreme, the minimum distance apart? Well, they could be right next to each other. The 180-foot rope would just dangle down, 90 feet on each side, and 10 feet off the ground. We've found the answer by going to extremes: the two flagpoles have to be right next to each other.

Here's another one. We have an eyewitness to shoplifting in a coin shop. The witness says that the suspect had six pockets and that he saw him steal precisely 20 silver dollars. The report: "He left with exactly 20 silver dollars in his pockets, at least one dollar in each, and a different number in each pocket." Should we trust that eyewitness? Press pause if you want to think about it. We can use the strategy of going to extremes to examine the credibility of our eye witness. With a different number of dollars in each pocket, the least number of total dollars has to be, one in pocket one, two in pocket two, three in pocket three, four in pocket four. The least number of silver dollars that can be divided between six pockets that way is, therefore, $1 + 2 + 3 + 4 + 5 + 6$. That totals to 21 silver dollars, not 20. The suspect couldn't have divided only 20 dollars between six pockets in the way the witness said. He would have needed at least 21 silver dollars.

We just refuted a theory by going to extremes. The quickest way to refute a theory is often to find a counter example, and the quickest way to find a counter example is often to go to the extremes What we want in our theories—in both science and in philosophy—are universal generalizations, natural laws that apply in all circumstances. You can see how powerful universal generalizations are, if true. That's why we work so hard to try to get them. Their Achilles heel is that it takes only one counter case to refute them. If I say that something holds for all of a class, it will take just one exception to prove me wrong.

Galileo dropped a single cannonball and musket ball from the top of the Leaning Tower of Pisa. The physics of the time was Aristotle's physics, and what Aristotle had said was that heavy objects will fall faster. A universal law. That theory fits most everyday experience. Drop an egg and an empty eggshell off the top of a building and the egg will hit first. Even Galileo started out thinking Aristotle must be right. But when Galileo dropped a cannonball and a musket ball, they hit at the same time. All it took was a single experiment. One simple case and Aristotle's physics was soundly refuted.

Galileo's was a physical experiment. But Galileo also offered a pure thought experiment for the same conclusion, without actually dropping cannon balls. It went like this: Aristotle says that heavier objects fall faster than lighter objects. So take an eight-pound cannonball. If you drop that, it will be falling pretty fast. If you drop a four-pound cannon ball, Aristotle says it will travel slower. Here's the thought experiment. What will happen if you tie them together and drop both? Aristotle says that the four-pounder will be traveling slower. So it seems it will hold the eight-pound cannonball back, like a little kid holding back his big brother. On this reasoning, the two cannonballs tied together should move slower than the eight-pound cannonball alone. But wait a minute. Tied together, the two cannonballs make a 12-pound complex. So the two together should move faster than the eight-pound cannonball alone. Aristotle can't have it both ways. Without dropping anything—with just a thought experiment—Galileo put all of Aristotelian physics into doubt.

Someone makes a sweeping and general claim. The strategy is to look for a simple counter-example that might put that claim in doubt. Here is a famous philosophical thought experiment regarding the claims of artificial intelligence. In the flush of early success in artificial intelligence, Roger Schank at Stanford claimed that with the right program a computer could understand English. By instantiating the right computer program, he claimed, a machine could do precisely what we do. It could understand a natural language—English or Chinese. The philosopher John Searle calls that claim Strong AI, and he regards it as pure hype. His ammunition against it is a thought experiment. Searle builds a counter example in which Schank's program is running, in which any program you like is running, but no real understanding is going on. Schank had said that running the right program would amount to understanding. But if the program can be running when no understanding is going on, running the program can't amount to understanding. That's the strategy.

Here's the Chinese Room. Searle imagines himself as a man in a room with a very large rule book. The book has a column of input symbols on one side of page—they just look like squiggles to him, Searle says—with a set of output squiggles on the other. The book has rules for which output squiggles to use given certain input squiggles. The room has an in slot and an out slot. Pieces of paper with squiggles come in on one side. Searle consults the book, copies out another set of squiggles, and drops those through the out slot. Suppose, Searle says, that the rule book is Schank's ideal program. Unbeknownst to the man in the room, the squiggles coming in the door are Chinese questions. The squiggles he puts out the other door are Chinese answers. Given the right rule book, input and output would match inputs and outputs of any program that Schank might want to invent.

But wait a minute, Searle says. The rule book is just a rule book. It doesn't understand Chinese. The room is just a room, no understanding there, and the man is just following the rules. He doesn't understand Chinese. Schank's program is running, but nothing in the Chinese room really understands Chinese. Searle points out that what a computer does isn't anything more than the guy with the rule book does. Nothing in the thought experiment case deserves to be called understanding, so nothing in the computer case does either. Searle's thought experiment took the debate into new and important channels, but it didn't end it. If following internal rules isn't enough to understand a language, exactly what is it that we do beyond that?

Let me give you another philosophical thought experiment along the same lines, a single imaginative case intended to throw doubt on a grand general claim. This is Black-and-White Mary, it's a thought experiment from the Australian philosopher Frank Jackson. Jackson is out to show that there is something about consciousness that goes beyond the facts that science can give us. Mary is raised in a black and white world. Everything she sees is colored black, white, or some shade of gray, including her own image in a mirror, no color, just black and white. Despite that impoverished upbringing, Mary has become the world's leading expert on color vision. She knows all the scientific facts regarding color vision, derived, of course, from her extensive reading in black and white. One day the black and white door is opened, and Mary walks out into the realm of color. "Aha," she says. "That's what red looks like!" At that moment Mary has learned something new. She now knows what red looks like. But she already knew all the scientific facts regarding color vision. The simple fact that red looks like that must, therefore, be something beyond scientific fact—something that science alone can't give us. Or so Jackson argues. Black and White Mary is famous for channeling discussion in certain directions, but Jackson's thought experiment can't be said to have settled the argument any more than Searles's did.

Next to philosophers, those who love thought experiments most seem to be physicists. Maxwell's Demon was a thought experiment that played a major role in the history of thermodynamics. Shrödinger's cat was a thought experiment central to debates regarding quantum mechanics. Here is a thought experiment crucial in the development of Einstein's theory of relativity. Just as he was a master of visualization, Einstein was a master of the thought experiment. Suppose a train car is moving along a track, with Mike in the middle of the car. He passes Fred, who is standing on the platform. We assume that the speed of light is a constant. Bolts of lightning hit both ends of the train car. The light from the two bolts reaches Fred at the same time. He sees them hit simultaneously.

What does Mike see in the middle of the car? Here it's crucial that the car is moving. The car is moving toward one lightning bolt, running up to and meeting the light coming from it. So Mike in the middle of the car will intersect light from that bolt first. The car is running away from the other bolt, adding distance for the light from it to travel. That light won't reach Mike until a little later. From Mike's perspective in the moving train, the bolt at the front of the train hit before the one at the back. It's called the Theory of Relativity, but it is in large part a series of deductions from the assumption that some things are not relative. What Einstein's thought experiment shows is that if the speed of light is not relative, observable simultaneity—whether two things are seen as happening at the same time—will be relative.

Einstein also followed a firm methodological principle: that theories must be written in terms of what is observable. When you rewrite this thought experiment purely in terms of what is observable, you can see how time and simultaneity will be relative to the comparative motion of two observers. That is the intuition at the core of Einstein's Special Relativity.

Visualize. Go to Extremes. If you're out to refute a theory, look for a counter-example. In all the examples I gave you there is also something else at play. All of those involved using your imagination to simplify a problem. There are lots of ways to simplify. One is the strategy of analogy. If the problem you face seems resistant to your best attempts, think of a problem that's like it, but simpler. Some problems are hard just because they have a lot of parts. For those, a strategy of divide and conquer can prove effective. Solve the parts independently, then put them back together: a complex solution for a complex problem but made of simple solutions to simple ones.

Both strategies can be illustrated with the classic Tower of Hanoi problem. This is a hard one. You have three posts in a row. The right and middle posts are empty. On the leftmost post are stacked a series of six disks, ordered from largest at the bottom to smallest at the top. Your task is to end up with all six disks in the same order as they are now but on either of the two other poles. But there are restrictions. You can only move one disk at a time. You can only move a disk from one pole to another—you can't just put it to the side, for example. And you can never put a larger disk on top of a smaller one. So how would you proceed? How many moves do you think it would take you to get all six disks from one pole to another? Gosh, where should we start? Well, we can take the top disk and move it to pole number two. then the second disk has to go to pole number three because it can't go on top of a smaller disk. Now I guess that the one on number two has to go to three. Oh, it's all so confusing. I'm lost. Cool it. Use a few conceptual strategies.

First, visualize. You want to be able to see the problem. Maybe you were already visualizing the setup mentally as I described the problem. It's hard for me not to do that. Even better, draw a sketch of the three poles and the six disks. Then you can really see it and not merely in your mind's eye. The best option of all is to get a three-dimensional version. You could construct that yourself—all you need are six disks, each smaller than the one before it. You could see the problem in 3-D even if your poles were just three spots on a table in front of you. Each step, from mind's eye, to two-dimensional drawing, to the realm of three-dimensional manipulation is a giant step in visualization. We want to know how many moves are required to move six disks from one post to another inside the rules. One disk at a time from pole to pole, never a larger one on top of a smaller one.

The first strategy: Visualize. The second strategy: Simplify. If a complex problem has you blocked, try an analogous problem at a simpler extreme. It may not be what you were asked to do, but it may give you a clue as to how to do what you were asked to do. So, what if the problem wasn't about six disks; it was just about two disks? How many moves to transfer two disks from one post to another? That's easy. The top one goes to an empty pole, the bigger one goes to the other empty pole, the top one goes on top of it. That was three moves. But note that the top had to go somewhere that it didn't end up to allow the one underneath it to move. That is clearly going to be the case when we're dealing with more disks too. Smaller disks have to go someplace to wait until we move a larger disk underneath them.

So the two-disk problem is easy, three steps. Now suppose you had three disks to move from one pole to another; call them little, medium, and big. Well, little goes to pole number three, say. Medium goes to two. So they're laid out big, medium, little on the posts. Now, little can climb on top of medium there. That leaves this pole empty, and big can move there. That looks good. Now one can go to this pole, temporarily, letting two climb on top of three, and little can finally go on top of that. That wasn't so tough. If you count it, that was seven steps for moving three disks. Perhaps you can see a strategy in the simple cases that we can scale up to six disks. When we went from two disks two three, we really moved the top two like when we were doing just two disks, then transferred the big disk to the bottom, and then moved the top two again like when we were doing just two. I'll bet that doubling up of previous move patterns is going to happen with every additional disk.

What if we try four disks? We'll go through the moves for the top three like we just figured, seven steps, then move the super-big disk on the bottom, then repeat the shuffle of the top three. That will be seven moves, plus one, plus seven. so 15 moves for four disks. You see the pattern? The problem with two disks took three moves. The three-disk problem took that many moves, plus moving a bigger disk, plus that many moves again, seven steps. The problem with four disks was really the three-disk solution, plus moving super big, plus the three-disk solution. So for four disks it's seven steps, plus one, plus seven, giving us a total of 15 steps. Scale that up and you can see the pattern for six disks. The minimum number of steps for five disks is the number of steps to move four, plus a move for the fifth one, plus the number of steps to move the four on top of it again, $15 + 1 + 15 = 31$. Six disks is going to be that, plus one, plus that again. $31 + 1 + 31 = 63$ steps for six disks. Here are the strategies we used, visualization, then, going to extremes. In this case that just meant going to the simplest versions: two and three disks. From those analogous problems the basic pattern started to emerge. Divide and conquer was at work too. The six-disk problem had the five-disk as a component, the five disk had the four disk, and so forth on down.

We started with an extremely simple form of the problem and bootstrapped up from there. There is some psychological evidence that that often works as a strategy. The psychologist John Sweller gave some college students a starting number, eight. That was their number in hand. They were told to convert eight to 15 in exactly six moves. At each move they had to multiply their number in hand by two, or subtract seven. One group got just that complex problem. The other group got three problems. The first two problems were simpler versions, parts of the whole. For example, start with eight as your number in hand. At each move, multiply your number in hand by two, or subtract seven. Can you get from eight to nine in two moves?

Not too tough, $2 \times 8 = 16$, $16 - 7 = 9$, from eight to nine in two moves. But how about getting from eight to fifteen in six moves? The solution here is to alternate operations. You can get from eight to fifteen by multiplying eight by two to get sixteen, subtracting seven to get nine, times two to eighteen, minus seven to eleven, times two to twenty two, minus seven to fifteen. So Sweller gave one group just the complex problem. He gave the other group three problems, two simple ones, like the move from 8 to 9, and the complex one. So who got the hard problem fastest? On average, those who got just the complex problem needed an average of over five minutes to solve it. Those who got that problem after two simpler ones needed only about a minute and a half to solve the complex one.

This statistic is even more telling. The group presented with only the complex problem took over five minutes to solve it. The three-problem group took an average of only a little over three minutes to solve all three problems. My colleague Marvin Levine, who gave me several of the examples in this lecture, says that someone faced with just the six-disk problem of Hanoi will take, maybe, ten minutes to solve it. He predicts that someone who spends five minutes thinking about the 2- and 3-disk case will need only another two minutes to solve six disks, for only seven minutes or so in total.

Not all thought experiments are out to prove something, or to find the right answer. Sometimes thought experiments are explorations in a very different sense. There are imaginative explorations in epidemiology. If you didn't have vaccine for the whole population, where would it do the most good? There are imaginative explorations in economics. What changes in the financial structure would best stimulate employment? In business, many a startup triumph began in thought experiment. What if you could access digital information worldwide? What if you could social network online?

Here is a thought experiment in the ethical realm, the Runaway Trolley. A trolley is running out of control down a track. The bad news is that it is hurtling toward five people tied to the track by a mad philosopher. The good news is that you are standing next to a rail switch. All you have to do is pull that lever and the trolley will go down a second track instead. The bad news is that the mad philosopher has also tied one person to the track on that side. What do you do? Do you do nothing, allowing the five people to die? Or do you pull the switch to save them, thereby sacrificing the one? This thought experiment was originally presented by the philosopher Philippa Foot. It was designed merely to point up differences in ethical theory. As it turns out, it may also reveal differences in the way different people approach ethical problems. In ethics, Utilitarians hold that the right thing to do is that which results in the greatest good for the greatest number. They would clearly pull the switch, sacrificing one in order to save five. Would you?

We can change the thought experiment at will. What if that one person on the second track is someone very special to you, your mother perhaps, or your child. Now would you pull the switch? Is that the ethical thing to do? A utilitarian would say it was. Here is another variation. This time there is no branch in the tracks. You are on a bridge above it as the runaway trolley rushes toward the five people. Only by stopping it with a large object can you save them. The good news is that there is a very large man sitting on the railing next to you. All you have to do is to give him a shove. When he falls in front of the train he will be killed instantly, but his bulk will save the five.

Should you push him off? Would that be the ethical thing to do? A utilitarian would still say yes. But perhaps you're now less inclined to be a utilitarian.

It's not just philosophers who are interested in ethical thinking. The runaway trolley has been used with fMRI scans to explore the parts of the brain that different people use when thinking about the problem. Joshua Greene is a psychologist at Harvard who has done much of this work. The first conclusion he draws is that there is no single part of the brain devoted to moral decision alone. There is no ethics central. Moral judgment recruits various portions of your brain. Greene claims that people who arrive at different decisions in the Runaway Trolley use a different balance of brain areas. If you refused to pull the switch when someone important to you was on the track, and you refused to push the very large man off the bridge, emotional centers of your brain played a major role. Those emotional centers play a smaller role for Utilitarians, dominated by more cognitive centers, focusing purely on the numbers of lives lost. The implications of Greene's work remain controversial. Can any brain scan tell us whether something is ethically right? That looks like a question we might explore with further thought experiments.

We've looked at thought experiments in the history of physics as well as the history of philosophy, in ethics and neuro-ethics. There's no rule book for imaginative thinking. But there are some general strategies, visualize, go to extremes, simplify by analogy, divide and conquer. In all these ways and more, give your imagination time to roam. The role of thought experiments in philosophy, physics, epidemiology, economics, and business make it clear that imagination isn't merely an escape from reality. Imagination is one of our finest tools for exploring reality. Imagination is one of our best routes in.

In the next lecture we'll return to simple visualizations that give us all the logic of Aristotle.

Thinking like Aristotle
Lecture 6

In this lecture and the next, we will concentrate on the logic of Aristotle. The first attempts to analyze thinking systematically started with Aristotle, around 350 B.C.E. Aristotle's logic was the gold standard through the Middle Ages and the Renaissance—and continued until about 150 years ago. What we've emphasized so far are techniques for logical thinking, with visualization as an important component. Aristotle himself lays out his logic using a visualization—one that takes up concepts and propositions. What we're working toward is an understanding of the flow of rational argument. Aristotle is a major step in that direction.

A Brief History of Greek Philosophy
- The genealogy of Greek philosophy runs from Socrates, to Plato, to Aristotle. That transition of great minds is also a transition from oral to written culture.

- Socrates wrote nothing down. Indeed, he insisted that writing things down negatively affects our capacity for memory and even for wisdom. Plato came next, transmitting the Socratic tradition to us by making Socrates a character in his dialogues.

- Aristotle studied with Plato for 20 years. He wrote on physics, astronomy, biology, and the soul and is credited with the invention of logic. At the core of his logic—indeed of all logic—is this breathtaking idea: Maybe we can systematize thought.

Aristotle's Goal: To Systematize Thought
- Aristotle's theory was that if we could systematize thought, we could perhaps make it less difficult, more accurate, more effective, and faster. Aristotle's work remained the core of logic all the way through the Middle Ages and the Renaissance. At the beginning of the 19th century, Immanuel Kant said that logic (meaning

Aristotelian logic) was a perfected science, in which no more progress could be made.

- Like all such predictions, Kant's turned out to be wrong. By the end of the 19th century, we get the beginnings of what we consider contemporary logic, and the 20th century was one of the great centuries in the history of logic.

- The computer age is grounded in logic. The idea of formalizing, systematizing, even mechanizing thought stretches as one unbroken chain from Aristotle to the logic that drives your laptop.

The Magic of Aristotle's Visualization

- A few lectures ago we talked about concepts or categories—the atoms of thought. We used Venn diagrams to visualize how two categories link together into what are called categorical propositions.

- Logic is about structure. Aristotle's first insight was that the structure of these categorical propositions isn't tied to their content. Thus, we can forget the Venn diagrams and represent the structures themselves using just an S for the subject term of the proposition—any subject term—and P for the predicate. With that **abstraction**, we are looking at four basic types of categorical propositions:

No S are P:

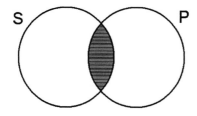

All S are P:

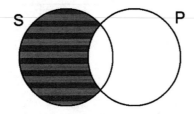

Some S are P:

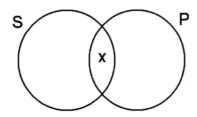

Some S are not P:

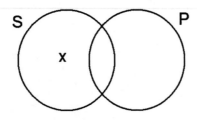

- That visualization leads to a second one—Aristotle's **square of opposition**. At the core of Aristotle's logic is a picture of the logical relations between those kinds of propositions.

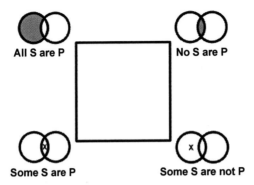

- The four types of categorical propositions are on specific corners of the square. Across the top of the square are **universal propositions**: those that are all or nothing. Across the bottom are **particular propositions**: the ones that are about "some." Running down the left side are the positive propositions; running down the right side are the negative propositions.

- But the following is the real magic of Aristotle's visualization.

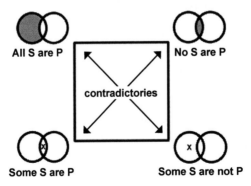

- First, look at the propositions on the diagonals. Those are **contradictories**. They can't both be true, and they can't both be false. No matter what S and P are, the propositions on the diagonals will contradict each other.

- Now look at the relationship between the top and bottom on the left and the top and bottom on the right. In each case, the top proposition implies the one below. That means that if the top one is true, the bottom one must be true, too.

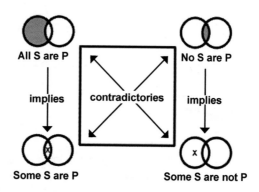

- Now think of it from the bottom to the top on each side. And think in terms of false rather than true. If the bottom one is false on either side, the top must be false, too.

- Here's another way to express that relation. True at the top on either side gives us true at the bottom. That's the direction of **implication**. False at the bottom on either side gives us false at the top. You can think of that as counterindication.

- Two other aspects of the square are a bit more obscure. They're called **contraries** and **subcontraries**.
  - ○ Consider the propositions across the top. Those can't both be true: For example, it can't be true that both all tickets are winning tickets and no tickets are winning tickets. But they could both be false—if some are winning tickets and some aren't. The propositions across the top are called contraries: both can't be true, but both could be false.

- The propositions across the bottom have exactly the opposite relationship. Both could be true, but they can't both be false. Those are called subcontraries.

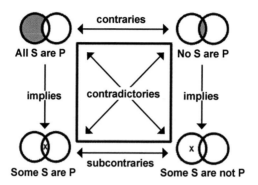

Reasoning Better Through Aristotle's Logic
- How does all this help in everyday reasoning? If you visualize propositions as located on the square of opposition, you will instantly know the logical relations between them.
  - Consider these two propositions: All bivalves are pelecypods. Some bivalves are not pelecypods.

  - Now, drop them onto the square. They're diagonals. Therefore, they are contradictories that can't both be true and can't both be false.

- We can imagine Aristotle drawing his vision with a stick in the sand of Athens. He was probably amazed at the result, convinced that he was looking at the logical structure of the universe. Laid out in the square of opposition are all the relations between our four types of categorical propositions: universal and particular propositions, positives and negatives, **contradictions** and implications, and contraries and subcontraries.

Transforming Propositions

- To sum up, what the square of opposition gives us is a visualization for all of these: contradictions, contraries, implications, and counterindications. Those are the logical relations fundamental for rationality, in all the ways that rationality is important to us.

- The visualizations we've developed also allow us to understand the logic of certain ways of transforming propositions. We'll be talking about transforming a proposition into its **converse**, or its **contrapositive**. For certain types of propositions, on specific corners of the square, these transforms give us immediate **inferences**: If the original sentence is true, so is its transform. But it will be important to remember what positions on the square these work for.
  - Starting with the converse, that transform holds for propositions in the upper right and lower left. If one of those is true, it will still be true when we switch the subject and predicate. If it's true, it will still be true. If it's false, it will still be false. If we go back to our Venn diagrams, we can see exactly why the converse works for the upper right and lower left.

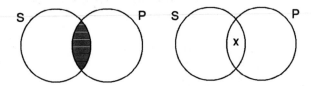

  - Both diagrams are symmetrical. One says there is something in the overlap area and puts an *x* there. One says there is nothing in the overlap area and blanks it out. But the overlap is the same whether we think of it as the overlap between S and P or the overlap between P and S. It doesn't matter which one we make the subject term and which we make the predicate term. What the proposition says is the same in each case. They're equivalent.

- The other forms of proposition—the upper left and lower right—are not symmetrical. The converse is not guaranteed to hold for those. But there is a transform that does hold for those: the contrapositive.

- In order to understand the contrapositive, we have to introduce one more concept: the **complement**. Let's take the category "senators." The complement is the category of everyone who isn't a senator—that is, non-senators.

- We take the two categories in a proposition and use their complements in order to arrive at the contrapositive. While leaving everything else the same, we switch subject and predicate, but this time, we also replace each by its complement.

- If we're dealing with the upper-left or lower-right corners of the square, that again gives us an equivalent proposition, an immediate inference.
  - All senators are congressmen.

  - Switch subject and predicate and replace them with their complements: All non-congressmen are non-senators.

- The inference is immediate and guaranteed. But try this on your friends and you'll see how difficult it is for people. Next time a friend says something like "All the widgets are defective," say, "Do you mean that everything that isn't defective isn't a widget?" The logical inference is immediate. But you'll be amazed how long it takes your friend to process it.

- Contrapositive works for the lower right, too: If some liquids are not flammable, some nonflammable things are liquids.

- In summary, then, for upper right and lower left, the converse gives us an immediate inference. We can just switch subject and predicate. Converse isn't guaranteed for the other two.

- For upper left and lower right, the contrapositive gives us an immediate inference. We can switch subject and predicate and replace each by its complement. Contrapositive isn't guaranteed for the other two.

- We'll return to Aristotle in the next lecture, when we take our logic well beyond him.

Terms to Know

abstraction: The thought process that allows us to derive general concepts, qualities, or characteristics from specific instances or examples.

Aristotelian logic: Aristotle's attempt to systematize thought by outlining a set of formal relations between concepts and propositions. These relations can be visualized by his square of opposition and his treatment of arguments as syllogisms.

complement: In logic, given any category, the complement comprises all those things that do not fall in that category. For example, "senators" and "non-senators" are complements.

contradiction: A statement that both asserts and denies some proposition, P, often represented in the form "P and not-P." If either part of a contradiction is true, the other cannot be true, and thus, a contradiction P and not-P is treated as universally false.

contradictories: The relationship between propositions on the diagonals of Aristotle's square of opposition. It is a contradiction for both propositions on a diagonal to be true; if one proposition of the diagonal is true, the other must be false.

contrapositive: A way of transforming categorical propositions by switching subject and predicate and replacing each with its complement. For some categorical propositions, the result is an immediate inference: the truth or falsity of the proposition is not altered. The contrapositive transformation preserves equivalence only for propositions in the upper left and lower right on the square of opposition: the universal positive ("All S are P") and particular negative ("Some S are not P").

contraries: The relationship between propositions at the top left ("All S are P") and right ("No S are P") of Aristotle's square of opposition. If two propositions are contraries, it is not possible for both propositions to be true, but it is possible for both propositions to be false.

converse: A way of transforming categorical propositions by switching subject and predicate. For some categorical propositions, the result is an immediate inference: the truth or falsity of the proposition is not altered. The converse preserves equivalence only for propositions in the upper right and lower left on the square of opposition: the universal negative ("No S are P") and the particular positive ("Some S are P").

implication: In Aristotelian logic, the relationship moving from the top to the bottom left corners or the top to the bottom right corners of the square of opposition. If the proposition on the top left corner is true, then the proposition on the bottom left corner is also true; if the proposition on the top right corner is true, then the proposition on the bottom right corner is also true. Expressed as "if all S are P, then some S are P" for the left side of the square of opposition and "if no S are P, then some S are not P" for the right side of the square of opposition.

inference: In logic, the derivation of a conclusion from information contained in the premises.

particular proposition: In logic, a proposition about "some" rather than "all": "Some S are P" (a particular positive, e.g., some cleaning products are poisons) or "Some S are not P" (a particular negative, e.g., some cleaning products are not poisons). The particular positive occupies the lower-left corner of Aristotle's square of opposition; the particular negative occupies the lower-right corner.

square of opposition: Aristotle's visualization of the logical relations between categorical propositions.

subcontraries: The relationship between propositions at the bottom left ("Some S are P") and right ("Some S are not P") of Aristotle's square of opposition. If two propositions are subcontraries, it is possible for both propositions to be true, but it is not possible for both propositions to be false.

universal proposition: In logic, a universal proposition refers to a claim either in the form "All S are P" (universal affirmative, e.g., all snakes are reptiles) or in the form "No S are P" (universal negative, e.g., no snakes are reptiles). The universal affirmative occupies the upper-left corner of Aristotle's square of opposition; the universal negative occupies the upper right.

Suggested Reading

Aristotle (Ackrill, trans.), *Categories and De Interpretatione.*

Groarke, "Aristotle's Logic."

Questions to Consider

1. Are there ideas you think you could have come up with if you had been present at the right time and place? Are there ideas you think wouldn't have thought of, even if you had been present at the right time and place?

2. Why do you think it is that Aristotle's "exoteric" works—popular pieces for a wide popular audience—have disappeared completely, while his "esoteric" works—lectures for a small body of students—have survived?

3. One of the claims in this lecture is that the attempt to "systematize, formalize, and mechanize thought" is a goal that runs all the way from Aristotle's logic to contemporary computers. If that is true, what do you think will be the next step in the trajectory?

Below are elements of a square of opposition that are not in the Aristotelian positions:

All philosophers
are idealists.

Some philosophers
are not idealists.

No philosophers
are idealists.

Some philosophers
are idealists.

Fill in the Venn diagrams appropriately for each categorical proposition.

Draw solid lines between contradictories ———————————

Draw dashed lines between contraries ------------------

Draw dotted lines between subcontraries ••••••••••••••••••••••••••

(See "Answers" section at the end of this guidebook.)

Thinking like Aristotle
Lecture 6—Transcript

Professor Grim: The first attempts to analyze thinking systematically start with Aristotle, around 350 BCE. Aristotle's logic was the gold standard through the Middle Ages and the Renaissance, indeed until about 150 years ago. In this lecture and the next I want to concentrate on the logic of Aristotle. That will serve as a link between what we've discussed so far and what we're working toward. What we've emphasized so far are techniques for logical thinking with visualization as an important component. Aristotle himself lays out his logic using a visualization, a visualization that takes up concepts and propositions right where we left off. What we're working toward is an understanding of the flow of rational argument. Aristotle is a major step in that direction. Later on I'll have you put that understanding to work analyzing arguments in the Great Debate.

Let me start with a little history. The genealogy of Greek philosophy runs from Socrates to Plato to Aristotle That transition of great minds is also a transition from oral to written culture. Socrates wrote nothing down. Indeed, he insisted that writing things down negatively impacts our capacity for memory and even for wisdom. Plato came next, transmitting the Socratic tradition to us in the half-way house of theater. Plato made Socrates a character in the plays we know as Dialogues. In Plato's theatrical performances, an actor spoke as Socrates, and indeed spoke against writing things down, but the plays themselves were written.

Aristotle studied with Plato for 20 years. He produced a range of popular works including dialogues, the exoteric works, but none of those survive. The Aristotle we know is a scholarly writer, a writer of discursive treatises. Those are Aristotle's esoteric works, based on lecture notes, either his own notes or his students', or some combination of the two. He wrote on physics, astronomy, biology, the soul, and logic. I want you to try to visualize Aristotle. What did he look like? We know he had small eyes and spindly legs. The earliest visual representations show him bearded and balding. We know he either stammered or spoke with a lisp. I also want you to try to visualize the texts we are dealing with. Papyrus rolls, written in Greek, containing all the logic of Aristotle. I'll talk about the history of those texts a little later in the lecture.

Aristotle is credited with the invention of logic. At the core of his logic—indeed of all logic—is this breathtaking idea: Maybe we can systematize thought. One of our greatest sources of power is our ability to think, but there are problems with thinking. It's often difficult, and it often seems to go only so far. But if we could systematize thought, we could perhaps make it less difficult. If we could formalize thought we might be able to do the same amount of thinking with, well, with less thinking. If we could schematize thought, we might be able to make our thinking faster and more accurate and more effective. That attempt to systematize thought starts with Aristotle.

There are some ideas I think that I could have had if I had been at the right place at the right time. Maybe this is just conceit, but if I had been at the right place at the right time, I think I could have come up with the internal combustion engine. The mechanical elements were there in the steam engine: expansion in pistons, regulation by valves. The chemical elements were there in the combustion of petroleum. The internal combustion engine? That seems like an idea that I might have had. But there are some ideas that I think I would never have had. One of those is the very idea of logic. For that we needed an Aristotle. Aristotle's work remains the core of logic all the way through the Middle Ages and the Renaissance. At some point Kant says that Logic—and he means Aristotelian logic—is a perfected science in which no more progress can be made. That was at the beginning of the 19th century. Like all such predictions, it turned out to be wrong. By the end of the 19th century we get the beginnings of what we consider contemporary logic, and the 20th was one of the great centuries in the history of logic.

The computer age is grounded in logic. The idea of formalizing, systematizing, even of mechanizing thought stretches as one unbroken chain from Aristotle to the logic that drives your laptop. So let me give you the fundamentals of Aristotle's logic.

A few lectures ago we talked about concepts or categories, the atoms of thought. We used Venn diagrams to visualize our two categories, how they link together into what are called categorical propositions. Let me remind you. The categorical proposition "No frogs are philosophers" looks like this.

The proposition tells us that no frogs are philosophers. There is no overlap—nothing is both a frog and a philosopher—so that middle area is empty. We black it out. "All neurologists are MDs." There aren't any neurologists that aren't in the overlap. So we black out that left area. No neurologists over there. "Some laws are unjust." For this we used an x to indicate that there is something in the overlap. Finally, "some inventions are not patentable." We use an x over in the inventions area, outside of the patentable area, to indicate that some inventions are not patentable.

Logic is about structure. Aristotle's first insight was that the structure of these categorical propositions isn't tied to their content. We can forget about frogs, philosophers, and patentable inventions. We can represent the structures themselves using just an S for the subject term of the proposition—any subject term—and P for the predicate. With that abstraction, we are looking at four basic types of categorical propositions: No S are P. All S are P. Some S are P. Some S are not P.

That visualization leads to a second one, Aristotle's Square of Opposition. At the core of Aristotle's logic is a picture of the logical relations between those kinds of propositions. The four types of categorical propositions are on very specific corners of the square. All S are P is up there on the left. No S are P is up here on the right. Some S are P is lower left. Some S are not P is lower right. Across the top of the square are what we call the universal propositions, those that are all or nothing: All S are P, or No S are P. Across the bottom are what are called particular propositions, the ones that are about some S—not all, not none—just some. Some S are P. Some S are not P. That's not all. Running down the left side are the positive propositions, All S are, some S are. Running down the right side are the negative propositions, some are not or none are. That's the setup. Universal positive, upper left: All S are P. Universal negative, upper right: No S are P. Particular positive, lower left: Some S are P. Particular negative, lower right: Some S are not P.

Ok. I'll rattle off a few categorical propositions. As I do so, I want you to mentally position them on the square of opposition. For each one ask yourself: Is this universal positive on the upper left, universal negative on the upper right, particular positive on the lower left, or particular negative on the lower right?

All snakes are reptiles.

No snakes are reptiles

Some snakes are reptiles

Some snakes are not reptiles.

Those positions in order were: left to right across the top, then left to right across the bottom. All snakes are reptiles is universal affirmative, upper left. No snakes are reptiles is universal negative, upper right; some snakes are reptiles, particular positive, lower left; some snakes are not, particular negative, lower right.

I'll give you four more. As you hear them, drop them into place in your mental square of opposition.

Some philosophers are not logicians.

No philosophers are logicians.

All philosophers are logicians.

Some philosophers are logicians.

Did you get those? They should have moved counter clockwise from the lower right. With the positions of each type of proposition firmly in your head, you are ready to see the real magic of Aristotle's visualization.

First, look at the propositions on the diagonals. Those are contradictories. They can't both be true, and they can't both be false. No matter what S and P are, the propositions on the diagonals will contradict each other. It can't be true that all snakes are reptiles and also true that some aren't, no way, that's a contradiction. It can't be false that no philosophers are logicians—that would mean that some are—and also be false that some philosophers are logicians. That too is a contradiction.

Now look at the relationship between top and bottom on the left and the top and bottom on the right. All S are P, top left; Some S are P, below it; No S are P, top right; Some S are not P, below it. In each case, the top proposition implies the one below. That means, if the top one is true, the bottom one has to be true too. If all tickets are winning tickets, some have to be; that's moving down on the left. If no tickets are winning tickets, some tickets have to not be winning tickets; that's moving down on the right.

Now think of it from the bottom to the top on each side and think in terms of false rather than true. If the bottom one is false on either side, the top one has to be false too. If it's false that some S are P, it must also be false that all S are P. If it's false that Some S are not P, it must also be false that No S are P. Here's another way to express that relation. True at the top on either side gives us true at the bottom. That's the direction of implication. False at the bottom on either side gives us false at the top. You can think of that as counter indication.

Let's sum up. Across the diagonals, both propositions can't be true and both can't be false. Those are contradictories. Going down the sides, if the top one is true the bottom one has to be true. Those are implications. If the bottom is false, the top has to be false too, counter indication.

Two other aspects of the square are a bit more obscure. They're called contraries and subcontraries. Consider the propositions across the top. Those can't both be true: It can't be both true that all tickets are winning tickets and no tickets are winning tickets. But they could both be false. How? Well, if some are winning tickets and some aren't. The propositions across the top are called contraries: both can't be true, but both could be false.

The propositions across the bottom have exactly the opposite relationship. Both could be true, but they can't both be false. Let me give you an example. Both of these propositions could be true: Some men are professional football players. Some men are not. But both couldn't be false. This is a little harder to see, but it's there. Suppose it's false that some men are football players. That must mean that no men are football players. As long as we have men, it follows that some men are not football players. But that gives us true on the lower right. Both lowers can't be false. But both could be true. Those are called subcontraries.

How does all this help in everyday reasoning? If you visualize propositions as located on the square of opposition, you will instantly know the logical relations between them. I'm going to give you two propositions. I want you to tell me their logical relationship. Are they contradictory? Does one imply the other? Are they contraries? Start with these two: All bivalves are pelecypods; some bivalves are not pelecypods. Drop them onto the square. They're diagonals. That's all you need to know. Contradictories can't both be true, can't both be false. You don't have to know anything at all about pelecypods or bivalves. From their form alone the square of opposition tells you they're contradictories.

How about these? All legal drivers are insured; some legal drivers are insured. What's their relationship? We're going down the left-hand side of the square. So the first one implies the second one. If the first one is true, the second one has to be true too. If the second one is false, the first one has to be false too. One more. No absentee votes are counted; all absentee votes are counted. What is their relationship? Those are contraries. Both couldn't be true. but they could both be false.

You can imagine Aristotle drawing his vision with a stick in the sand of Athens. My guess is that he was amazed. My guess is that he was convinced he was looking at the logical structure of the universe. Laid out in the square of opposition are all the relations between our four types of categorical propositions: universal propositions, particular, positive, negative, contradictions, implications, contraries, and subcontraries. It is still a powerful image. I don't expect you to have gotten all the details in this brief presentation, but take some time to study the square of opposition, and you'll have the logic of Aristotle at your fingertips.

I'll let some of that logic sink in while I fill in a little background history, just for fun. Aristotle's logic appears in a series of works collected as *The*

Organon, or the "Instrument," a set of lectures that outlined instruments for thought. Aristotle's *Organon* was the first Philosophical Toolkit lecture series. *The Organon* starts with the categories, just as we started with concepts. It includes the Prior Analytics, the Posterior Analytics, On Interpretation. Those are the sacred texts of logic. Earlier in the lecture I asked you to visualize the real man that was Aristotle and the real texts that were those lecture notes, written in Greek on papyrus. What happened to the originals? Can you see them in the Athens museum? No. The history of Aristotle's texts is much stranger than that.

We said that the intellectual genealogy went from Socrates, to Plato, to Aristotle. It actually goes one step farther to one of Aristotle's students. Aristotle's father had been court physician to the court of Macedonia. Aristotle eventually became tutor to the son of Philip of Macedonia. The boy's name was Alexander, and he went on to conquer most of the known world. You know him as Alexander the Great. After Plato died Aristotle founded his own school, the Lyceum. Funding came from his patron and former student, Alexander the Great. Aristotle died in 322 BC, at the age of 62. The ancient biographer Diogenes Laertes tells us the text of his will: "All will be well, but in case anything should happen, Aristotle has made these dispositions. " In the will he names his executor and makes provision for his second wife and his two children. He asks that he be buried next to his first wife, in accordance with her wishes. His books were undoubtedly the core of the Lyceum's library. Those he leaves to Theophrastus, his successor as Director. It is Theophrastus who takes charge of the Aristotelian manuscripts.

We also have a transcript of Theophrastus's will. The important line is: "I leave all my books to Neleus." Neleus had known Aristotle. In fact, he was, by that time, probably the only living man who had been taught by Aristotle. It was natural for Theophrastus to assume that Neleus would become the new Director of the Lyceum. But the scholars of the Lyceum didn't choose Neleus. They chose Strato. Neleus left in a tiff and took Aristotle's manuscripts with him. When Epicurus came to Athens as a 20 year old, all the Aristotle he could get were the popular works. The Exoteric works, those have now disappeared entirely.

The Aristotelian manuscripts? Neleus took them to his hometown, where they were buried in a cellar and passed down through his family for generations. The Library at Alexandria tried to get them but failed. If they had succeeded, Aristotle's logic might have disappeared when the library burned. Generations later, one of Neleus's descendants dug up the scrolls and sold them to a book collector named Appelicon of Teos. He took them to Athens, but they were decayed and full of holes by that time. But Apellicon thought of himself as a philosopher as well as a book-collector. So he filled in the holes. That's been a problem for scholars ever since. Some of what we're reading may be Aristotle. Some of what we're reading may be Appelicon of Teos.

In 86 BC the Roman general Sulla sacked Athens. Apellicon was one of the first to fall, and Sulla took his library back to Rome. There, after dinner, he took pride in letting visitors see and copy out portions of Aristotle's manuscripts. Sulla bequeathed the original Aristotelian manuscripts to his son Faustus. Faustus was a spendthrift and eventually had to sell off everything he owned. It's at that point that the original manuscripts disappear from history. What we have are copies of copies, probably of manuscripts seen at Sulla's after dinner, with holes of decay filled in by Appellicon of Teos.

Let's get back to the logic. What the Square of Opposition gives us is a visualization for all of these: contradictions, contraries, implications, counter-indications. Here's why those relations are important. Those are the logical relations fundamental for rationality, in all the ways that rationality is important to us. Think of debate, an attempt to decide between conflicting views. That's what political debate is all about. That's what legal argument in a court of law is all about. That's what negotiation is all about. And when do two views conflict? Precisely when they are either contradictories or contraries. Think of the kinds of calculations we do in mathematics and engineering. Think also of the predictions we make on the basis of scientific theory. Those are all about implication. If this mathematical fact is true, so is this one. Given this scientific theory, and given these initial conditions, this is what we can expect to happen. Think finally of how we test scientific theories. A theory predicts an eclipse at 12:30, but it doesn't happen. That theory has to be wrong. That is disconfirmation, crucial to all scientific rationality, and it follows the structure of counter indication. Rational debate, calculation and prediction, empirical tests and disconfirmation: The logical relations crucial to those are precisely those laid out in Aristotle's Square of Opposition.

The visualizations we've developed also allow us to understand the logic of certain ways of transforming propositions. We'll be talking about transforming a proposition into its converse or its contrapositive. For certain types of propositions, on specific corners of the square, these transforms give us immediate inferences, If the original sentence is true, so is its transform. But it will be important to remember what positions on the square these work for. Let's start with the converse. That transform holds for propositions in the upper right and the lower left, the universal negative, No S are P and the particular positive, Some S are P. If one of those is true, it will still be true when you switch subject and predicate. If it's true, it will still be true. If it's false, it will still be false. But remember, that holds only for upper right and lower left. So take the proposition No fish are reptiles, universal negative, upper right. And if no fish are reptiles, no reptiles are fish.

Some experimental failures are scientific breakthroughs. Switch subject and predicate and you get: Some scientific breakthroughs are experimental failures. If one is true, the other is. If one is false, the other is too. An immediate inference. If some cleaning products are poisons, it follows immediately that some poisons are cleaning products. If some primates are philosophers, it follows immediately that some philosophers are primates. If you go back to our Venn diagrams, you can see exactly why converse works for the upper right and lower left. Look at our Venn diagrams for those. Both diagrams are symmetrical. One says there is something in the overlap area and puts an x there. One says there is nothing in the overlap area and blanks it out. But the overlap is the same whether you think of it as the overlap between S and P or the overlap between P and S. It doesn't matter which one you make the subject term and which one you make the predicate term. What the proposition says is the same in each case. They're equivalent.

The other forms of proposition—the upper left and lower right—are not symmetrical. The converse is not guaranteed to hold for those. But there is a transform that does hold for those. It's called the contrapositive. In order to understand it, we have to introduce one more concept: the complement. Let's take the category Senators. The complement is the category of everyone who isn't a Senator. The non Senators. Or consider all the poisonous substances, and consider the substances that aren't poisonous. Those two categories are complements. You take the two categories in a proposition and you use their complements in order to arrive at the contrapositive. While leaving everything else the same, you switch subject and predicate, but this time you also replace each by their complement. If you're dealing with the upper left or lower right corners of the square, that, again, gives you an equivalent proposition. That, again, gives you an immediate inference. All Senators are Congressmen. Switch subject and predicate and replace them with their complements. All non-Congressmen are non-Senators. All people are primates. Switch subject and predicate and replace them with their complements, all non-primates are something other than people. The inference is immediate and guaranteed. But try this on your friends and you'll see how difficult it is for people. Next time they say something like, "All the widgets are defective," say "Do you mean that everything that isn't defective isn't a widget?" The logical inference is immediate. But you'll be amazed how long it takes them to process it. Contrapositive works for the lower right, too. But this does get a little tangled. If some liquids are not flammable, Some non-flammable things are liquids. I cancelled a couple of double negatives there.

In summary, then: For upper right and lower left, the converse gives you an immediate inference. You can just switch subject and predicate. Converse isn't guaranteed for the other two. For upper left and lower right, the contrapositive gives you an immediate inference. You can switch subject and predicate and replace each by its complement. Contrapositive isn't guaranteed for the other two.

Let me try you on these. These are immediate inferences. With a square of opposition clearly in mind, you should be able to give me these answers without even thinking very hard. Everyone on the committee supports the measure. Does that mean that everyone who doesn't support the measure isn't on the committee? Yes. Immediate inference. That's the contraposition of the upper left corner, and that guarantees equivalence. Some registered voters support the measure. Does that mean that some people who support the measure are registered voters? Yes. Immediate inference. That's the converse of the lower left, equivalence guaranteed. Some registered voters support the measure. Does that mean that some who do not support the measure are not registered voters? No. That's a contrapositive transform, but in this case it was applied to the lower left, no guarantee. That might be false. Contrapositive is only guaranteed for upper left and lower right.

What I hope you will take away from his lecture isn't so much a conclusion as an image. I want you to take away that image at the core of Aristotle's logic: the Square of Opposition. Picture the categorical propositions laid out that way—universal across the top, particular across the bottom. positive to the left side, negative to the right. With that image, the core logical relations are laid out before you. Contradictions on the diagonals, they can't both be true; they can't both be false. Implications down the sides; if the first is true, the second is. That gives us counter indications, too. If the second is false, so is the first. Contraries across the top, contradiction's weaker sister. They couldn't both be true, but they could both be false. Sub contraries across the bottom, they couldn't both be false, but they could both be true. We threw in Converses and Contrapositives for good measure.

We'll return to Aristotle in the next lecture. With a break for creative thinking, we'll then take our logic well beyond it.

Ironclad, Airtight Validity
Lecture 7

In the last lecture, we were introduced to the logic of Aristotle in terms of a central visualization: the square of opposition. This lecture will cover what makes an argument valid and what defines ironclad, airtight validity. Here is how an argument is valid in that strongest sense: If the premises are true, the conclusion absolutely must be true. If the premises are true, it's impossible for the conclusion to be false. We can't get a much tighter connection than that.

What Is a Valid Argument?

- Propositions don't just stand alone; they go together to form arguments. In arguments, we see the dynamics of thought. In the philosophical sense, an argument is a connected series of statements intended to establish a proposition. An argument in our sense represents the conceptual dynamic from proposition, to proposition, to **conclusion**. It is that movement of thought that is central to all rationality.

- But of course, it's not just the intended transition that is important. We want to be able to evaluate arguments. When does a set of reasons give us a solid inference? When does evidence really support a hypothesis? When do **premises** really entail a conclusion? What guarantees that a transition of thought is a rational transition? Those are questions about **validity**.

- That's what Aristotle was after. That's what all of logic is after. The whole purpose of logic is to capture validity: the crucial category of argument evaluation. Perhaps we can systematize thought by capturing what makes it rational. Perhaps we can formalize validity.

Ironclad, Airtight Validity

- For an argument to be valid: If the premises are true, the conclusion must be true. That is important information, but it's conditional information. What we know is that the conclusion must be true if the premises are true. Following is an example; determine whether it is valid or invalid:

○ All purple vegetables are poisonous.

○ Carrots are a purple vegetable.

○ Conclusion: Carrots are poisonous.

- The conclusion is perfectly ridiculous, but the argument is also perfectly valid. Not all purple vegetables are poisonous. But suppose they were. Carrots aren't a purple vegetable. But suppose they were. If those two premises were both true—if carrots were purple vegetables and all purple vegetables were poisonous—it would also have to be true that carrots were poisonous.

Validity Depends on Structure

- Suppose we have an argument in which all premises and the conclusion are true. Is that a valid argument? The answer is: not necessarily. It's the connection that matters.

- We might have a valid argument in which all premises and the conclusion are true. That would be great, but validity alone doesn't tell us that. We might have a valid argument in which all premises and the conclusion are false. The "purple carrots" argument above was one of those. We might, in fact, have a valid argument in which the premises are false but the conclusion is true. For example:
  ○ Everything poisonous is orange.

  ○ Carrots are poisonous.

  ○ Therefore, carrots are orange.

- That is a valid argument. If the premises are true, the conclusion must be true. Here, the premises are false, but they entail a conclusion that is true anyway. Sometimes, two wrongs do make a right.

- There is only one pattern of truth and falsity that validity rules out. We will never find a valid argument that goes from true premises to a false conclusion. A valid argument is one in which, if the premises are true, the conclusion must be true.

Syllogisms
- Aristotle worked out the structure of ironclad, airtight validity in terms of three-step arguments called **syllogisms**. A syllogism takes us from categories, to propositions, to steps in an argument. It uses three categories in all: A, B, and C. It links those categories in pairs to make categorical propositions: "All A's are B's"; "Some B's are not C's." The syllogism itself is an argument in three propositional steps, such as the following:
  - All A's are B's.

  - Some B's are not C's.

  - No C's are A's.

- Some syllogisms are valid. Some are not. The one above is not a valid syllogism.

Using Visualization to Determine Validity
Visualization is the key to sorting the valid syllogisms from the invalid ones. We can do all of Aristotle's logic with the Venn diagrams we already have at our disposal.

All S are P:

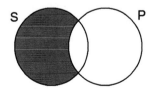

No S are P:

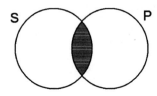

Some S are P:

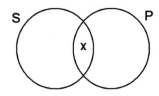

Some S are not P:

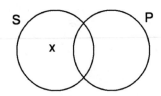

- Those are the propositional elements of syllogistic logic. But syllogisms involve three propositions. Each proposition of the syllogism—each step in the argument—uses just two of our three categories. In order to use Venn diagrams to track validity, we need to overlap three circles into a triangular Venn diagram.
  ○ Our first proposition, linking categories A and B, will look something like this:

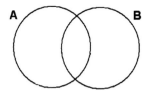

○ Our second, linking B and C, will look like this:

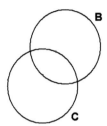

○ Our third, linking A and C, will look like this:

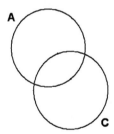

○ All three categories, and all three propositions, come together in the syllogism. We put them together like this:

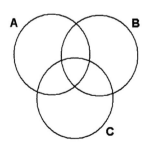

- The result is an overlap of three categories, just as the syllogism always involves three categories. The triangle is formed from three categorical propositions: pairs of linked circles.

How to Visualize Logic: An Example
- Following is an easy syllogism that is also a valid one:
  - All proteins are polypeptides.

  - All polypeptides function by elaborate folding.

  - Therefore, all proteins function by elaborate folding.

- Here, we have three categories: proteins, polypeptides, and "function by elaborate folding." We label the three circles in the diagram accordingly:

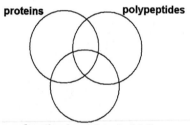

function by elaborate folding

- We now enter each of our premises in Venn diagram style. First, all proteins are polypeptides.

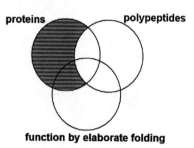

function by elaborate folding

- All polypeptides function by elaborate folding.

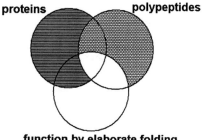

function by elaborate folding

- And the last step—the crucial step in telling whether a syllogism is valid: We want to look at our third proposition—the conclusion—and ask whether its information is already in the diagram we've constructed for our premises. If it is, the argument is valid. If it's not, the argument is invalid. If the concluding information is already provided by the premises, it has to be true when they are, and it's valid. If the premises don't give us that information, the conclusion doesn't follow from the premises, and it's invalid.

- In order to test for validity, we want to know whether the concluding information is already there in the logical space of the premises. As you look at the diagram, ask yourself what would it mean to draw in the conclusion: "All proteins function by elaborate folding." That would mean looking at just "proteins" and "elaborate folding"—those two circles on the left.

- For it to be true that "All proteins function by elaborate folding," the "proteins" area outside of the overlap would be blacked out. You will see from the diagram that that area is already blacked out. There aren't any proteins in our diagram that aren't already squeezed into some part of the "function by elaborate folding" area.

- That means that the information in step 3—the information in the conclusion—was already there in the premises. That's the sign of validity. If the premises are true, the conclusion must be true. The information in the premises is sufficient to give us the information in the conclusion. The argument is valid.

The Syllogism: No New Information
- A valid argument in the deductive sense is one that squeezes out in the conclusion some information that you already had. Immanuel Kant speaks of an "analytic" proposition as one in which the predicate is already "contained" in the subject term. A valid deductive argument is one in which the information in the conclusion is already contained in the premises.

- It therefore makes sense that we can show an argument to be valid or invalid depending on whether the information in the conclusion is already there in the premises. But it also means that a syllogism will never give us any really new information.

The Three-Circle Technique Embodies Ironclad, Airtight Validity
- We can do all of Aristotle's logic, all the logic of the medieval philosophers, all the logic that we've had for 2,000 years, with those three circles. The three-circle technique embodies the concept of ironclad, airtight validity. With that visualization, we can test validity for any three-step syllogism. However, the concept of ironclad, airtight validity sets a standard for inference in which all the information in a conclusion must already be contained, somehow, in the premises.

- For 2,000 years, ironclad, airtight validity was held up as the gold standard for all knowledge. But that meant all information had to flow from above—from something you already knew. Mathematics is about the only discipline that works that way. It's no surprise that **axiomatic** geometry was the standard for all knowledge.

Getting to Reasoning That Offers New Information
- Aristotle's legacy, therefore, had a downside, too. It was only with the rise of empirical science—in such figures as Galileo—that an alternative became clear. Perhaps an inference from observed

cases to universal laws is what was needed, even if it wouldn't be ironclad, airtight, guaranteed in Aristotle's sense.

- If we weaken the demands on validity, we get the possibility of rational information expansion for the first time—reasoning, together with observation, that offers new information.

- Ironclad won't give us new information, and anything that gives us new information won't be ironclad. We have bought the benefits of science by weakening our demands on validity, by giving up Aristotelian certainty. We'll start talking about reasoning and validity in that wider sense a few lectures down the line.

- In the next lecture, however, we'll take a break for something equally important but a little more fun: a workshop on creative thinking.

Terms to Know

axiomatic: Organized in the form of axioms and derivations from them. Euclidean geometry is an example of an axiomatic system.

conclusion: The endpoint of an argument; in a logical argument, the claim to which the reasoning flows is the conclusion.

premise(s): The proposition(s) or claims that are given as support for a conclusion; in a rational argument, the reasoning flows from the premises to the conclusions.

syllogism: An argument using three categories (A, B, and C) that are linked in pairs to make categorical propositions (e.g., "All A's are B's," or "Some B's are not C's"), which are then combined into a three-step argument. Some syllogisms are valid and some are not.

validity: An argument is valid if the conclusion follows from the premises, if the premises offer sufficient logical or evidential support for the conclusion.

An argument is deductively valid if it is impossible for all premises to be true and the conclusion to be false.

Suggested Reading

Kelley, *The Art of Reasoning*.

Questions to Consider

1. Aristotle sometimes seems to suggest that *all* reasoning can be analyzed in terms of syllogisms. Can you think of a counterexample?

2. Aristotle's notion of ironclad, airtight validity is that it if the premises are true, it is absolutely *impossible* for the conclusion to be false. This lecture makes the point that that notion of validity was the gold standard for inference for 2,000 years.

   (a) In what ways do you think that was a positive force in the history of thought?

   (b) In what ways might it not have been so positive?

3. Knowing whether an argument is valid demands knowing whether the premises and conclusion are *actually* true. It demands knowing something about *possibility*: Could the premises be true and the conclusion be false? If we live in the actual world, how can we know things about mere possibility?

Exercises

More on validity and truth:

   (a) If we know that the premises of an argument are true and its conclusion is true, what do we know about validity?

   (b) If we know that the premises of an argument are false and the conclusion is false, what do we know about validity?

(c) If we know that the premises are false and the conclusion is true, what do we know about validity?

(d) If we know that the premises are true and the conclusion is false, what do we know about validity?

More on Aristotelian syllogisms:

Determine validity or invalidity for each of these using Venn diagrams:

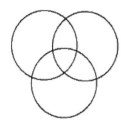   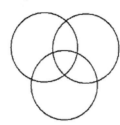   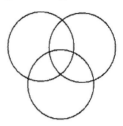

No M are P.	All M are P.	All M are P.
All Q are P.	All Q are P.	Some P are Q.
No M are Q.	Some M are not Q.	Some M are Q.

(See "Answers" section at the end of this guidebook.)

Ironclad, Airtight Validity
Lecture 7—Transcript

Professor Grim: In the last lecture I introduced the logic of Aristotle in terms of a central visualization, the Square of Opposition. I want you to call that image to mind. Here's a quick review. In Venn diagrams we use overlapping circles S and P for the two concepts involved. We black out areas to indicate that there's nothing in that area; we use an x to indicate there is something there. All S are P is in the upper left of the Square of Opposition. We black out the area of S that isn't in the overlap with P. If all S are P, there aren't any that aren't P. No S are P is on upper right. In this case there is nothing in the overlap of S and P, so we black that out. In the lower left is Some S are P. We use an x in the intersection to show that there's at least something in the overlap; something is both S and P. Some S are not P is lower right with an x in the outer S area to indicate there is at least one S outside the overlap with P. Some S's are not P. Then the magic. Contradictories are on the diagonals. They can't both be true, and they can't both be false. If one is true, the other has to be false. If one is false, the other has to be true. Implications run down the sides. If the upper one is true, the lower one is true. In that direction, truth is contagious; it flows from the top to the bottom. We called the correlate movement from the bottom up counter-indication. If the lower one is false, the upper one has to be false too. Contraries face each other across the top. Both can be false, but both can't be true. Subcontraries face each other across the bottom. Both can be true, some are, some aren't, but both can't be false.

This time I want to give you the rest of Aristotle's logic using another visualization. All of Aristotle's logic in two lectures. That's not bad, considering that Aristotle's logic remained the gold standard for over 2,000 years. Aristotle invented logic. At the core of that logic—and of all logic since—is one breathtaking idea: Maybe we can systematize thought. Concepts are the atoms of thought. In combination they form the molecules we call propositions. It's the fundamental relations between propositions that shine out in the Square of Opposition.

But it's the next step that is really the most important. Propositions don't just stand alone. Propositions go together to form arguments. In arguments you see the dynamics of thought. Reasoning involves moving from one proposition to another and another beyond that. That's the movement of argument, from premises to a conclusion. There was a Monty Python skit called The Argument Clinic. I'm sure you can find it online. In that skit, Michael Palin plays the client. John Cleese is the guy behind the desk. Michael Palin enters the room. "Is this the right room for an argument?" "I've told you once," says Cleese. "No you haven't." "Yes I have." It goes downhill from there. The whole joke turns on the term "argument." What Cleese means by "argument" is the everyday sense of bickering conflict. At one point the Client says: "An argument isn't just contradiction. An argument is a connected series of statements intended to establish a proposition." That is our sense of argument, the philosophical sense. A connected series of statements intended to establish a proposition. An argument in our sense represents the conceptual dynamic from proposition, to proposition, to conclusion. It's that movement of thought that is central to all rationality.

But of course it's not just intended transition that is important. We want to be able to evaluate arguments. When do a set of reasons really give you a solid inference? When does evidence really support a hypothesis? When do premises really entail a conclusion? What guarantees that a transition of thought is a rational transition? Those are questions about validity. When is the dynamics of thought rational? When is an argument valid? That's what Aristotle was after. That's what all of logic is after. The whole purpose of logic is to capture validity, the crucial category of argument evaluation. Maybe we can systematize thought by capturing what makes it rational. Maybe we can formalize validity.

Let me emphasize what that means. An argument is valid if the link between premises and conclusion is logically tight. I'll talk more about truth in a bit. The important first point, however, is that validity and truth are importantly different. The validity of an argument—the fact that there is a tight logical connection between premises and conclusion—doesn't guarantee that the premises are true. It doesn't guarantee that the conclusion is true. What you know when you know that an argument is valid is just this: If the premises are true, the conclusion must be true. That is important information, but it's conditional information. What you know is that the conclusion has to be true if the premises are true. In this lecture I want to talk about ironclad, airtight validity. I'll leave lesser shades to a later lecture. We're starting with Aristotle, and Aristotle took as his target validity in the highest degree.

Here's what it means for an argument to be valid in that strongest sense: If the premises are true, the conclusion absolutely must be true. If the premises are true, it's impossible for the conclusion to be false. You can't get a much tighter connection than that. If the premises are true the conclusion is absolutely guaranteed. Let me give you three short arguments, each with two premises and a conclusion. You tell me which ones are valid:

First argument:

> Rice is grown in every county of Louisiana.
>
> Any county that grows rice also harvests crawfish.
>
> Conclusion: Crawfish are harvested in every county of Louisiana.

Valid or invalid?

Second argument:

> All purple vegetables are poisonous.
>
> Carrots are a purple vegetable.
>
> Conclusion: Carrots are poisonous

Valid or invalid?

Third argument:

> The Declaration of Independence lays down a principle of self-determination: a right to "life, liberty, and the pursuit of happiness."
>
> "One man, one vote" is a core principle in democratic decision making.

Conclusion: "Presumed innocent until proven guilty" is a major principle of American law.

Valid or invalid?

Which of those are valid arguments? If you said the first two are valid, and that the third one is not, you're absolutely right. But look at how that differs from truth. The propositions of the first argument may well be true. But when the question is validity, their truth doesn't really matter. If they're true, the conclusion has to be too. Validity is about the logical connection between premises and conclusion.

All purple vegetables are poisonous.

Carrots are a purple vegetable.

Conclusion: Carrots are poisonous.

That's perfectly ridiculous. But it's also perfectly valid. Not all purple vegetables are poisonous, but suppose they were. Carrots aren't a purple vegetable, but suppose they were. If those two premises were both true—if carrots were purple vegetables and all purple vegetables were poisonous—it would also have to be true that carrots were poisonous.

The first was a valid argument with true claims. The second was a valid argument with silly claims. In the third argument we have all true claims again, but it's not valid. The Declaration of Independence lays down a principle of self determination: a right to "life, liberty, and the pursuit of happiness." "One man, one vote" is a core principle in democratic decision making.

Conclusion: "Presumed innocent until proven guilty" is a major principle of American law. Those are all propositions about principles, they're all true, but that's not enough to make a valid argument. They lack the logical connection that validity demands.

So let me see if you've got this. Suppose you've got an argument in which all premises and the conclusion are true. Is that a valid argument? Not necessarily. It's the connection that matters. You might have a valid argument in which all premises and the conclusion are true. That would be great, but validity alone doesn't tell you that. You might have a valid argument in which all the premises and the conclusion are false. The purple carrots was one of those. You might, in fact, have a valid argument in which the premises are false but the conclusion is true. Here's one:

Everything poisonous is orange.

Carrots are poisonous.

Therefore carrots are orange.

That a valid argument. If the premises are true, the conclusion has to be true. The premises are false, but they entail a conclusion that is true anyway. I guess sometimes two wrongs do make a right. There is only one pattern of truth and falsity that validity rules out. You will never find a valid argument that goes from true premises to a false conclusion. A valid argument is one in which, if the premises are true, the conclusion must be true.

Aristotle's goal—the goal of all logic—is to capture what it is about an argument that makes it valid. The core insight, validity doesn't turn on what an argument is about. Validity is a matter of structure. Here, for example, are two arguments about very different things:

Argument one. Another silly one:

All carrots are legumes.

All legumes are delicious.

Conclusion: All carrots are delicious.

Argument two, not so silly:

All countries in the Middle East are in turmoil.

All states in turmoil have unpredictable futures.

Conclusion: All countries in the Middle East have unpredictable futures.

If one of those is valid, the other one is valid. Why? Because validity depends purely on the structure of the argument. Both of them have this structure:

All As are Bs.

All Bs are Cs.

Conclusion: All As are Cs.

Validity depends not on content but on structure. I don't think I would have ever seen that. Reasoning is so thoroughly human, as context-bound as the things it's about, as variable as the humans who do it. I don't think I would ever have seen that what makes it real reasoning—that what makes something a valid argument—is a matter of abstract structure. For that we needed an Aristotle.

Aristotle worked out the structure of ironclad validity in terms of three-step arguments called syllogisms. A syllogism takes us from categories, to propositions, to steps in an argument. It uses three categories over all, A, B, and C. It links those categories in pairs to make categorical propositions, All As are Bs, Some Bs are not Cs. The syllogism itself is an argument in three propositional steps with maybe a structure like this:

All As are Bs.

Some Bs are not Cs.

No Cs are As.

Some syllogisms are valid. Some are not. The one I just gave you is a syllogism, sure enough, but it's not a valid syllogism.

So is there some way of sorting the sheep from the goats? Is there some way of sorting the valid syllogisms from the invalid ones? There are lots of ways. The Medieval monks memorized an exhaustive list using a code in terms of women's names to remember which syllogisms were valid and which weren't. Yep. Those celibate Medieval monks learned their logic by memorizing women's names. Barbara is virtuous, and valid. Celise is not. But there is a much easier way. Here, again, visualization is the key. We can do all of Aristotle's logic with the Venn diagrams you already have at your disposal. Remember the Venn diagrams for the categorical propositions: All S are P. No S are P. Some S are P. Some S are not P.

Those are the propositional elements of Syllogistic logic. But we said that syllogisms involve three propositions. Each proposition of the syllogism—each step in the argument—uses just two of our three categories. In order to use Venn diagrams to track validity we need to overlap three circles into a triangular Venn diagram. Our first proposition, linking categories A and b, will look something like this. Our second, linking B and C, will look something like this. Our third, linking A and C, will look like this.

All three categories and all three propositions come together in the syllogism. We put them together like this. The result is an overlap of three categories, just as the syllogism always involves three categories. The triangle is formed from three categorical propositions, pairs of linked circles. You should learn to see each of those pairs individually in that diagram.

Let's take an easy syllogism. The Medieval monks called it Barbara:

All proteins are polypeptides.

All polypeptides function by elaborate folding.

Therefore all proteins function by elaborate folding.

That one is obviously valid. I'll use it to introduce the technique, which can then be used for cases that aren't so obvious. We have three categories, proteins, polypeptides, and function by elaborate folding. We label the three circles in the diagram accordingly. We now enter each of our premises in

Venn-diagram style. So enter our first premise: All proteins are polypeptides. In order to do this, we act as if that bottom circle isn't even there. Fill in the shaded area just like we did before for All S are P. That will mean shading in that whole left portion of proteins, all proteins are polypeptides.

Now take the second proposition: all polypeptides function by elaborate folding. That premise is just about polypeptides and folding. We, therefore, concentrate on that part of the diagram, ignoring the proteins circle. How do you represent "all polypeptides function by elaborate folding"? By striking out any polypeptides that aren't in the overlap with "function by elaborate folding."

Now step back and look at the diagram as a whole. Philosophers talk of logical space, and what we have just done is to map the logical space of our first two propositions. If those are both true, the diagram shows what the categorical universe must look like. Here comes the last step, the crucial step, in telling whether a syllogism is valid. Some people find it the trickiest. We want to look at our third proposition—the conclusion—and ask whether its information is already in the diagram we've constructed for our premises. If it is, the argument is valid. If it's not, the argument is invalid. If the concluding information is already provided by the premises, it has to be true when they are—valid. If the premises don't give us that information, the conclusion doesn't follow from the premises—invalid.

In order to test for validity, we want to know whether the concluding information is already there in the logical space of the premises. So look at the diagram. What would it mean to draw in the conclusion "all proteins function by elaborate folding"? That would mean looking at just protein and elaborate folding, those two circles on the left. For it to be true that all proteins function by elaborate folding, the protein area outside of the overlap would be blacked out. Take a look. That area is already blacked out. There aren't any proteins in our diagram that aren't already squeezed into some part of the "function by elaborate folding" area. That means that the information in step three—the information in the conclusion—was already there in the premises. The sign of validity. If the premises are true, the conclusion has to be true. The information in the premises is sufficient to give us the information in the conclusion. The argument is valid.

Think about that a minute, and let me emphasize two points. The first point is that what I've just told you actually does make sense. Let me convince you of that. I'll then deal with some lingering disappointment. Here is the makes-sense part. If we graph the information in the premises, and the information offered by the conclusion is already there, the argument is valid. Remember, we're talking about the highest degree of validity. If the premises are true, the conclusion has to be true. That is what is happening here. If the premises are true, the conclusion has to be true precisely because the information in the conclusion is already there in the premises. If that weren't the case, if somehow the conclusion was telling you something new, then the information in the conclusion wouldn't already be in the premises. It would be independent of the premises, a rogue conclusion from somewhere else, not from the premises, and thus the argument wouldn't be valid. So it does make sense. A valid argument in the deductive sense is one that squeezes out in the conclusion some information that you already had. Kant speaks of an analytic proposition as one in which the predicate is already contained in the subject term. A valid deductive argument is one in which the information in the conclusion is already contained in the premises. It, therefore, makes sense that you can show an argument to be valid or invalid depending on whether the information in the conclusion is information already there in the premises. But that's a little disappointing. It means that a syllogism will never give you any really new information.

I have a memory along those lines that I can date to when I was five. Somehow I had gotten the idea of a syllogism, or at least of stepped reasoning, though I'm not sure where. I would try to deduce conclusions as I imagined Sherlock Holmes would. Walking up the street to Wendy's house, I remember thinking something like this:

If Wendy's brother is home, Brian is home.

If Brian is home, Wendy won't be the only one home.

So, if Wendy's brother is home, Wendy won't be the only one home.

But that's obvious! I knew that already! I wanted ironclad, airtight, valid reasoning that would give me some genuinely new information. But whenever it was air tight, that kind of reasoning never seemed to give me anything new. I found that very disappointing when I was five. Maybe I'm still a little disappointed.

You can do all of Aristotle's logic, all the logic of the Medievals, all the logic that we had for 2,000 years, with those three circles. It gets a little trickier with the x's in "some are" or "some are not," but not a whole lot. Consider, for example:

Anything alkaline turns red litmus paper blue.

Some poisons are alkaline.

Therefore, all poisons turn red litmus paper blue.

Valid or invalid? Use the three circles again, this time labeled alkaline, poison, and turns litmus paper blue.

Now, using your ability to concentrate on just two circles at a time, put in the information of the first premise, anything alkaline turns red litmus paper blue. That means nothing alkaline is outside of the turns-red-litmus-paper-blue area. We black out that upper left portion.

Now what about the second premise, some poisons are alkaline? That means that there is something—we'll say at least one thing—that is in the intersection of the left two circles. When we go to put it there, of course, we see that part of that area is blacked out already, nothing there, so our x can't go there. It will have to wedge into that middle area.

There, right in front of us, we have all the information contained in our premises. Step three, we test the syllogism by seeing whether the conclusion would tell us anything new. What does the conclusion say and how would we graph that? All poisons turn litmus paper blue. That would mean that there are no poisons outside of the turns-litmus-paper-blue area. Is that information there already? Is that information we have already? No. If it were, that lower area would be blacked out already, but it isn't. We have a rogue conclusion, an addition to logical space beyond our premises. It doesn't follow. That argument is invalid.

There were actually two little tricks in that last example that are worth noting as we go by. One is this: I quickly went from "some poisons are alkaline" to "at least one…" That is in fact a standard move in logic, to read "some" as meaning "at least one." It is so standard, in fact, that it has become the technical meaning for "some" in standardized tests. If you ever have to take the GREs, the LSATs, the MCATs, and there is a question about "some," just re-read it as "at least one." That makes things easier. But even logicians have to admit that it doesn't fit very well with how we ordinarily talk. In normal usage, the word "some" at least implies a plurality. If I just got married, I have a wife. Count them, one. So I guess it's true that I have at least one wife. But it would certainly be peculiar to say "I just got married, so I have some wives." The other trick in the last example is a visualization trick. We drew in the universal premise first: Anything alkaline turns red litmus paper blue. That told us, when our x arrived, that it had to squeeze over into the remaining white area.

What if we had done the "some" proposition first? Where would we have put the x? We would know it would go in the overlap of poison and alkaline, for example. But on which side of that other line that's in there?

In that case we would put it on the line in the relevant area, as a way of reminding us that it might be on either side. A further universal premise might tell us that we have to nudge it one way or the other.

That's how it works. In order to really incorporate Aristotle's logic, one needs to practice with lots of cases, but the technique is guaranteed. Every time, if you do it right, you will be able to distinguish a deductively valid categorical syllogism from an invalid one. Here's another one for you to try:

All absentee ballots have been counted

Some counted ballots were illegible

Therefore some absentee ballots were illegible

Stop the tape, take a minute, and work it out. Valid or invalid? Here's what your diagram of the premises should look like. The conclusion is some absentee ballots were illegible. Is that information already there? No. It would be there if we knew that x that is somewhere in the overlap between counted and illegible were in that central area that is also the overlap between absentee and counted ballots. But we don't know that. We only know it's somewhere in that larger overlap. It could be in either portion. That makes sense. The first premise tells us that all absentee ballots were counted. But we don't know that only absentee ballots were counted. The second premise tells us that some counted ballots are illegible. It doesn't tell us which ones. The possibility is therefore left open that the counted and illegible ballots are not the absentee ones. In that case it wouldn't be true that some absentee ballots were illegible. In that case the premises would be true but the conclusion false. The truth of those premises doesn't guarantee truth of conclusion. The argument is invalid.

What I want you to remember is that three-circle technique and the concept of airtight, ironclad validity that it represents. With that visualization you can test validity for any three-step syllogism. With that visualization you have all of Aristotle's logic at your fingertips. But let me return for a minute to that aspect of disappointment. The concept of airtight, ironclad validity sets a standard for inference in which all the information in a conclusion has to already be contained, somehow, in the premises. For two thousand years, ironclad, airtight validity was held up as the gold standard for all knowledge. But that meant all information had to flow from above from something you already knew. Mathematics is about the only discipline that works that way. It's no surprise that axiomatic geometry was the standard for all knowledge.

Aristotle's legacy, therefore, had a downside too. It was only with the rise of empirical science in figures like Galileo that an alternative became clear. Maybe an inference from observed cases to universal laws is what we need. Maybe it's legitimate. Maybe it's rational even if it isn't ironclad guaranteed in Aristotle's sense. If you weaken the demands on validity, you get the possibility of rational information expansion for the first time: reasoning, together with observation, that offers new information. Iron clad won't give you new information. Anything that gives you new information won't be ironclad. We have bought the benefits of science by weakening our demands on validity, by giving up full Aristotelian certainty. I think that was worth the price. We'll start talking about reasoning and validity in that wider sense a few lectures down the line. Next time, however, I want to take a break for something equally important but a little more fun. Next time I want to offer a workshop on creative thinking.

Thinking outside the Box
Lecture 8

In this lecture, we'll take a break from Aristotle—a break from visualizing argument structure—to talk about what may be the most important kind of thinking of all and, perhaps, the most fun: creative thinking... lateral thinking...thinking outside the box. Here, we'll discuss a number of problems or puzzles and how they can be solved through creative thinking.

The Phenomenon of "Mental Set" or "Expectancy"

- A cowboy rides into town on Tuesday. He stays in town for exactly three days. On the first day, he works in the general store. On the second day, he works in the livery stable. On the third day, he hangs out in the sheriff's office. He leaves promptly the next day and rides out of town on Tuesday.

- How can that be? The answer is that his horse is named Tuesday. He rides into town on Tuesday, stays three days, and rides out again on Tuesday.

- This story illustrates the phenomenon of **mental set**, or expectancy. You are led into thinking about things in a certain way. You aren't able to change perspectives quickly enough to solve the puzzle. The whole setup is intended to get you to think in terms of days—one day in the general store, one day in the livery stable—blinding you to the fact that Tuesday could also be the name of a horse.

- Mental set or expectancy is a well-established psychological phenomena. We learn to anticipate particular patterns or continuations of patterns. We get stuck in a pattern rut.

Breaking Out of the Pattern Rut

- Creative thinking demands the ability to break out of the mental set that a problem brings with it. That mental set is the box. Creative thinking is thinking outside the box.

- Often, real-life problems, when they can be solved at all, can be solved only by looking at things in an entirely different way. The same holds for life's real opportunities. Opportunities may be invisible unless you can look at a situation differently—sideways—until you can think outside the box. Real problems and real opportunities demand more than standardized thinking. For real life, we often need some real creativity.

- Here's an example based on real events: A woman is driving across the desert, without a cell phone, and has a flat tire in the pouring rain. She manages to jack up the car and takes off the five nuts on the wheel. Just then, a flash flood washes the nuts away beyond recovery deep in the sand. How can she get the car back on the road?
  - Here's a creative solution: She has lost the five nuts from one wheel but still has five nuts on each of the other three.

  - Borrow a nut from each of the others, use those three nuts on this wheel, and she can be on her way.

- In another case, some friends are swimming together in a pond. The area is residential; the pond is ringed with lawns and gardens, but it is deep and has a tangle of water plants at the bottom. One of the friends becomes entangled in the pond plants, underwater, just a foot from the surface of the pool. He is struggling desperately, but his friends realize they won't be able to free him in time.
  - What else can the friends do?

  - The story is that they grabbed a garden hose from a yard nearby, cut a two-foot segment with a pair of garden shears, and gave him a makeshift snorkel to use until they could get help to work him free.

Cultivating Creative Thinking: Change Your Expectations

- Creative thinking can't be taught with a template, but it can be cultivated. You can learn to broaden the horizon of possibilities you consider. You can learn to first recognize what the box is and then deliberately look for a way to think outside it.

- For example, take 12 toothpicks (or pencils) and arrange them into four squares. Here's the challenge: By removing only two toothpicks and without moving any other toothpicks, create not four squares but only two. They must be squares, not rectangles. And there should be only two squares: no floating toothpicks that aren't part of any square.

  ○ Here's the solution: Remove two neighboring toothpicks from the original cross. If you do that, you'll be looking at one large square with a smaller one tucked into one corner. By removing two toothpicks and moving none of the others, you end up with two squares, no leftovers.

  ○ Of course, those squares aren't the same size, and they're not two of the original four. If there was a mental set imposed by the problem, it was the expectation that the two remaining squares would be two of the four you started with. That was the mold you had to break in order to think about the problem in a different way.

- Here's another exercise: Use a pencil to connect all nine dots below by drawing four straight lines, without lifting your pencil from the paper. Draw a line, then start the next one from where that line stopped, and so on, for a total of just four straight lines. Those four lines must go through the middle of all nine dots.

  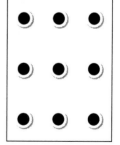

  ○ The key here is that your lines don't all have to start and stop at a dot.

  ○ In order to solve the problem, your lines will have to extend beyond the box of nine and turn some corners outside of it.

Cultivating Creative Thinking: Break Habits of Thought

- Here's another problem: Arrange 10 pennies in such a way so that you have five straight lines of pennies—five rows—each of which has exactly four pennies in the row. The solution is to draw a five-pointed star and place a penny at each of the five points and each of the five inside corners. When you do that, you'll find that you have five lines, each with four pennies, for a total of 10.

- If you didn't solve that one, what mental set got in the way? Was it that you expected the lines to be parallel? Was it that you expected them to cross only at right angles? If you analyze your own thinking on problems such as these, you'll get a feeling for how you can step outside the box.

- What we're doing is trying to develop an ability to think beyond the expected...to look for options that aren't anticipated in the question or the situation. We're trying to develop an ability to think creatively and perhaps even develop a habit of breaking habits of thought.

Cultivating Creative Thinking: Overcome Conceptual Limitations

- For the last problem, work side by side with a friend: You have 12 toothpicks. Give six to your friend. Lay three of those toothpicks down in front of you to form a triangle. Now, by adding just three more toothpicks, form four triangles, of precisely the shape and size as the one you just made.

- How do you get four triangles with just three more toothpicks? Here, creative thinking takes you literally into another dimension. On top of your original two-dimensional triangle, form a three-dimensional pyramid with your three additional toothpicks. There you have it: four triangles of precisely the shape and size as the original. If you want it to stand on its own, of course, you'll need to stick the toothpicks together. (Try miniature marshmallows.)

- In order to get that problem, you had to move conceptually beyond two dimensions and into three. There is a wonderfully creative book on that theme, written in 1884 by Edwin Abbott, called *Flatland:*

A Romance of Many Dimensions. Abbott's narrator is "A Square," who lives in a two-dimensional plane, unable to conceive or understand what happens when things enter that two-dimensional world from a three-dimensional outside. It's a charming story in its own right and a wonderful metaphor for the kinds of conceptual limitations we inevitably carry with us.

Thinking Better

- We've been talking all along about thinking outside the box. That's a metaphor, of course, but some recent experiments indicate that it may be a particularly apt metaphor. One hundred undergraduates were given a word task designed to measure innovative thinking. Half of them had to do the task inside a five-foot-square box made of plastic pipe and cardboard. Half did the task outside and next to the box. Interestingly, those who were doing their thinking "outside the box" came up with 20 percent more creative solutions.

- The same kinds of differences showed up between students who had to walk a particular path as they worked on a problem, as opposed to those who could wander at will. Those who could "leave the beaten path" came up with 25 percent more original ideas.

- In the first lecture, we emphasized the idea that physical context can make a mental difference, and these experiments illustrate that point. Sometimes, creativity demands that you really do get away from it all.

- One final creative thinking puzzle: It is a dark and stormy night, and you're driving in your sports car, which has only two seats. Suddenly, by the side of the road, you see three people stranded at a bus stop.
  - One is a stranger who is having a heart attack at that very moment. Another is a childhood friend who has often saved your life and has long been begging to ride in your sports car. The third person is the man or woman of your dreams—your soul mate—whom you may never see again. You have just one empty seat in your car. Who do you pick up?

○ The solution requires you to stop thinking that you have to pick up just one person and that only you can be behind the wheel. Ask your friend to drive the heart attack victim to the nearest hospital, then wrap your coat around the person of your dreams.

Term to Know

mental set: In perception, the background expectation that may influence what is perceived; that is, when one is expecting a normal deck of cards, mental set may lead one to ignore altered cards, such as a red ace of spades.

Suggested Reading

Abbott, *Flatland: A Romance of Many Dimensions.*

Burger, *Sphereland: A Fantasy About Curved Spaces and an Expanding Universe.*

Weston, *Creativity for Critical Thinkers.*

Questions to Consider

1. Below are a few more easy "lateral thinking" problems.
   (a) 1985 pennies are worth almost $20.00. Why is that?

   (b) Is it legal for a man in Tennessee to marry his widow's sister?

   (c) A clerk in the butcher shop is 5 feet, 10 inches tall. What does he weigh?

   (d) Can you rearrange these letters to make one word? new door

   (e) A woman has two sons, born at the same minute of the same hour of the same day of the same year. But they are not twins. How can that be?

(See "Answers" section at the end of this guidebook.)

2. Are there aspects of creativity that you think can be taught? Are there aspects you think cannot be taught?

Exercise

Write down 25 things you can do with a bucket of water. Time yourself.

- Can't come up with 25? Put the exercise aside and come back to it another time. Did you include water as a cleaning fluid? As something to put fires out with? Did you consider its weight? Did you think of using the bucket by itself?

- Did you come up with 25 in less than 4 minutes? A very respectable showing.

- Less than 3 minutes? Outstanding.

- Less than 2 minutes? Phenomenal.

Now can you come up with 40 things to do with two paper bags full of sand?

Thinking outside the Box
Lecture 8—Transcript

Professor Grim: In this lecture I want to take a break from Aristotle, a break from visualizing argument structure, in order to talk about what may be the most important kind of thinking of all. Also the most fun, Creative thinking, lateral thinking, thinking outside the box. In order to deal with creative thinking I'm going to have to use a different kind of lecture. This will be more of a workshop than a lecture, a workshop in creative, sideways thinking. Because of that, you will have to participate in a different way. I'll need you to stop the lecture occasionally in order to think about a problem and manipulate some objects. When you have a solution, start the lecture again and move on.

You'll need a creative thinking kit for this workshop. Your kit will have to include each of these things. As they say, familiar items from around the house. 12 toothpicks, pencils would work, too.10 small circular objects: pennies, buttons, or the like. If you have 10 of those miniature marshmallows, those would be great. Two wrapped lollipops, and a picture that appears in your guidebook. You'll need a pencil to write with. You'll also need a friend. At the very end, I'll be asking you to bring in a friend for participation. If you can assemble that list of things, and you are in a situation in which you can start and stop the lecture at will, we're ready to go. If that's not possible right now, go on to another lecture and come back to this one. Ready? Let's go.

I'll start with an experiment. I'm going to spell a word. I want you to pronounce it. So how is this pronounced? M.A.C. That's right, mac. M.A.C.N.A.M.A.R.A. Quick now. MacNamara, that's right. M.A.C.H.I.N.E. Did you say MacHine? If so, think about it again. Machine spells machine.

Here's another experiment, this time in the form of a puzzle. A cowboy rides into town on Tuesday. He stays in town for exactly three days. On the first day he works in the general store. On the second day he works in the livery stable. On the third day he hangs out in the sheriff's office. He leaves promptly the next day, and rides out of town on Tuesday. How can that be? The cowboy puzzle is as old as the hills, what they call a chestnut. If you don't know the answer, pause the lecture for a few minutes to think about it, then come back. How can that be? The answer is that his horse is named Tuesday. He rides into town on Tuesday, stays three days, and rides out again on Tuesday.

Both of those experiments illustrate the phenomenon of mental set, or expectancy. You are led into thinking about things in a certain way. You aren't able to change perspectives quickly enough to solve the puzzle. In the machine case, the whole setup leads you to expect a hard c in the next mac word, not the soft ch of machine. In the Tuesday case, the whole setup is intended to get you to think in terms of days—one day in the general store, one day in the livery stable—blinding you to the fact that Tuesday could also be the name of a horse. Mental set, or expectancy, is a well-established psychological phenomena. We learn to anticipate particular patterns or continuations of patterns. We get stuck in a pattern rut. Creativity involves breaking that mental set, getting out of that rut, exploring unanticipated options. Interestingly, mental set can change with the culture over time. Here's a chestnut that no longer works. It stumps almost nobody. But it used to. A father and his son are both hurt in an automobile accident, and the son is rushed to the hospital for immediate surgery. The surgeon looks down at the boy and says, "I can't operate on him, he's my son." How can that be? The answer, of course, is that the surgeon is his mother. At one time the cultural presumption was that a surgeon must be male, and that made the puzzle hard. It isn't the problem that has changed, it's the mental sets of our cultural expectations.

Creative thinking demands being able to break the mental set, or expectancy, that a problem brings with it. That mental set is the box. Creative thinking is thinking outside the box. Don't get me wrong. For some problems you need a standard pattern, a template designed for that kind of problem. That holds, for example, for the kinds of problems that appear on standardized tests. Standardized tests often have a section with problems like this: There are five women at the fair. Mrs. A, Mrs. B, Mrs. C, Miss D and Miss E. On the basis of the following elimination, tell us who is wearing which color: red, gray, brown, or yellow. Two of the women are wearing the same color. The married women are wearing the brightest colors. Mrs. C hates her red dress and wishes she were wearing the same color as the other married women.

Miss E is wearing gray. The standard template for solving problems like that is the visualization we mentioned in an earlier lecture. Construct a matrix—a checkerboard of squares. Mrs. A, Mrs. B, Mrs. C, Miss D, and Miss E are written above the columns. Red, gray, brown, and yellow appear as labels on the rows. By process of elimination you can use the information given and fill in the squares to find out who is wearing what color.

There are only so many types of questions on standardized tests. What is taught in prep courses are a handful of templates for those kinds of questions. Once you have mastered the templates, the only challenge is to recognize which template to apply to a specific problem. Real life tends not to be like that. Most real-life problems don't come in a tidy format with a standard cookie-cutter answer. Often real life problems, when they can be solved at all, can be solved only by looking at things an entirely different way. The same thing holds for life's real opportunities. Opportunities may be invisible unless you can look at a situation differently—sideways—unless you can think outside the box. Real problems and real opportunities demand more than standardized thinking. For real life we often need some real creativity. How about this one? A man walks into a bar. The bartender pulls a gun from behind the counter and points it at him. The man says, "Thank you." and leaves. Why? Hmm. Why would the bartender pull a gun? And why on earth would the man say, "Thank you"? You may know the answer immediately. If not, think about it and then come back. You've undoubtedly heard that a surprise or a shock is a sure cure for hiccups. The man entering the bar has the hiccups. The bartender pulls a gun and points it at him to help with the hiccups. That's why he says, "Thank you."

Let me give you another puzzle before I move on to some real-world examples. A man lives in a penthouse apartment in New York City. When he goes to work, he takes the elevator down to the lobby. When he comes home he only takes it halfway up unless it's raining, when he takes it all the way up. Why? Press pause and take a little time to think about it. Then come back. Why does he only go halfway up to his apartment unless it's raining? Because he is very short. He can only reach the lower buttons in the elevator. But when it rains he carries his umbrella. With his umbrella he can punch the top buttons. What was the mental set, or expectancy, that you had to break in this case? Unlike the others, it wasn't something imposed by the language of the puzzle. Here the whole key was to think about something that wasn't mentioned at all—how tall he was.

Let me give you some stories that go beyond puzzles. I've heard these are based on real events. I can't vouch for that, but they certainly do make the point that creative thinking may not be just an entertaining add-on; it may be a necessity in a crisis. A woman is driving across the desert without a cell phone and has a flat tire in the pouring rain. She manages to jack up the car and takes off the five nuts on the wheel. Just then, a flash flood washes the nuts beyond recovery deep in the sand. How can she get the car back on the road? If you know the answer, stay with me. If not, press pause. Here's a creative solution. She has lost the five nuts from that wheel, but still has five nuts on each of the other three. Borrow a nut from each of the others, use those three nuts on this wheel, and she can be on her way.

Another case. Some friends are swimming together in a pond. The area is residential. The pond is ringed with lawns and gardens, but it's deep and has a tangle of water plants at the bottom. One of the friends becomes entangled in the pond plants underwater, just a foot from the surface of the pool. He is struggling desperately, but his friends realize they won't be able to free him in time. What else could they do? The story is that they grabbed a garden hose from a yard nearby, cut a two-foot segment with a pair of garden sheers, and gave him a makeshift snorkel to use until they could get help to work him free. That's creative thinking when it's needed most. I'm not even sure they needed the garden shears; the hose itself could work as an air supply.

Creative thinking can't be taught by handing you a template., but can be cultivated. You can learn to broaden the horizon of possibilities you consider. You can learn to first recognize what the box is and then deliberately look for a way to think outside it. So, here are some exercises along those lines. I'll talk in terms of toothpicks, though you could use pencils or any other 12 sticks of the same size. I want you to take your 12 toothpicks and arrange them into this pattern, four squares made of 12 toothpicks. Square one, square two, square three, square four.

Now, here's the challenge: By removing two toothpicks, only two, and without moving any other toothpicks, I want you to end up looking not at four squares but at only two. They have to be squares, not rectangles, and I want two squares with nothing else—no detritus, no floating toothpicks that aren't part of any square. Press pause and take a few minutes to figure out how to cut the figure down from four squares to two by removing just two toothpicks. When you think you've got it, come back.

Here's the solution. Remove two neighboring toothpicks like this. If you do that you'll be looking at one large square with a smaller one tucked into one corner. By removing two toothpicks, moving none of the others, you end up with two squares, just like I asked for, no leftovers, just like I insisted. Of course those squares aren't the same size and they're not two of the original four. If there was a mental set imposed by the problem, it was the expectation that the two remaining squares would be two of the four you started with. That was the mold you had to break in order to think about the problem a different way.

Let's try another exercise. In your guidebook is a picture with nine dots, three evenly spaced rows of three. In fact, they look just like the points at which our toothpicks met in the last case. With a pencil, I'm going to ask you to connect all nine dots by drawing four straight lines without taking your pencil off the paper. You'll draw a line, then start the next one from where that stopped, and so forth, for a total of just four straight lines. Those four straight lines will go through the middle of all nine dots. See if you can do that, four straight lines, passing through all the dots, without lifting your pencil from the page. Pause the lecture and come back when you think you've solved it.

The key is that your lines don't all have to start and stop at a dot. In order to solve the problem, your lines will have to extend beyond the box of nine and turn some corners outside of it. There are several ways to do this. Here's one. Draw your first line on the diagonal from the top left dot, through the middle dot, to the bottom-right dot. When you are at the dot at the bottom, draw your second line straight up the right side of the box but keep going beyond that corner of the box. Go up far enough so that drawing your third line at a 45 degree angle down to the left will catch those two highest unconnected dots.

You will cross your original line. Continue that third line outside the box until you can turn right to catch the bottom row of dots with your fourth line. Like I said, there are other ways of drawing the lines, but they all require extending lines beyond the box of dots. Is that something like what you did? If so, you're thinking creatively. If so, you're literally thinking outside the box.

Here's a third problem. Take out your 10 circular objects, marshmallows, buttons, whatever. I'll call them pennies. I want you to arrange those pennies in such a way that you have five straight lines of pennies—five rows, each of which has exactly four pennies in the row; five straight lines of four pennies using just 10 pennies in all. This one is harder for most people. I know it was for me. Take a few minutes and see if you can get it.

Here's the solution. Draw a five-pointed star. It will have five points and five inside corners, the star's armpits. Make the picture big enough to put a penny at each outer point and each inner corner. If you drew your star with a flourish of overlapping lines, like people often do, put a penny at each outer point and at the inside corners where your lines cross. When you do that, you'll find that you have five lines, each with four pennies, for a total of 10. If you didn't solve that one, what mental set got in the way? Was it that you expected the lines to be parallel? Was it that you expected them to cross only at right angles? If you analyze your own thinking on problems like these, you'll get a feeling for the kinds of boxes you need to break out of. What we're doing is trying to develop an ability to think beyond the expected, to look for options that aren't anticipated in the question or the situation. We're trying to develop an ability to think creatively, maybe even develop a habit of breaking habits of thought.

For the next exercise take just eight of your round objects. I want you to imagine that these are eight coins, all of which look and feel alike. One of them is different, however, in ways indistinguishable by look or feel. All are exactly the same weight except one. One of the eight weighs just a little more than the others. You also have a balance scale. It won't tell you the precise weight of anything, but it will tell you whether the thing you put on one side is heavier or equal to the thing on the other side. It won't say three ounces, but if you put anything on one side and anything on the other side, it will tell you which side is heavier, or whether both sides weigh the same. Okay, each time you put things into the scale and weigh them, that counts as one weighing. You have eight coins in front of you. In just two weighings, how would you find the single coin that weighs more? I want you to tell me, first I'll weight this, then I'll weigh that, and that will tell me, precisely, which of the eight coins is heavier. Take a minute, think about that one, and then get back to me.

When most people first approach the weighing problem, they think of dividing the eight coins in half. Four on one side of the scale, four on the other side. That would tell you that the heavy coin is on one side, maybe this one. But then how do you get the four coins on the heavy side down to the single heavier coin in just one weighing? If you divide them in half, you'll only know that the heavy coin is one of two, and that's not what we want. Most people's initial expectation is that the problem will be solved with that kind of symmetry. Lots of problems are. But for this one you have to think asymmetrically. On the first weighing, put three coins on each side of the scale, six coins on the scale—three on this side, three over here, and two left out over here. With that first weighing one of two things might happen. It might come out that both groups of three weigh the same. That would be great. It means the heavier coin has to be one of the remaining two, and a single additional weighing will find it. The other thing that might happen is that one of the groups of three weighs more. You'll then know that the heavy coin is one of the three. So, do the same thing. Weigh just two of the three coins. If one of these weighs more, you've found your coin, your man. If they weigh the same, you know that the one that doesn't weigh the same is that one. Once again, you've found your man, or, well, you've found your coin.

This one's easier. Here's a picture of a cake. It's just a squat little cylinder with a dotted line across the diameter of the top. What that dotted line shows you is how to cut the cake into two equal parts. Now, just because a problem calls for creative thinking, doesn't mean it has to be hard. And maybe you're better at creative thinking already. If I ask you to make two straight cuts to cut the cake into four equal parts, I'm sure you could do that. I'll ask for something al little harder. Give me three straight cuts that will cut the cake into eight equal parts. Start the lecture again when you have the answer. The answer is to do two cross cuts across the top, like you would in dividing the cake into four pieces. Then do one sideways, all the way through the body of the cake. That's thinking sideways. You end up with eight equal pieces, equal by volume, anyway. Though, if you really like frosting, you still want one of the top pieces.

The last item on your list was a friend. For the last two problems I want you to work side by side with your friend. We'll see who solves these problems first. You have 12 toothpicks. Give six to your friend. I want each of you to lay three of those toothpicks down in front of you to form a triangle. Here's the challenge: By adding just three more toothpicks, I want each of you to form four triangles of precisely the shape and size as the one you just made. That's right, just six toothpicks, but four triangles of precisely the size and shape as the one you started with. Let's see who gets it first. Here's our original triangle. How do you get four triangles of exactly that size and shape with just three more toothpicks? Here creative thinking takes you, literally, into another dimension. On top of your original two-dimensional triangle, form a three-dimensional pyramid with your three additional toothpicks. There you have it, four triangles of precisely the shape and size as the original. If you want it to stand on its own, of course, you'll need those toothpicks to stick together. That's where the miniature marshmallows work great.

In order to get that problem you had to move conceptually beyond two dimensions and into three. There is a wonderfully creative little book on that theme written in 1884 by the English schoolmaster Edwin Abbott. It's called *Flatland: A Romance of Many Dimensions*. Abbott's narrator is "A Square," who lives in a two-dimensional plane, unable to conceive or understand what happens when things enter into that two-dimensional world from a three-dimensional outside. It's a charming little story in its own right and a wonderful metaphor for the kinds of conceptual limitations we inevitably carry with us.

It's important that your friend is along for the ride on the last exercise, too. Here, again, we'll see who finds the answer first. Give your friend one of the wrapped lollipops. I want each of you to put that lollipop, still wrapped, on a table in front of you. Now, you can't touch your lollipop with your hands or elbows. Without getting your nose closer than six inches to the table, I want you to get that lollipop unwrapped and into your mouth. Press pause and come back to the lecture when you're happily sucking on your lollipop.

Did you get it? I hope so. You couldn't have picked it up with your mouth, because you weren't supposed to get your nose closer than six inches to the table. But you could use your mouth in another way. You could use your mouth by turning to your friend and saying, "Will you please unwrap that lollipop and put it in my mouth?" Sometimes the best way to solve a problem, sometimes the most creative way, is not to try to solve it alone. Sometimes the best way to solve a problem is to ask a friend to help.

The whole idea of creative problem solving is to break the mold, change the mental set, and approach the issue from an entirely new direction. That can't be taught by handing you a template, but it is an ability that can be developed with practice. Here's an exercise for still more practice. Take an ordinary object, a brick, for example. How many things can you think to do with a brick? The more you can think of, the more creative your thinking. Most people can think of using a brick as a doorstop, perhaps a bookend, but there is no limit to the things you can do with a brick. Put it on your head and learn to walk like a lady, break a window, get into your car when the keys are locked inside,   save water in your toilet tank, use it as weapon, wrap it as a present, grind it on the pavement to make some war paint, use it to prop up a table, open nuts with it,   kill bugs, step on it to reach the high shelf, use it as a weight measure, use it as a ruler, warm it up and put it in bed in the cold weather. See if you can figure out 50 things to do with a brick.

We've been talking all along about thinking outside the box. That's a metaphor, of course, but some recent experiments indicate it may a particularly apt metaphor. One hundred undergraduates were given a word task designed to measure innovative thinking. Half of them had to do the task inside a five-foot-square box made of plastic pipe and cardboard. Half did it outside of and next to the box. You guessed it. Those who were doing their thinking outside the box came up with 20 percent more creative solutions. T same kinds of differences showed up between students who had to walk a particular path as they worked on the problem, as opposed to those who could wander at will. Those who could leave the beaten path came up with 25 percent more original ideas. In the first lecture we emphasized that physical context can make a mental difference. That's what's happening here. Sometimes creativity demands that you really do get away from it all.

I'll leave you with one final creative thinking puzzle. I got this from my one-time colleague Anthony Weston. It is a dark and stormy night, and you're driving in your sports car—a good looking little car—but with only two seats, like an old MG. Suddenly, by the side of the road, you see three people stranded at a bus stop. One is a stranger who is having a heart attack at that very moment. Another is a childhood friend who has often saved your life. He has long been begging to ride in your sports car. The third person is the man or woman of your dreams. It's love at first sight. You recognize that you've suddenly found your soul mate, who you may never see again. You have just one empty seat in your car. Who do you pick up? Well, you certainly owe a debt of gratitude to your childhood friend, but you should also pick up the stranger and save his life. And what of romance? As I say, you may never see the person of your dreams again. What do you do? There's a very nice solution. But you have to stop thinking as if you have to pick up just one person. You have to stop thinking as if you have to be behind the wheel. Hand the car keys to your friend and ask him to drive the heart-attack victim to the nearest hospital. Then wrap your coat around the person of your dreams. That's creative thinking.

In the next lecture we'll return to the flow of argument but take it far beyond the logic of Aristotle.

The Flow of Argument
Lecture 9

In a previous lecture, we used Aristotle's syllogisms to emphasize the central concept of validity. Visualizing syllogisms in terms of three-circle Venn diagrams gave us a picture of validity in the strongest Aristotelian sense: airtight, ironclad validity. In this lecture, we will go beyond Aristotle to look at validity much more broadly. In general, validity refers to the degree of support between premises and conclusion. Does this fact support this claim? Do these premises render the conclusion highly plausible? Does this evidence give us good reason to believe the conclusion? The reasoning may not be airtight, but is it solid enough to act upon?

Beyond the Syllogism

- Here's an example of the kind of argument we have to deal with every day. You'll notice that it's far more complicated than a syllogism:

  1. Spending $50 billion in annual economic aid to foreign countries would be justified only if we knew that either they would genuinely benefit from the exchange or we would.

  2. We don't know that we would benefit from the exchange.

  3. It might further weaken our already vulnerable economy.

  4. We don't know that they would genuinely benefit.

  5. The track record of foreign aid that has been misdirected, misappropriated, or lost in foreign corruption is all too familiar.

  6. With $50 billion a year, we could offer college scholarships to every high school graduate and major health benefits for every man, woman, and child in the United States.

7. Our obligations are first and foremost to our fellow citizens.

_____

8. We should spend the $50 billion here rather than in foreign aid.

- There are lots of premises in that argument, far more than in Aristotle's syllogisms, and lots of transition steps. Together, they are intended to drive us toward the conclusion. But how good is the argument? And how should we analyze an argument such as this? Trying to deconstruct it into syllogisms is nearly impossible. We will see how to analyze it later in this lecture.

Flow Diagrams
- The best way to analyze complex arguments is with a simple visualization: a **flow diagram**. Such a diagram will help us see the validity of a complex argument. Breaks, disconnects, and weak logical links can show us invalidity in an argument.

- The basic rule is simple: When one claim is intended to support another, we put an arrow from the first to the second. The problem, however, is that propositions don't come labeled as premises or conclusions. They can function in either role. It all depends on the context, on the structure of the argument. Another difficulty is that the first sentence you hear doesn't always come first logically.

- Consider this argument, for example: (1) If the governor is impeached, we might be no better off. (2) Impeaching the governor would require another election. (3) But there is always the chance that people would then vote for someone equally corrupt.

- The propositions are numbered in the order of presentation, but what we want to know is something different. We want to know where the reasoning flows from and where it flows to. It will help if we can identify the conclusion. Which is the conclusion: (1), (2), or (3)?

- It's proposition (1) that is the conclusion, right at the beginning. Everything else is offered in support of that conclusion. The logic of the argument starts from (2), which leads to (3). And that leads to the conclusion: If the governor is impeached, we might be no better off.

$$2 \longrightarrow 3 \longrightarrow 1$$

Branching Flow Diagrams
- Of course, arguments get more complicated than that, so we need to add complications in the flow diagrams. First of all, arrows can branch. A set of propositions can lead to multiple conclusions or parallel conclusions.
  - Think about how to graph the following: (1) We can get only so much money from taxes; taxation resources have to be balanced among different social needs. (2) Taxation for prisons must, therefore, be balanced against taxation for education. (3) If we build more prisons, we'll have less for education. (4) If we spend more on education, we'll have less for the prisons we may need.

  - That argument has branching conclusions. The first proposition leads directly to the second. From (1) we graph an arrow to (2), but at that point, our arrows branch.

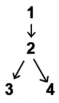

- Further, just as arrows can branch out, they can branch in. Sometimes several propositions lead to the same place.
  - Consider this example: (1) We are dangerously reliant on foreign energy sources. (2) Our oil comes primarily from the Middle East. (3) Most of our natural gas does, as well. (4) Even the elements in our batteries come from such places as Zambia, Nairobi, and China.

- First, find the conclusion. The conclusion is the first sentence. Each of the other propositions is an argument for that conclusion in its own right. We graph it by having arrows converge on a single conclusion.

Independent and Dependent Reasons

- **Independent reasons** function independently. But sometimes reasons have to function together in order to lead to a conclusion. **Dependent reasons** only function together. In a case where all three propositions work together as dependent reasons, we can mark them like this:

- How do we know whether propositions are working independently or dependently toward a conclusion? The answer is argument **stress testing**. If we have independent reasons and one of them fails, the argument should still go through. If we knock out a reason and the argument is still standing, it must be an independent reason. However, that won't hold for dependent reasons.

Graphing a Complex Argument

- We now have the basic elements of any argument flow diagram. But when we start to graph real arguments, we can see how those elements can combine into an immense variety of structures.

- Consider the argument we started with above, about spending $50 billion in foreign aid or for college scholarships here. The conclusion is (8). How are the rest of the propositions meant to support that conclusion? Here's a sketch of that argument

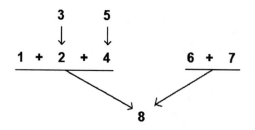

- The argument uses two dependent clusters, functioning independently of each other. One cluster uses (1), (2), and (4): Together, they offer an independent argument. If a stress test showed that the information about the college education in (6) and (7) was false, then (1), (2), and (3) together would still stand as an independent argument for the conclusion. Propositions (3) and (5) are backups for (2) and (4). All those together make one argument for the conclusion. Another argument comes from (6) and (7) working together.

Data and Warrants

- We can expand on flow diagrams for arguments using the work of the philosopher Stephen Toulmin. We have talked about premises that function together, representing them as numbers with a plus sign between them.

- What Toulmin points out is that there is often more structure just beneath that plus sign. Some of those numbers may stand for premises that function as what he calls data. Some function instead as what he calls **warrants**. They function together but in a very specific way.

- Any argument starts with some kind of data. But an argument often doesn't just go from data to conclusion. It has a middle step—a step that says how the data are supposed to lead to the conclusion. Thus, instead of a plus sign between premises (1) and (2), it might be better to represent them like this:

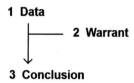

1 Data

2 Warrant

3 Conclusion

- Here's an example: The CT scan shows a shadow on the lungs. When there is that kind of shadow, cancer is a real possibility and further tests are in order. We should do a biopsy. The fact that the CT scan shows a shadow on the lungs is the data in this case. The conclusion is that we should do a biopsy.

- The warrant is the general principle that takes us from the data to the conclusion. It is not so much an additional piece of information as an inference: Given these data, we should draw this conclusion. In this case, the warrant is "When there is that kind of shadow, cancer is a real possibility and further tests are in order."

Different Kinds of Warrants

- Just as there are different kinds of arguments—scientific, legal, ethical—there are different kinds of warrants appropriate for those arguments.

- For some arguments—in scientific contexts, for example—a probability warrant is appropriate, which is a strong inductive or empirical support. Sometimes, the argument is over the use of a term; in that case, the warrant may be a definition. Sometimes, a legal principle or an ethical rule may serve as a warrant.

- Graphing arguments in terms of data and warrant can help in analyzing both individual arguments and the structure of debates. We all know that there are often two sides to an issue. The distinction between data and warrant lets us see that there are two different ways those sides may be in conflict.

Thinking Better

- Here's a real challenge: Clip out a newspaper editorial. Isolate just one or two paragraphs and graph out the argument. If you do that, you'll discover a number of things:

  - Just as we said, the conclusion doesn't always come last. Sometimes, it takes real work to figure out what the conclusion is supposed to be.

  - You will find that the flow of reasons can be very complex. Often, two claims support a third, which is a subconclusion. Much like a subtotal, it is an intermediate conclusion, just part of the way toward a final conclusion.

  - You will find that some claims in an editorial are very important. Some will be minor. Some will be widely accepted. Some will be accepted only in very specific belief structures.

  - The most important thing you'll notice is how much goes unsaid in normal arguments. What goes unsaid is often the most important part. An argument is often most vulnerable at the points at which major assumptions are made but are not made explicit.

  - In the next lecture, we'll move to aspects of our reasoning that don't need graphing: the conceptual heuristics that make us smart.

Terms to Know

dependent reasons: Premises that support the conclusion only when they are both present; propositions or claims that function together but are insufficient alone as support for the conclusion.

flow diagram: A systematic sketch of a train of thought illustrating the lines of support between premises and conclusions in a rational argument; when one claim is intended as support for a second claim, an arrow is drawn from the first to the second.

independent reasons: A group of premises, or reasons, that are given as support for a conclusion, each of which could support the conclusion on its own.

stress test: A technique used to examine the strength or stability of an entity under operational conditions that are more extreme than what is expected normally. In analyzing argument flow, a technique for detecting dependency between reasons by eliminating them individually in order to see whether the argument still goes through.

warrant: A general underlying principle that licenses an inference from data to a conclusion. In a probability warrant, the strength of the link between premise and conclusion is expressed in terms of probabilistic connection (e.g., 90 percent of the time, premise A is linked to conclusion B). In a definitional warrant, the premises are linked to conclusion as a matter of definition (e.g., whales are mammals by definition because they breathe air and give live birth). A legal warrant relies on a point of law as the link between the premise and conclusion (e.g., a contract requires a signature; thus, this unsigned document is unenforceable). An ethical warrant relies on an underlying ethical belief (e.g., if there is a shared belief that one should not deceive, then the conclusion that a deliberately deceitful act was wrong is warranted).

Suggested Reading

Kelley, *The Art of Reasoning.*

Toulmin, *The Uses of Argument.*

Questions to Consider

1. In Aristotle's sense of validity, it is impossible for the conclusion to be false if the premises are true. This lecture uses a broader notion of validity: An argument is valid if the premises support the conclusion, even if they don't make it impossible for the conclusion to be false. What do we stand to gain with this broader sense of validity? What do we stand to lose?

2. All things being equal, which kind of argument is stronger: one that relies on dependent reasons or one that relies on independent reasons?

3. Both Mr. Able and Mrs. Brightman agree that Colonel Mustard deliberately misled his stockholders about the imminent bankruptcy. Mr. Able argues that what Colonel Mustard did was permissible because the laws of contract do not demand full disclosure in such a case. Mrs. Brightman argues that what he did was impermissible because lying is always wrong. Do these people disagree on data in Toulmin's sense or on warrant? What difference might that make as to where the argument can go from this point?

Make up an argument of your own that follows this pattern of reasoning:

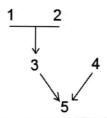

Pick a letter to the editor from the local newspaper.

    (a) Draw a flow diagram of the argument. Note that you may need to add "hidden premises" or assumptions.

    (b) Do the premises really support the conclusion? How solid is the reasoning from premises to conclusion?

The Flow of Argument
Lecture 9—Transcript

Professor Grim: In our last lecture, we used Aristotle's syllogisms to emphasize the central concept of validity. We need to know when the logic of an argument is solid, a genuinely logical transition from premises to conclusion. Visualizing syllogisms in terms of three-circle Venn diagrams gave us a picture of validity in the strongest Aristotelian sense: air-tight, iron-clad validity. Now it's time to go beyond Aristotle. His is a wonderful little logic, but it's a little logic. The structure of human reasoning goes much farther.

Structure is the key to logic, just as Aristotle said. But we need to look at the structure of argument much more broadly than he did. We also need to look at validity much more broadly. We emphasized that Aristotle's validity is validity in the highest degree. It's what we now call deductive validity. Given the truth of the premises, an argument is deductively valid if the conclusion absolutely has to be true. Given the truth of the premises, it's impossible for the conclusion to be false. Iron-clad, air-tight validity is great when you can get it. But it's actually pretty rare outside of the realm of pure mathematics or perfect little syllogisms. If we restricted our reasoning to cases in which arguments had to be absolutely iron-clad, our reasoning wouldn't get us very far. The full concept of validity is wider than that. In general, validity refers to the degree of support between premises and conclusion; all kinds of support of all different degrees. Does this fact support this claim? Do these premises render the conclusion highly plausible? Does this evidence give us good reason to believe the conclusion? The reasoning may not be air-tight, but is it solid enough to act upon? Here we'll emphasize validity in this wider sense as a matter of logical support between premises and conclusion, even if not iron-clad.

Here is an example of the kind of argument we have to deal with every day. You'll notice that it's far more complicated than a syllogism:

> Spending 50-billion dollars in annual economic aid to foreign countries would be justified only if we knew that either they would genuinely benefit from the exchange or that we would.

We don't know that we would benefit from the exchange.

It might further weaken our already vulnerable economy.

We don't know that they would genuinely benefit.

The track record of foreign aid that has been misdirected, misappropriated, or lost in foreign corruption is all too familiar.

With 50-billion dollars a year we could offer college scholarships to every high school graduate and major health benefits for every man, woman, and child in the United States.

Our obligations are first and foremost to our fellow citizens.

We should spend that 50-billion here rather than in foreign aid.

There are lots of premises in that argument; far more than in Aristotle's syllogisms; lots of transition steps. Together they are intended to drive us toward the conclusion. But how good is the argument? And how should we analyze an argument like that? Trying to decompose it into syllogisms seems a little crazy. Is there some other tool we can use to analyze more complex arguments?

Yes. A very simple visualization. It doesn't give us an automatic answer, like Venn diagrams did. We've left the comfy realm of automatic answers behind. But in gauging the validity of an argument, we still have to look systematically at logical flow. A great tool for that is a flow diagram. The idea is to use a flow diagram to sketch the train of thought. That will let us see how the premises are intended to support the conclusion. We can then look at how strong those lines of support really are. A flow graph of the argument will help us see how valid it is.

As in the case of syllogisms, of course, even a solidly valid argument isn't enough to guarantee truth; validity doesn't address the truth of the premises, for example. But it can tell us whether the truth of those premises would be enough to assure us of the conclusion, or whether we wouldn't be justified in believing the conclusion even if the premises were true. Breaks, disconnects, and weak logical links can show us invalidity in an argument. So, how to analyze complex arguments? Let's give flow diagrams a try.

The basic rule is simple. When one claim is intended to support another, we put an arrow from the first to the second. That sounds pretty simple, but it can be tricky. Here's one problem. Let me give you a proposition. You tell me whether we should enter it as a premise or a conclusion.

Gladys's marriage is on the rocks.

Is that a premise, or a conclusion? The answer is that it might be either. In one context it might appear as a premise. In another context it might appear as a conclusion. It all depends on the context of argument. Here, for example, it appears in the conclusion:

Gladys is no longer interested in Bill.

Bill seems to find no time for Gladys.

Gladys's marriage is on the rocks.

But the same proposition might appear as a premise in a different argument. Like this one:

Gladys's marriage is on the rocks.

Her sister Ethel has mounting gambling debts.

Their mother is in the hospital.

The whole family is having trouble.

There Gladys's marriage comes in as just one premise toward a different conclusion, that the whole family is having trouble. Propositions don't come labeled as premises or conclusions. They can function in either role. It all depends on the context, on the structure of the argument.

Here is another difficulty. The first sentence you hear doesn't always come first logically. Consider this little argument, for example:

(1) If the Governor is impeached, we might be no better off.

(2) Impeaching him would require another election.

(3) But there is always the chance that people would then vote for someone equally corrupt.

We can number the propositions in the order of presentation. But what we want to know is something different. We want to know where the reasoning flows from and where it flows to. It will help if we can identify the conclusion. Which is the conclusion, the first, second, or third sentence?

(1) If the Governor is impeached, we might be no better off.

(2) Impeachment would require another election.

(3) We would therefore have the possibility that people would vote for someone equally corrupt.

It's the first, proposition (1), that is the conclusion, right there at the beginning. Everything else is offered in support of that conclusion. The logic of the argument starts from:

(2) Impeachment would require another election.

That leads to

(3) We would therefore have the possibility that people would vote for someone equally corrupt.

And that leads to the conclusion. (1) If the Governor is impeached, we might be no better off.

Draw it in a flow diagram, and you have an arrow from 2 to 3 and another from there to 1. Arrows for logical flow. Simple argument, simple flow diagram.

Of course arguments get more complicated than that. We'll need to add complications in the flow diagrams. First of all, arrows can branch. A set of propositions can lead to multiple conclusions, parallel conclusions. Listen to this argument and think about how you would graph it.

(1) We can only get so much money from taxes. Taxation resources have to be balanced across different social needs.

(2) Taxation for prisons must therefore be balanced against taxation for education.

(3) If we build more prisons, we'll have less for education.

(4) If we spend more on education, we'll have less for the prisons we may need.

Number those (1) through (4), and figure out what the argument flow diagram will look like. Press pause if you want, and then come back.

That argument has branching conclusions. The first proposition leads directly to the second. From (1) We can only get so much money from taxes; taxation resources have to be balanced across different social needs we graph an arrow to (2) Taxation for prisons must, therefore, be balanced against taxation for education. But at that point our arrows branch. If taxation for prisons must be balanced against taxation for education, it follows that (3), If we build more prisons, we'll have less for education. If taxation for each has to be balanced against the other, it also follows that (4), If we spend more on education, we'll have less for the prisons we may need.

Notice that I haven't said anything about the truth of those premises. Here, as in Aristotle, we're just examining the flow of the argument. We want to trace the intended train of thought, a logical line that is supposed to step us from one proposition, to another, to the conclusion, or, as in this case, to multiple conclusions.

Here's a second complication in argument flow diagrams. Just as arrows can branch out, they can branch in. Sometimes several propositions lead to the same place. We'll start with the simple case in which there are a set of different premises that lead independently to the same conclusion. "Independent" reasons are reasons that don't need each other. They are, each, little arguments in their own right. So here's an example.

We are dangerously reliant on foreign energy sources. Our oil comes primarily from the Middle East. Most of our natural gas does as well. Even the elements in our batteries come from places like Zambia, Nairobi, and China.

In order to analyze the argument, start by numbering its propositions. Then draw a flow diagram of how the logic is supposed to go. First, find the conclusion. It's not the last sentence; this isn't an argument about batteries. It's an argument for a general conclusion about our reliance on foreign energy sources. The conclusion is the first sentence. Each of the other propositions is a little argument for that conclusion in its own right. In an advertising layout this is called bullet type. You're given three independent reasons to buy a Hyundai, for example. It gets good gas mileage. It looks great. It comes with an extended warrantee. All independent reasons. The reasons in this argument work the same way. Our oil comes primarily from the Middle East. Conclusion: we are dangerously reliant on foreign energy sources. Our natural gas does too. Conclusion: we are dangerously reliant on foreign energy sources. Even the elements in our batteries come from places like Zambia, Nairobi, and China. Conclusion: we are dangerously reliant on foreign energy sources. Three independent lines of support. We graph it by having little arrows converge on a single conclusion.

Independent reasons function independently, but sometimes reasons to have to function together in order to lead to a conclusion. Dependent reasons only function together. Here's what an argument based on dependent reasons looks like.

If both partners are indifferent, any marriage is guaranteed to fail. Gladys no longer cares about the relationship. Her husband Bill is far too busy with work to think about his relationship with Gladys. This marriage is doomed.

The context indicates that Gladys and Bill are married. We can number the first proposition (1): If both partners are indifferent, any marriage is guaranteed to fail. Note that it says both partners. Along with (1) we, therefore, need both (2) that Gladys is indifferent, and (3), that Bill is indifferent in order to get our conclusion. Here, all three propositions work together. Those are dependent reasons. In order to mark them as a set that has to work together, I graph them with a plus between them, a bar under all three, and a single arrow from there to the conclusion.

So how do you know whether propositions are working independently toward a conclusion or working dependently? Argument stress testing. Stress testing in a machine, or an organization, involves asking what will happen if a certain component fails. We do the same thing here. If you have independent reasons and one of them fails the argument should still go through.

Let's stress test our example about energy reliance.

(1) We are dangerously reliant on foreign energy sources.

(2) Our oil comes primarily from the Middle East.

(3) Most of our natural gas does as well.

(4) Even the elements in our batteries come from places like Zambia, Nairobi, and China.

Put it through a stress test. The conclusion is (1). Propositions (2), (3), and (4) are given in support. But what if (3) turns out to be false? "Excuse me, Mr. Speaker, but the fact is that most of our natural gas does not come from the Middle East." Is Mr. Speaker lunched? Is the argument over? No. He may be a bit embarrassed, but the argument is still standing. "Ahem. Well, all right. But be that as it may, it is still true that we are dangerously reliant on foreign energy. Our oil comes primarily from the Middle East. We use batteries for energy, but even the elements in our batteries come from places like Zambia, Nairobi, and China." You can test each of the other premises the same way. If you knock out a reason and the argument is still standing, it must be because that was an independent reason.

That won't hold for dependent reasons. Consider the second example.

> If both partners are indifferent, any marriage is guaranteed to fail.

> Gladys no longer cares about the relationship.

> Her husband Bill is far too busy with work to think about his relationship with Gladys.

> This marriage is doomed.

"Excuse me, Mr. Speaker, but you're wrong about Gladys. She still cares deeply about the relationship." Unless he changes one of the other claims, the argument is now dead in the water. The basic premise was that if both partners are indifferent, a marriage is guaranteed to fail. His conclusion was that this marriage is, therefore, doomed. But if he's wrong about Gladys, the other parts aren't enough to give him that conclusion. That's the sign of propositions functioning dependently.

You now have the basic elements of any argument flow graph, but when you start to graph real arguments you see how those elements can combine into an immense variety of structures. Take the argument we started with, for example. We'll number the propositions in order:

> (1) Spending 50-billion dollars in annual economic aid to foreign countries would be justified only if we knew that either they would genuinely benefit from the exchange or that we would.

> (2) We don't know that we would benefit from the exchange.

> (3) It might further weaken our already vulnerable economy.

(4) We don't know that they would genuinely benefit.

(5) We're all too familiar with the fact that our foreign aid has very often been misdirected, misappropriated, or lost in foreign corruption.

(6) With 50-billion dollars a year we could offer college scholarships to every high school graduate and major health benefits for every man, woman, and child in the United States.

(7) Our obligations are first and foremost to our fellow citizens.

(8) We should spend that 50 billion here rather than in foreign aid.

The conclusion is (8). How are the rest of the propositions meant to support that conclusion? Press pause and make a graph of how you think this argument is supposed to flow. Take a few minutes. This is a pretty complicated argument. Then come back.

Here's my sketch of that argument. The argument uses two dependent clusters functioning independently of each other. One cluster uses (1), (2), and (4). Spending 50-billion dollars in economic aid to foreign countries would be justified only if we knew either A or B. We don't know A. We don't know B. So spending the money that way isn't justified.

Those work together. That's why I represent them with a plus between them and a bar beneath them. Together they offer a little independent argument. If a stress test showed that the stuff about the college education in (6) and (7) was false, (1), (2), and (4) together would still stand as an independent argument for the conclusion.

Where do (3) and (5) appear in that argument? They're really little backups for (2) and (4). Proposition (2) says we don't know that we would benefit from the exchange. Why? Because (3), it might further weaken our economy. Proposition (4) says that we don't know that they would benefit. Why? Because (5), foreign aid has so often gone astray. All of those together make one argument for the conclusion. There is another argument in there that comes from (6) and (7) working together:

With that money we could offer college scholarships to every High School graduate and major health benefits for every man, woman, and child in the United States.

and

Our obligations are first and foremost to our fellow citizens.

Those together offer their own little argument for spending the money here rather than in foreign aid.

As real arguments go, even that one is fairly simple. Here's the real challenge: Clip out a newspaper editorial. Take just one or two paragraphs and graph out the argument. If you do that, you'll discover a number of things: Just as we said, the conclusion doesn't always come last. Sometimes it takes real work to figure out what the conclusion is supposed to be. You will also find that the flow of reasons can be very complex. Often two claims support a third, which is a sub-conclusion. It's like a sub-total. It's an intermediate conclusion just part of the way toward a final conclusion. You will find some claims in an editorial that are very important. Some will be minor. Some will be widely accepted. Some will be accepted only in very specific belief structures. The most important thing you'll notice is how much goes unsaid in normal arguments. What goes unsaid is often the most important part. An argument is often most vulnerable at the points at which major assumptions are made but are not made explicit.

With an understanding of the flow of argument, we have the tools for looking at validity and vulnerability. Argumentative strengths and weaknesses. How valid is the argument? That means, how strong is the logical flow from premises to conclusion? If the premises are true, is the conclusion extremely plausible? Highly probable? Downright inescapable? If the answer to any of these is yes, we have a valid argument. This is still an internal evaluation of logical strength. It doesn't tell us whether the premises are in fact true. If they are true, we have something even better, a sound argument. A sound argument is one that is both valid and does have true premises. Because they're true, and because the reasoning is valid, we have solid reasons to believe the conclusion is true too.

I'm going to expand a little on flow graphs for arguments using the work of the philosopher Stephen Toulmin. We've talked about premises that function together, representing them as little numbers with a plus between them.

What Toulmin points out is that there is often more structure just beneath that plus. Some of those little numbers may stand for premises that function as what he calls data. Some function, instead, as what he calls warrants. They function together, just like we've said, but function together in a specific way. Here's what I mean. Any argument starts with some kind of data. But an argument often doesn't just go from data to conclusion. It has a middle step, a step that says how the data is supposed to lead to the conclusion. So instead of just a plus between premises 1 and 2, it might be better to represent them like this.

Here's an example:

The CT scan shows a shadow on the lungs.

When there is that kind of shadow, cancer is a real possibility and so further tests are in order.

We should do a biopsy.

The data in this case? The fact that the CT scan shows a shadow on the lungs. The conclusion? That we should do a biopsy. The warrant is the general principle that takes us from the data to the conclusion. It's not so much an additional piece of information as an inference-ticket. Given this data, we should draw this conclusion. In this case the warrant is, when there is that kind of shadow, cancer is a real possibility and so further tests are in order.

Here's another example. You tell me what the conclusion is, what's playing the role of data, and what's acting as warrant:

He can't be guilty of first degree murder.

The defendant was heavily intoxicated at the time and acted in the heat of emotion.

First degree murder demands premeditation.

Press pause and think about it for a minute if you want to. The data? The claim that the defendant was intoxicated at the time and acted in the heat of emotion. The conclusion? That he can't be guilty of first degree murder. What's the warrant? What greases the track from data to conclusion? The principle that first degree murder demands premeditation. Data: emotion and intoxication at the time; warrant: first degree demands premeditation; conclusion: he can't be guilty of first degree murder.

There are, of course, very different kinds of arguments: scientific arguments, legal arguments, ethical arguments. It's important to note that very different kinds of warrants are appropriate for those different kinds of arguments. For some arguments, in scientific contexts, for example, it's a probability warrant that's appropriate. Ninety percent of the experiments showed a latency effect. That's strong inductive or empirical support. Sometimes the argument is over the use of a term. In that case the warrant may be a definition. Whales breathe air and give live birth, so they're mammals. Sometimes it's a legal principle that serves as warrant. The contract wasn't signed. It is therefore unenforceable. The warrant? A legal principle that unsigned contracts aren't enforceable. Sometimes it's an ethical rule that serves as warrant. It was deliberately deceitful, so what he did was wrong. Here the warrant is our belief that deception is unethical.

Graphing arguments in terms of data and warrant can help in analyzing both individual arguments and the structure of debates. We all know that there are often two sides to an issue. The distinction between data and warrant lets us see that there are two different ways those different sides may be in conflict. Sometimes what's at issue in a controversy is disagreement about the data. Any argument will go from data to conclusion. If one side doesn't recognize the other side's starting point as genuine data, there is going to be no meeting of the minds. That is a problem in disputes regarding abortion. Opponents of legalized abortion may start the argument with "A fetus is a person." Precisely because that claim is not accepted as genuine data by the opposition, the argument won't convince the other side. Proponents of legalized abortion, on the other hand, may start the argument with "A woman's rights regarding her own body are absolute." Precisely because that is not accepted as data on the other side, their argument fails to convince the other side as well. But controversies often continue even when both sides do agree on the data. Why? Because the two sides don't agree on what the data warrants.

Controversies regarding gun control seem to be like this. The statistics on gun ownership, crime, and accidents in the home may be conceded on both sides. Both sides may agree that there is a right to self-defense. They may even agree on what the second amendment says. But one side thinks the statistics warrant restrictions on gun ownership despite individual rights. The other side thinks that individual rights trump the social statistics, that the statistics alone are insufficient warrant for further restriction. When you graph that controversy in terms of data and warrant, it's clear that the core dispute is over the prioritization of values. Individual rights vs. the prevention of harm. The two sides may agree on the data, but emphasis on different background principles takes them from that same data to different conclusions. In that case you have a dispute over warrant.

The next time you find yourself in apparent controversy with someone, ask whether you disagree about the data or the warrant. That may help to keep people from talking past each other. Graphing logical flow can only take you so far, of course. It alone can't resolve disputes like this, but it can help in deciding where the dispute really lies.

Here's what I want you to take home from this lecture. Rational argument is a matter of logical flow. To understand an argument is to be able to see that logical flow, to see lines of argument that branch to multiple conclusions, to see how different reasons may work independently to support a conclusion, or to see how they have to work dependently together. If we can sketch out the logical flow, we have a better grasp on validity, on how strong the reasons offered really are for the conclusion. We can also identify stress points of vulnerability. Analysis in terms of data and warrant offers a further fine tuning of the graphing approach, particularly useful in analyzing what's really in dispute in long-standing controversies.

In the next lecture we'll move to aspects of our reasoning that don't need graphing. Aspects of reasoning that are quick and dirty but often very effective. We'll be talking about conceptual heuristics that make us smart.

Simple Heuristics That Make Us Smart
Lecture 10

A heuristic is a rule of thumb, such as "You should always have three to six months' pay in the bank in case of emergency" or "To approximate a 15 percent tip, double the sales tax." Heuristics are guides to action that are simpler than a complete calculation or a logical examination of all possibilities. In this lecture, we'll learn about heuristics that can be simpler, cruder, faster, and just as good as or even better than a complete rational calculation. Simple heuristics come with a warning label, however. Although they're often useful, they're not guaranteed. Calculating a 15 percent tip by doubling the sales tax works only if the tax rate is 7.5 percent.

The Recognition Heuristic
- Two groups of students, one at the University of Chicago and one at the University of Munich were asked the following question: "Which city has a larger population: San Diego or San Antonio?" Which group do you think did better?

- The University of Chicago students are a smart bunch, and of course, they're American. Sixty-two percent of them got the right answer (San Diego), but 100 percent of the German students answered correctly.

- In fact, the Germans got the answer right precisely because they know less. They seemed to have used this **heuristic**—this rule of thumb—"I've certainly heard of San Diego. I'm not sure I've heard of San Antonio. That must be because San Diego is a more prominent city, which probably means it's larger."

- Recognition is a fast-and-frugal heuristic. It's frugal because it works with extremely limited information. In fact, it demands that your information be extremely limited.

- Suppose you have three groups of people. One is extremely knowledgeable about an issue. One knows about half as much as the first, and one is totally ignorant. For some questions, because of the **recognition heuristic**, it's the middle group that will do the best. Those who know absolutely nothing can't use it—they don't recognize any of the alternatives. Those who know too much can't use it—they recognize them all. It is those in the middle, who recognize some but not all of the alternatives, who can put the recognition heuristic to work effectively.

The Wisdom of Crowds

- When it comes to advice or data, what we need to be able to do is to tell the real experts—those who actually know—from the so-called experts. But it may be hard to pick those out. The **wisdom-of-crowds heuristic** indicates that we may not need to know who the real experts are.

- Sir Francis Galton was a distinguished British scientist. He also happened to be a cousin of Charles Darwin. Galton is remembered positively for his studies of heredity and intelligence. He is remembered negatively for founding **eugenics**.

- Galton was convinced that some people are inherently smarter than others. He thought we should breed the smart ones and prevent reproduction by those who were not as smart. He considered the intelligence of the general public to be deplorable. That fact about Galton makes this next case particularly interesting.

- In the fall of 1906, the 85-year-old Galton attended the annual West of England Fat Stock and Poultry Exhibition, which had a weight-judging competition. A prize ox was put on display, and people put wagers on how much the ox would weigh after it had been butchered and dressed. There was a prize for the person whose guess was the closest.

- Galton examined all the tickets and then took the **median**—the guess in the very middle, with half of the guesses greater and half of the guesses less. The median guess was 1,207 pounds. The actual weight of the ox, when butchered and dressed, was 1,198 pounds. That's an error of less than 1 percent. For the simple average of the guesses, the result was even more amazing. The average of the

The studio audience for the show *Who Wants to Be a Millionaire?* gave the right answer 91 percent of the time.

guesses for the weight of the ox was 1,197 pounds: an error of less 0.001 percent.

- What is the heuristic here? In this case it's social. When in doubt, go with the **mean** judgment. When in doubt, rely on the wisdom of crowds.

Statistics Explains the Wisdom of Crowds

- The statistical phenomenon behind the wisdom of crowds is the fact that the ignorant answers randomize over all possibilities. When we go for the answer that got the most votes, the ignorant answers effectively cancel each other out. That leaves the knowledgeable answers to tip the scale.

- Statistics shows when you can rely on this heuristic and when you had better not.
  - In order for this to work, you must have a crowd in which some people really do know the answer. A crowd of schoolchildren, for example, won't outguess Warren Buffett on stock choices.

  - The other condition is that the wrong answers must really be random. The heuristic won't work if all the ignorant people have the same kind of ignorance, for example, if they all give the answer "Mississippi" to all questions about rivers because it's hard to spell. Lots of people with lots of different biases is fine—those biases will cancel each other out. What you don't want is to have all the biases in the same direction.

The Heuristic of "Satisficing"

- In decision making, the ideally rational setup is one in which you examine all courses of action, calculate consequences, compare benefits and losses, and then decide. If you are making an important decision and you have the time, that's what you should do. But the demands of action often preclude that kind of extended consideration.

- Often, the alternatives are too many and the demands are too immediate. If you are a firefighter deciding how to fight a basement fire, you don't have time to be a "perfect maximizer." You don't have time to look up the blueprints, map out all possible ways of entry, analyze the probabilities that certain fire-retardant materials were used in construction, access medical records in order to determine the probability that a child on the third floor is asthmatic. Lives are at risk and property is going up in flames.

- The heuristic to use in situations where we don't have time to be perfect maximizers is **satisficing**. Pick a realistic set of goals. Now, start looking for a course of action that will achieve those goals. Consider alternatives in any order they happen to occur to you. You're not looking for perfection; you're looking for good enough. You're trying to be a "satisficer." The heuristic: Take the first course of action that meets your goals and go.

- There is biological evidence that satisficing is built into a wide range of animal behavior. Believe it or not, sastisficing is often used in computer programs, as well. Even computers don't have time to consider all the options in every case.

Rational Calculation or "Go with Your Gut"?

- Heuristics are rational shortcuts. Given the selective pressure imposed by the environment, it is not surprising that animals should have some heuristic shortcuts built in. The animal has to do something to shake off the predator and has to do it now. If there is a way of doing that without all the time delay of careful logical calculation, evolution will select for it.

- People also have an evolutionary history. When we look at ourselves, we can expect to find some generally useful heuristics built in.

- Should you try to calculate things rationally, or should you "go with your gut"? The answer seems to be both. Neither emotions nor gut reactions should be ignored. They have the experience of our ancestors built in. When a decision is of the same type as those made by our ancestors, and it has to be made fast—such as getting out of the way of an approaching stampede— extended calculation is clearly a bad idea. Go with your gut.

- On the other hand, our rationality is also a major part of our evolutionary inheritance. If it hadn't been good for anything, we wouldn't have it. Rational calculation is particularly useful for long-range decisions. In that case, we should use all our resources, survey all available information, and weigh all options carefully. To rely on gut reactions and emotions alone might well be to make the wrong decision.

Cultivating Heuristics

- Not all heuristics are built in. Some can be deliberately acquired. Here's a lesson on acquired heuristics from one of the great thinkers of our time: Neumann János Lajos.
  - Born in Hungary in 1903, he changed his name to John von Neumann by the time he came to America. He made major advances in set theory, number theory, and computer science and basically invented game theory. Von Neumann was a major mover in the Manhattan project, responsible for the atomic bomb in World War II, and in the later hydrogen

bomb, and he participated in the early years of the Atomic Energy Commission.

- o Along with Einstein, he was one of the original members of the Institute for Advanced Studies at Princeton. All our computers embody what is known as the von Neumann architecture.

- o The interesting thing about von Neumann is that he quite deliberately built his own kit of heuristics. And it was that kit of heuristics that let him cross fields so easily and productively, from physics and mathematics, to economics, to game theory.

- o We've all seen technical textbooks with lists of formulae at the back. Although they may seem a little forbidding to us, those were von Neumann's favorite part of any book. He would deliberately memorize the formulae in the back of textbooks: formulae in matrix arithmetic, in physics, in theoretical biology, in economics.

- o He would then use those patterns as heuristics. Von Neumann's strategy was to familiarize himself with those formal patterns in order to be able to see connections and analogies when he saw something like them somewhere else. He stocked up on mental patterns in one place so he could apply them elsewhere.

- • In this lecture, we've emphasized the sunny side of conceptual patterns. But there's a dark side, too—a dark side we can see in some of the mistakes we make. We'll talk about those in the next lecture.

Terms to Know

eugenics: A social movement that advocates practices and direct interventions aimed at changing, or ostensibly "improving," the genetic characteristics of a population.

heuristics: Simple guides to action or rules of thumb that allow us to act or make a decision without calculation or deliberation.

mean: The average of a set of numbers.

median: The midpoint of a set of numbers that has been ordered from lowest to highest; that point at which there are as many below as above.

recognition heuristic: A fast-and-frugal rule of thumb for decision making in which one picks the alternative or orders alternatives in terms of recognition; often most effective when one is working on very limited information about alternatives.

satisficing: A heuristic, or rule of thumb, for decision making in which one picks a realistic goal or set of goals, then selects the first course of action that meets the goal or goals. This heuristic emphasizes that one is not seeking perfection, just something that is "good enough."

wisdom-of-crowds heuristic: A rule of thumb in which one bases one's decision on the most popular answer selected by a random group of people. In order for this heuristic to be effective, certain conditions of randomness must be met. The reason it works can be explained statistically by the fact that in an appropriately random group of people, most will be ignorant about the topic, but a few will actually know the answer. The answers of the "ignorant" will distribute randomly and cancel each other out, allowing the answers of those who really know to tip the scale.

Suggested Reading

Gigerenzer, Todd, and the ABC Research Group, *Simple Heuristics That Make Us Smart*.

Surowiecki, *The Wisdom of Crowds*.

Questions to Consider

1. For what kinds of decisions are you a "satisficer"—taking the first option that satisfies your basic goals? For what kinds of decisions are you a "perfect maximize"?

2. From your own experience, give a case where you "went with your gut" and ended up with a good decision. Give a case where you "went with your gut" and ended up with a bad decision. Was there a difference in the kind of decision at issue?

Exercise

Try the recognition heuristic on these:

(a) Which is the larger South American city, Bogota or Rio de Janeiro?

(b) List these people in order of wealth: Larry Ellison, Bill Gates, Christy Walton.

(c) Which is the larger African country, Libya or Sudan?

(d) List these pitchers in terms of the most career home runs: Wes Ferrell, Don Drysdale, Walter Johnson.

(See "Answers" section at the end of this guidebook.)

Simple Heuristics That Make Us Smart
Lecture 10—Transcript

Professor Grim: A heuristic is a rule of thumb. Like: You should always have three to six months' pay in the bank in case of emergency. Or: To approximate a 15% tip, double the sales tax. Heuristics are guides to actions that are simpler and cruder than a complete calculation or a logical examination of all the possibilities. Simpler, cruder, and faster. Albert Einstein published a paper in 1905 entitled "On a Heuristic Point of View Concerning the Generation and Transformation of Light." He used that title to indicate that the approach in the paper was sketchy and incomplete, maybe even false by the lights of a complete theory, but useful nonetheless. That's what heuristics are like.

In this lecture I want to introduce you to heuristics that can be simpler, cruder, faster, and just as good, or even better, than a complete, rational calculation. Some of these heuristics seem to generate knowledge from ignorance in an almost magical way. In the phrase used by Gerd Gigerenzer, the German psychologist who did much of the background research, these are "simple heuristics that make us smart." Simple heuristics also come with a warning label, however. While they're often useful, they're not guaranteed. Calculating a 15% tip by doubling the sales tax only works if the tax rate is 7 1/2%.

Let's start with a simple but obscure question. Which city has a larger population, San Diego or San Antonio? That question was posed to two groups of students, to students at the University of Chicago and to students at the University of Munich. Who do you think did better? The University of Chicago students are a smart bunch, and of course, they're American. Sixty-two percent of them got the answer right. San Diego has the larger population. What about the Germans? This may surprise you. One-hundred percent of the German students got the question right. All of them said that San Diego was larger. How could that be? Are the German students better informed about American demographics than University of Chicago students? No, in fact, they generally know significantly less about American cities and their population. One-hundred percent of the German students got the answer right despite the fact that they know less about American cities.

In fact, they got the answer right precisely because they know less. The German students seem to have used this heuristic, this rule of thumb: "I've certainly heard of San Diego. I'm not sure I've heard of San Antonio. That must be because San Diego is a more prominent city. That probably means it's larger. Which has the larger population? I'll go with San Diego." This is a recognition heuristic. In general, you are asked about the value of two objects—whatever value, size of population, number of department stores, tax revenues, it doesn't matter. If you recognize one of them and not the other, pick the one you recognize as having the higher value.

The German students could use that heuristic because they had a significantly clearer recognition of the one city than the other. They used their ignorance to good effect. The Chicago students didn't have that advantage; they didn't have that valuable ignorance. They were familiar with both San Diego and San Antonio and so perhaps had to fall back on thinking of the position of the cities on a map, or "San Diego is near the ocean, and probably crowded with suburbs. San Antonio has lots of space. Things in Texas are always bigger." That's why only 62% of the Chicago students ended up with the right answer. Recognition is a fast and frugal heuristic. It's frugal because it works with extremely limited information. In fact, it demands that your information be extremely limited.

Let's try you on the recognition heuristic. So here are two German cities. Which do you think has the larger population? Munich, Berlin. Do you have that? If you said Berlin is larger—perhaps because it's more familiar—you're right. The more detailed we get, the riskier it is, but let me try you on one more. Here are five German cities. How would you rank them in terms of population? Stuttgart, Bremen, Cologne, Berlin, Munich. If you're like me, you don't really know anything about relative population sizes. But if we list them in terms of recognition, they might go like this: Berlin, Munich, Cologne, Stuttgart, Bremen. That is in fact their order in terms of population size.

Of course this rule of thumb also has important limitations. Will it always work? No. A recognition heuristic works only when there is a real correlation behind it—a correlation between familiarity and population size, for example. When you don't have any further information, that correlation can sometimes be a good bet. It's interesting that it demands limited knowledge of the topic in order for this work. If you know too much, you can't use it. So suppose you have three groups of people. One is extremely knowledgeable about the issue, one knows about half that much, one group is totally ignorant. For some questions, because of the recognition heuristic, it's the middle group that will do the best. Those who know absolutely nothing can't use it—they don't recognize any of the alternatives. Those who know too much can't use it—they recognize them all. It's those in the middle, who recognize some but not all of the alternatives, who can put the heuristic to work effectively.

The group that ran the experiment with German and American students put the results to the test by gauging success on the stock market. They made a list of companies from the New York Stock Exchange and from several German exchanges. They asked four groups of people to indicate which stocks they recognized. The four groups: American experts on the stock market, German experts, American laypeople—they actually used pedestrians off the street, and German laypeople. They asked each group, "Which of these companies do you recognize?" German laypeople recognized German companies like Allianz AG, Daimler Benz, and Siemens, for example. U.S. laypeople recognized U.S. companies like Amoco, Chrysler, Maytag, Sears Roebuck. Both groups recognized Lufthansa. Experts on each side, of course, recognized more. The experimenters then put together mock portfolios of the stocks recognized by over 90% of each group. They made a portfolio of the German stocks that German laypeople recognized, a portfolio of German stocks that German experts recognized. Similarly for Americans and American stocks. They also made a portfolio of American stocks that German laypeople recognized and one of American stocks that German experts recognized, and similarly for Americans. So, eight different pretend stock portfolios. They then tracked how well each did, compared with the market index and mutual funds.

Could it actually be possible that mere recognition beats the stock market index? When it comes to bets across borders, the answer is yes. Take the recognition portfolio of American laypeople on German stocks. That did better than both mutual funds and the market index. It even did a little better than the portfolio of German stocks based on recognition by American experts. Returns for German laypeople and German experts on American stocks did better than either the market index or mutual funds. When the bets were on stocks within countries, the results were mixed. German laypeople and German experts still did better by recognition alone. American laypeople and American experts betting on American stocks didn't. That was the only area in which the recognition heuristic lost. Maybe Americans weren't ignorant enough to use it. Maybe they knew too much.

The portfolios constructed of stocks recognized by experts didn't do particularly well. That doesn't mean they don't know their stuff. It's natural that the experts will recognize more stocks, both good and bad. But there are plenty of cases in which the experts do turn out to be wrong. Harry Warner was one of the founding brothers of Warner Brothers, but in 1927, just as movies changed from silent to sound, he predicted that movies with sound would be a flop. "Who the hell wants to hear actors talk?" In 1943, Thomas Watson, the father of IBM, said, "I think there is a world market for maybe five computers." Neils Bohr defined an expert as "A man who has made all the mistakes which can be made in a very narrow field." Carl Sandberg characterized an expert as "A damn fool a long way from home." What we need to be able to do is to tell the real experts—those who actually know—from the so-called experts. The suspicion is that many so-called experts may not in fact be particularly knowledgeable. Real experts are going to be those who really do know, but it may be hard to pick those out.

This next heuristic indicates that you may not need to know who the real experts are. This is the rule of thumb known as the wisdom of crowds. Sir Francis Galton was a distinguished British scientist. He also happened to be a cousin of Charles Darwin. Galton is remembered positively for his studies of heredity and intelligence. He is remembered negatively for founding eugenics. Galton was convinced that some people are inherently smarter than others. He thought we should breed the smart ones and prevent reproduction by the dumb ones. He considered the intelligence of the general public to be deplorable. That fact about Galton makes this next case particularly interesting.

In the fall of 1906, the 85-year-old Galton attended the annual West of England Fat Stock and Poultry Exhibition. They had a weight-judging competition. A prize ox was put on display, and people put wagers on how much the ox weighed, or more precisely, how much it would weigh after it had been butchered and dressed. You bought a ticket for sixpence, and you wrote down your estimate. There was a prize for the person whose guess was the closest. Eight-hundred people entered the competition—all kinds of people. There were farmers and butchers whom you might expect to know this kind of thing. Those were the experts, but there were also all kinds of rabble and riffraff. With the attitude of a typical aristocratic snob, Galton said, "The average competitor was probably as well fitted for making a just estimate of the dressed weight of the ox as an average voter is of judging the merits of most political issues on which he votes." Because of his interest in statistics, and because of his prejudices regarding intelligence, Galton decided to conduct a little experiment. When the contest was over, he asked for the all the tickets that had been submitted, and there were exactly 800 tickets, but 13 were illegible. Galton cut the sample down to the remaining 787. He laid them out in order of the guessed weight. He then took the median—the guess in the very middle, with half of the guesses greater, half of the guesses less. The median guess was 1,207 pounds. The actual weight of the ox, when butchered and dressed, was 1,198 pounds. That's an error of less than 1%.

The result hit Galton's anti-democratic, aristocratic, eugenic-prone beliefs pretty hard. He said, "The result seems more creditable to the trustworthiness of a democratic judgment than might have been expected." Galton reported the median of the guesses, the point at which there are as many above as there are below. But he saved all the data, and from that we can figure out something even simpler, the mean, the simple average. And that result is even more amazing. The real weight of the ox, when butchered and dressed, was 1,198 pounds. The average of the guesses for the weight of the ox—the average guess of the assorted rabble, with all those non-experts—was 1,197 pounds. That's an error less than 1/10th of 1%. What's the heuristic here? In this case it's social. When in doubt, go with the mean judgment. When in doubt, rely on the wisdom of crowds.

Here is another case along those lines. Remember Regis Philbin and *Who Wants To Be A Millionaire?*? The idea of that show was that people were asked multiple-choice questions that got progressively more difficult. If you answered 15 questions in a row, you got a million dollars. If a contestant didn't know the answer to a question, they could do one of three things, they could have two of the four options removed, increasing the odds of guessing right; they could call a friend whom they had singled out in advance as one of the smartest people they knew, that's our expert; the third option was to ask the studio audience, with a majority answer polled immediately and electronically. So how well did the experts do in *Who Wants to be a Millionaire?*? How often did they choose the right answer? They actually did pretty well. The experts picked the correct answer 65% of the time. How well did the studio audience do? They picked the correct answer 91% of the time. Those two examples come from James Surowiecki, a business columnist for The New Yorker. It's not technically the mean, though it is the majority vote. When in doubt, rely on the wisdom of crowds. There are more than just stories behind this heuristic, however. The wisdom of crowds has a simple statistical explanation. The explanation shows why it works, when it works, and when you shouldn't expect it to work. So suppose you're a contestant on *Who Wants to Be a Millionaire?* and you get this question: Which river did Caesar cross?

    a. The Amazon

    b. The Danube

    c. The Mississippi

    d. The Rubicon

You have no idea of the answer. You could call your expert. The experts were right 65% of the time. Or you can ask for a poll of the studio audience. The audience was right 91% of the time. How can that be? The audience contains all kinds of people, many of whom have totally distorted ideas of who Caesar even was. Many of them are as ignorant as you are. How can a majority vote of that rabble be accurate 91% of the time? Here's how. The audience probably does contain at least a few people who do know the answer. I'll simplify this by taking this to extremes. Suppose that the audience had just two kinds of people: (1) people who know the answer, and (2) those who are totally ignorant. Take the people who know the answer. They'll pick d. The

Rubicon. Now take those who are totally ignorant. If they're totally ignorant, their answers will spread randomly across all four alternatives. That is going to be the key.

So let's start by tallying the answers of the ignorant. Those will spread randomly across all four alternatives. Then add the answers from the people who know. Take that answer that got the most votes. What is it? The Rubicon. The statistical phenomenon behind the wisdom of crowds is the fact that the ignorant answers randomize over all possibilities. When we go for the answer that got the most votes, the ignorant answers effectively cancel each other out. That leaves the knowledgeable answers to tip the scale.

In that simplification I assumed that people divided neatly between the ignorant and the knowledgeable. But the explanation stands even if we don't assume that. Suppose that there are people who are more or less knowledgeable and people who are more or less ignorant. Those who are more or less knowledgeable will hit the right answer more or less. Those who are more or less ignorant will guess randomly more or less, and the effect will be the same: the wisdom of crowds will reflect the degree of wisdom of even a few people in the crowd. That same explanation holds for Galton's ox. There were some people in that crowd who really did know how to estimate the weight of an ox. The rest of the people didn't. But precisely because they didn't, they were as likely to guess too high as to guess too low. When you line up the answers from low to high, the answers of those who really know will end up approximately in the middle. The highs and lows will cancel each other out. That's Galton's median. When you average all the answers, the random highs and lows cancel out, allowing the more knowledgeable answers to shine through.

So, statistics explains the effect. They also show when you can rely on it and when you'd better not. Here are the conditions. In order for this to work, you have to have an audience in which there are some people who really do know the answer. A crowd of school children won't out-guess Warren Buffet on stock choices. But it's interesting that you don't have to know who the people are who really know. You don't have to be able to identify the experts. All you have to know is that there are some people in that crowd who really do know. The other condition is that the wrong answers have to really be random. It won't work if all the ignorant people have the same kind of ignorance. It won't work if they all think that Mississippi is going to be the answer to all questions about rivers, for example, just because it's hard to spell. Lots of people with lots of different biases is fine—those are the biases that will cancel each other out. What you don't want is to have all the biases in the same direction.

Let's move on to another heuristic, important not for picking answers but for deciding what to do. In decision making, the ideally rational setup is one in which you examine all courses of action, calculate consequences, compare benefits and losses, and then decide. If you are making an important decision and you have the time, that's what you should do. But the demands of action often preclude that kind of extended consideration. Often the alternatives are too many and the demands are too immediate. If you are a fireman deciding how to fight a basement fire, you don't have time to be a perfect maximizer. You don't have time to look up the blueprints, map out all possible ways of entry, analyze the probabilities that certain fire retardant materials were used in construction, access medical records in order to determine the probability that a child on the third floor is asthmatic.. The fireman doesn't have time for all that. Lives are at risk and property is going up fast.

We often don't have time to be perfect maximizers. The heuristic to use in these kinds of cases is called satisficing. Pick a realistic set of goals. The fireman's may be these: Keep your comrades safe. Get everyone out of the building. Put the fire out before it spreads to the next building. Now start looking for a course of action that will achieve those goals—any course of action. Consider alternatives in any order they happen to occur to you. You're not looking for perfection. You're looking for good enough. You're trying to be a satisficer. So the heuristic here, take that first course of action that meets your goals. Go! There is evidence that satisficing is, in fact, the way firefighters think. There is also biological evidence that it's built into a wide range of animal behavior. Believe it or not, sastisficing is often used in computer programs as well. Even computers don't have time to consider all the options in every case.

Heuristics are rational shortcuts. Given the selective pressure imposed by the environment, it's not surprising that animals should have some heuristic shortcuts built in. The animal has to do something to shake off the predator and has to do it now. If there's a way of doing that without all the time delay of careful, logical calculation, evolution will select for it. Those organisms that have the technique built in will live to reproduce more consistently. So the result is a behavioral heuristic built into the species. We can expect generally successful heuristics to be chosen for even if they don't work all the time. Quick-and-dirty techniques that evade predators much of the time will be selected for too.

People have an evolutionary history. When we look at ourselves we can expect to find some generally useful heuristics built in. In an earlier lecture we looked at gut reactions and emotions in this light. Those are part of our behavioral inheritance. Fear is a heuristic for getting out of the way. Disgust is a heuristic for avoidance. The emotions involved in sexual attraction, on the other hand, are telling you something more positive. So, should you try to calculate things rationally, or should you go with your gut? The answer seems to be, both. Neither emotions nor gut reactions should be ignored. They have the experience of our ancestors built in. When a decision is of the same type as those made by our ancestors and it has to be made fast—like getting out of the way of an approaching stampede—extended calculation is clearly a bad idea. Go with your gut. On the other hand, our rationality is also a major part of our evolutionary inheritance. If it hadn't been good for anything, we wouldn't have it. Rational calculation is going to be particularly useful where a decision is long range, where it doesn't have to be made immediately. In that case we should use all of our resources, survey all available information, and weigh all options carefully. To rely on gut reactions and emotions alone might well be to make the wrong decision.

Rational deliberation is also going to be particularly important when the decisions we face are not the same as those faced by our ancestors. Evolution hasn't had time to catch up with the need to make presentations before one's business partners, for example. It hasn't had time to catch up with the internet and cell phones. Business partners aren't as hazardous as predators. They won't literally eat you. If your system is recruiting a heuristic appropriate to predators, then, it's recruiting the wrong heuristic. Rationality should override it. Cell phones and internet access are not as important as the food you need for tomorrow. If your system reacts to losing them as if it were a matter of literal survival, it's recruiting the wrong heuristic. Rationality should override.

Not all heuristics are built in. Some can be deliberately acquired. Here is a lesson on acquired heuristics from one of the great thinkers. Neumann János. I put him near the top of my list of the smartest people of the 20th century. Born in Hungary in 1903, he changed his name to John von Neumann by the time he came to America. He made major advances in set theory, in number theory, and in computer science. He basically invented game theory, which we'll talk about later. He was a major mover in the Manhattan project, responsible for the atomic bomb in World War II, in the later hydrogen bomb, and in the early years of the Atomic Energy Commission. Along with Einstein, he was one of the original members of the Institute for Advanced Studies at Princeton. And all of our computers embody what's known as the von Neumann architecture. The interesting thing about von Neumann is that he quite deliberately built his own kit of heuristics. It was that kit of heuristics that let him cross fields so easily and productively, from physics and mathematics, to economics, to game theory.

I'm sure you've seen technical textbooks with lists of formulae in the back—always a little forbidding. We go to them only when we have to. Not von Neumann. Those were von Neumann's favorite part. He would deliberately memorize the formulae in the back of textbooks: formulae in matrix arithmetic, formulae in physics, formulae in theoretical biology, formulae in economics. Why? So that he could use those patterns as heuristics; in particular, so that he could recognize similar patterns elsewhere. In the history of ideas it has repeatedly happened that an idea in one area, developed for one purpose, finds an unexpected application elsewhere. Concepts developed purely for philosophy of mathematics turned out to be just what you needed to build a computer. Statistical formulae developed for understanding genetic change in biology are now applied in both economics and in programming.

Von Neumann's strategy was to familiarize himself with those formal patterns in order to be able to see connections and analogies when he saw something like them somewhere else. He stocked up on mental patterns in one place so he could apply them somewhere else. Now I'm not arguing that we should all be von Neumanns, nor am I urging you to memorize the backs of technical textbooks. The lesson is much more general and applies to a well-rounded education in literature and the humanities as well. Pay attention to patterns and templates: patterns of reasoning, templates for decision-making, ways of life. Tuck them away for future reference. Those may come in handy in some other context.

What I want you to take home from this lecture are two fast-and-frugal heuristics that allow you to squeeze knowledge from ignorance. One is the recognition heuristic; remember San Diego and San Antonio. The other is the wisdom of crowds; remember *Who Wants to be a Millionaire?*. Remember satisficing, a decision heuristic where you need to move fast. Set reasonable criteria and take the first option that's good enough. Trust emotions for the short-term contexts that they're made for. For long-range decisions you'd better use your head as well as your gut.

In this lecture I've emphasized the sunny side of conceptual patterns, but there's a dark side too, a dark side you can see in some of the mistakes we make. We'll talk about those in the next lecture. Oh, yes. At the beginning I mentioned one of Einstein's early papers: "On a Heuristic Point of View Concerning the Generation and Transformation of Light." I noted that he used the term "heuristic" to indicate how sketchy and incomplete the work was. That was the paper for which he received the Nobel Prize.

Why We Make Misteaks

Lecture 11

H ow many times have you berated yourself for a stupid mistake? A friend may try to make you feel better: "You're only human. Everybody makes mistakes." It turns out there's a lot of truth in that. To err is human. Systematic error—not random but systematic and predictable—is built into some of the ways we think. Systematic error is even built into how we see things. In this lecture, we'll look at the ways that we as humans systematically go wrong and how we can correct for them. If we are aware of cognitive biases, we can compensate for them.

Perceptual Bias

- We have emphasized the power of visualization in a number of lectures. Because of our evolutionary heritage, people are particularly good at fast information processing in visual contexts. But our evolutionary heritage comes with certain built-in biases, as well. There are perceptual heuristics built into the system that systematically mislead us. That's what optical illusions are all about. A prime example is color vision.

- Although we may think we see shades of color the way they are, our perceptual system isn't built to see pure colors. It isn't built to detect the same wavelengths at different spots on a two-dimensional image; it's built to detect differences in a three-dimensional world—a world with shadows.

- How long in our evolutionary history have we been dealing with three-dimensional contexts and real shadows? Millions of years. That's what the system was built for. In contrast, we haven't been dealing with two-dimensional representations for very long at all.

- The heuristics built into the perceptual system were built for one kind of context: a three-dimensional world with real shadows. If we have to judge things in a different context—for example, for an artist who is judging color in a two-dimensional representation—we have to adjust for that built-in bias.

- We have similar built-in biases regarding judgments of length. For example, in an image of two tables, one seems to be long and narrow, while the other is closer to a square. Amazingly, though, the two shapes are precisely the same. The reason they don't look the same is that we are judging them as if we are viewing a three-dimensional scene. Our perceptual system is made to detect the way things really are in the world, so it sees them that way.

- The lesson here is that in a two-dimensional context, our perceptual system continues to carry a three-dimensional bias. If we're dealing with two-dimensional representations, we have to compensate for that.

- There is another surprising thing about these optical illusions. Even after we know the two shapes are the same, they still look different. People often talk about how belief influences perception—that we see what we want to see or what we expect to see. These optical illusions offer a counterexample.

- In cases like this, mere belief seems powerless in the face of our built-in **perceptual biases**. We know perfectly well that those two shapes are the same. But our built-in perceptual biases seem to ignore what we know.

Change Blindness
- Some of our mistakes are built into our information processing at a higher level. One source of error is **attention bias**. It is not true that we always see what we want to see or what we expect to see, and it's a good thing that we don't. But it is sometimes true that we don't see what we don't expect to see, particularly if we are concentrating on something else.

- There is a one-minute video from a famous experiment on attention bias that shows two groups of people—some in black shirts, some in white shirts—interweaving in a small room as they pass basketballs to one another. Subjects were told to count the number of times the basketball is passed from one white-shirted team member to another.

- At about the 30-second mark in the video, a person in a gorilla costume enters the room. The gorilla pauses, thumps his chest, and leaves after about 5 seconds. All this time, the subjects have been trying to count the number of basketball passes. A few minutes, later they are asked "Did you see anything unusual?"

- Amazingly, about half the experimental subjects failed to see the gorilla. They were concentrating on counting ball passes. Even when specifically asked, "Did you see anything unusual?" the answer was "No." They were victims of what is called **change blindness**, an attention bias.

Malleable Memory

- Our memories are far from perfect. We think of our memories as photographs of real events filed away for future reference. But our memories, however vivid, are often backward reconstructions from later reports. You may have a memory of something you did when you were a child. However, it's likely that what you remember is a reconstruction from family stories of that event, rather than the event itself.

© iStockphoto/Thinkstock.

Trying to name just three features on the side of a penny showing Abraham Lincoln reveals just how imperfect our memories are.

- The influence of verbal framing on memory has also been studied by researchers. Elizabeth Loftus is one of the world's foremost experts on eyewitness testimony.
  - In one of her classic studies, people were shown a film of a traffic accident and then were asked questions about it. The only difference was how the question was phrased. One group was asked, "About how fast were the cars going when they hit

each other?" The other group was asked, "About how fast were the cars going when they smashed into each other?"

○ One week later, the same people were asked a few more questions, such as "Did you see any broken glass?" There was, in fact, no broken glass in the accident. But those who had been asked about cars "smashing" into each other rather than "hitting" each other were twice as likely to say that there had been broken glass at the scene. The difference in a single word created different memories.

Availability Heuristic

- Some of our patterns of reasoning come mistake-prone. The previous lecture discussed heuristics that make us smart—fast-and-frugal intuitive rules of thumb that often seem to pay off. But there are also heuristics that make us not so smart. Those are patterns that we use all the time and that don't pay off—habitual patterns of thought or perception that systematically mislead us. One is called the **availability heuristic**.

- To illustrate this heuristic, answer this question: Are there more English words that start with the letter *K* or more English words that have the letter *K* in the third position?

  ○ If you're like most people, you would say that there are more English words that start with the letter *K*. Why? It's easy to rattle off words that start with *K*: *kite, kangaroo, kindergarten, Kool-Aid*. Those words are "available," and we follow an availability heuristic: If it's easier to think of, we tend to think it's more common.

  ○ But in fact, there are about three times as many English words that have a *K* in the third position: *bike, ask, acknowledge*. It's much harder to think of words in terms of a third letter rather than a first letter. But that has nothing to do with how many there are. The availability heuristic misleads us.

Anchor-and-Adjustment Heuristic

- Just as images can mislead us, so can our starting points. Those starting points are called anchors. Here, it's an **anchor-and-adjustment heuristic** that can mislead us.

- The starting point—the anchor—can come from many places, such as how a question is phrased. Memory can be influenced by how a question is phrased. It turns out that the same holds for rational calculation.

- Consider this: "What percentage of African nations do you think are members of the United Nations? Is it more than 10 percent?" People who were asked the question that way tended to give answers averaging about 25 percent.

- Other people were asked the question this way: "What percentage of African nations do you think are members of the United Nations? Is it more than 65 percent?" In that case, the average answer was 45 percent.

- The theory is that the figure given in the question is taken as an anchor. "Is it more than 10 percent?" Yes, it must be more than that; thus, you adjust up from that anchor. You end up in the 25 percent range, perhaps. When the question is phrased as "Is it more than 65 percent?" people use 65 percent as the anchor and adjust down, ending up in the 45 percent range.

- Of course, the real probabilities have nothing to do with how the question is asked. Something like 93 percent of African states are members of the United Nations. The anchor-and-adjustment heuristic has led us astray.

Cultural Patterns of Thought

- Heuristics are rules of thumb for patterns of thought. But the conceptual patterns don't have to be technical. Collecting patterns of thought across cultures can do the same thing, even if the patterns are something as simple as fairytales.

- Consider this puzzle: A treasure hunter is going to explore a cave on a hill near a beach. He suspects there might be many paths inside the cave and is afraid he might get lost. Obviously, he does not have a map of the cave; all he has with him are some common items, such as a flashlight and a bag. What can he do to make sure he does not get lost when trying to get back out of the cave? The answer is that he can take a bag full of sand from the beach and sprinkle it along his path.

- When that puzzle was given to American college students, about 75 percent gave the correct solution. But when the puzzle was given to Chinese students, only 25 percent solved it. The theory is that American students had a heuristic; they had a template of how to solve this kind of story. They had the story of Hansel and Gretel.

Terms to Know

anchor-and-adjustment heuristic: A common strategy used in calculating probabilities, but one that depends on how a question is phrased. Information given in the question is taken as a starting point, or anchor; individuals tend to adjust their responses upward if the anchor seems too low or downward if the anchor seems too high, arriving at an answer that is less extreme than the information given in the question but that may have little connection to the real answer.

attention bias: Overlooking the unexpected because we are attending to the expected.

availability heuristic: The tendency for individuals to assume that things that are easier to bring to mind must be more common or occur more frequently; the tendency to generalize from simple and vivid images generated by single or infrequent cases and to act as if these are representative.

change blindness: The propensity for individuals not to perceive unexpected changes, particularly when attention is focused on something else.

perceptual bias: A "hard-wired" tendency in our perceptual processing that forces us to perceive things in particular ways. Our color perception does not track pure wavelengths of light or actual lengths in a stimulus, for example, because our visual processing has evolved to interpret input immediately in terms of contextual cues regarding shadow and perspective.

Suggested Reading

Hallinan, *Why We Make Mistakes.*

Hoffman, *Visual Intelligence.*

Questions to Consider

1. Which line looks longer, the vertical or the horizontal?

   Adolf Fick first studied this illusion in 1851. We seem to have a built-in propensity to exaggerate the vertical. The division of the lower line contributes to the illusion.

   Measure the two lines.

2. You now know they are the same length. Does the vertical line still look longer? What does that say about the claim that our beliefs determine what we see?

3. The first lecture used a "bird in the thorn bush" example. We are much better at integrating images when they are moving. Can you offer an evolutionary hypothesis for that? Can you offer an evolutionary hypothesis for our propensity to exaggerate the vertical?

Watch the attention bias experiment on-line (http://www.theinvisiblegorilla.com/gorilla_experiment.html). When instructed to count the number of passes made by people in white shirts, almost half of those watching the video failed to see the gorilla.

The degree of our change blindness is astounding. Watch some of the examples posted online by the Department of Psychology at the University of British Columbia (http://www2.psych.ubc.ca/~rensink/flicker/download/) or search for "Change Blindness Experiments" on YouTube.

Why We Make Misteaks

Lecture 11—Transcript

Professor Grim: Why do we make mistakes? How many times have you berated yourself for what, looking back on it, was a really stupid mistake. "What was I thinking?" "How could I do that?" "That was a really stupid mistake." A friend, or a spouse, or a passerby may try to make you feel better. "You're only human. Everybody makes mistakes." It turns out there's a lot of truth in that. To err is human. Systematic error, not random, but systematic and predictable, is built into some of the ways we think. Systematic error is even built into how we see things. That's what this lecture is about: the ways that we as humans systematically go wrong and how we can correct for them. We don't have to be victims of our own systematic mistakes. If we are aware of cognitive biases, we can compensate for them.

Let's start with some built-in biases. I have emphasized the power of visualization in a number of lectures. Because of our evolutionary heritage, people are particularly good at fast information processing in visual contexts. We've been exploiting that ability at every opportunity. We used it, for example, in using Venn diagrams and flow diagrams to convert abstract relations into something we can see. But our evolutionary heritage comes with certain built-in biases as well. There are perceptual heuristics built into the system that systematically mislead us, and that's what optical illusions are all about. A prime example is color vision.

Look at squares A and B in this first image. Which is darker, patch A or patch B? Clearly A is; that's obvious. You can see that A is darker. But if we clip out those bits from the picture, move them over, and put them side by side, things look entirely different. Patches A and B are, in fact, exactly the same shade of gray. Why is it that they look so different in the original picture? Here's the theory.

In the context of that picture, square B is shown in shadow. Our perceptual system seems to compensate for that fact by making it appear lighter, as it takes its color to really be. It sees the shadow and compensates for the shadow by telling us that square B must really be lighter in color, and it tells us that by making B look lighter. We may think we see shades of color the way they are, but our perceptual system isn't built to see pure colors; it isn't built to detect the same wavelengths at different spots on a two-dimensional image. It's built to detect differences in a real three-dimensional world. A world with real shadows.

How long in our evolutionary history have we been dealing with three-dimensional contexts and real shadows? Millions of years. That's what the system was built for. How long have we been dealing with two-dimensional representations like the one you just saw? Not very long at all. It's only when we take bits of this two-dimensional picture out of that three-dimensional context and put them side by side that we see that our system has misled us as to the true wavelength of the patches in the picture.

The same thing happens with color. We can put dots of orange in squares A and B, dots of precisely the same wavelength. If we do, the one in the shadow will shine bright orange; the one out of shadow will look dull and darker.

The heuristics built into our perceptual system were built for one kind of context, a three-dimensional world with real shadows. If you have to judge things in a different context—for example, if you are an artist judging color in a 2-dimensional representation—you will have to adjust for that built-in bias.

We have similar built-in biases regarding judgments of length. Take a look at the tops of these two tables. Which one is narrower? Which one is closer to a square? Amazingly, the two shapes you see are precisely the same. Why don't they look the same? We are judging them as if we are viewing a three-dimensional scene. In order for a three-dimensional scene to project those shapes onto your retina, you'd need a narrow table on the left and a fat, square table on the right. Our perceptual system is made to detect the way things really are in the world, so it sees them that way. In a two-dimensional context, on the other hand, those shapes are exactly the same. The lesson: in a two-dimensional context our perceptual system continues to carry a three-dimensional bias. If we're dealing with two-dimensional representations, we will have to compensate for that. The table illusion persists even if we remove the clues of the legs.

There is another surprising thing about these optical illusions. Even after we know the colors are the same, they still look different. Even after we know the two shapes are the same, they still look different. People often talk about how belief influences perception. They say that you see what you want to see. Or that you see what you expect to see, and these optical illusions offer a counter-example. In cases like this, mere belief seems powerless in the face of our built-in perceptual biases. We know perfectly well that those two shades are the same. We know perfectly well that those two shades are identical. But our built-in perceptual biases seem to ignore what we know; they still look different.

Some of our mistakes are built into our information processing higher up. One source of error is attention bias. It's not true that we always see what we want to see or what we expect to see, and it's a good thing that we don't. But it is sometimes true that we don't see what we don't expect to see. Particularly if we are concentrating on something else. There is a one-minute video from a famous experiment on attention bias that shows two groups of people—some in black shirts, some in white shirts—interweaving in a small room as they pass basketballs to one another. If you were a subject participating in the experiment, you would have been told to count the number of times the basketball is passed from one white team member to another. At about the 30-second mark in the video, a person in a gorilla costume enters the room. The gorilla pauses, thumps his chest, and leaves after about five seconds. All this time the subjects have been trying to count the number of basketball passes. A few minutes later they're asked "Did you see anything unusual?" Amazingly, about half of the experimental subjects failed to see the gorilla. They were concentrating on counting ball passes. They weren't expecting a gorilla, and so they didn't see it. Even when specifically asked, "Did you see anything unusual?" the answer was no. If you go online to watch the video, of course, you will see the gorilla; you'll be looking for it. You may find it hard to believe that anyone would miss the gorilla. But then again, you may have been one of the people who failed to notice that the color of my shirt and tie changed after you saw the gorilla. If so, you were a victim of what is called change blindness.

There are people in the movie industry whose job is to assure a continuity between scenes—to make sure that Indiana Jones's hat doesn't change color between shots, for example. How many people will notice that kind of change? The answer is surprisingly few. The 1959 Hollywood classic *Ben Hur* won 11 Academy Awards. The most famous scene is the chariot race with Charlton Heston. It lasts 11 minutes on screen, though I'm told it took months to film. But even that famous scene has continuity problems. The race starts with nine chariots. Six crash in the scene. That should leave three, but if you watch it, you'll see that the race actually ends with four.

In videos constructed for psychological experiments, people often don't even notice that an actor has changed—that it's a completely different person carrying the action from one scene to the next. And it's not just changes on screen that we tend not to notice. Daniel Simons and Daniel Levin, then at Cornell, performed an experiment in which a stranger asks a person for directions. During their conversation, two people carrying a door walk between them. But there's a twist. The experimenters had one of the door carriers change places with the stranger as the door went by. So the conversation begins with one person, a door goes by, the conversation ends with someone else. In that experiment, only seven out of fifteen people noticed the switch. Attention, then, comes with its own biases. We tend not to notice things that aren't at the focus of our attention, even when that thing is a gorilla. We're change blind; we often don't notice when things change.

There are other gaps in our information processing. As you probably know, our memories are far from perfect. Just how imperfect? Here's an experiment for you. How many times have you seen a penny? Thousands. So tell me, tell me some things about pennies. Without looking—that would be cheating. Think of the side with Abe Lincoln in profile. Which way is he facing? Right or left? You probably got that one. He's facing to the right. Now name three other things on that side of the penny. Take your time. Just dip into your memory banks. Give me three other features on that side of the penny. If you want to hit pause and think about it a minute, do so, but no cheating. Then come back.

What are the other features on Abe's side of a penny? One of them is the date. Where is that? To the right of Lincoln, or the left? It's to the right. You've got two features left. What are they? If you've got a really good memory, you'll remember that it says "In God We Trust" across the top.

And the fourth feature? It's the one most people don't remember, even after handling pennies for their entire lives. The word "Liberty" appears just to the left of Lincoln's profile. Our memories are imperfect. They are also selective. Gamblers remember the times they won much more effectively than the times they lost. That's part of what keeps them gambling.

There's another distressing aspect about memory. We think of our memories as photographs of real events filed away for future reference. But they are often less a function of the original photograph than of later representations. Our memories, however vivid, are often backward reconstructions from later reports. You may have a memory of some crazy thing you did when you were a kid. Remember that Thanksgiving that Pat snuck into the kitchen and ate the whole chokecherry pie by himself? Chances are, however, however vivid your memory, that what you remember is a reconstruction from the family stories of that event, rather than the event itself. After I discussed this in one of my classes, one of my students went home for Thanksgiving and asked her family about a particularly vivid memory she had from her childhood. It turned out that the event she remembered so vividly as happening to her had in fact happened to her sister instead.

The influence of verbal framing on memory has also been studied in the lab. Elizabeth Loftus is one of the world's foremost experts on eyewitness testimony. This is one of her classic studies. People were shown a film of a traffic accident and were then asked questions about it. Immediately after viewing the film, half the people were asked, "About how fast were the cars going when they smashed into each other?" The other half were asked, "About how fast were the cars going when they hit each other?" The only difference was how the question was phrased:. About how fast were the cars going when they hit each other? About how fast were the cars going when they smashed into each other? One week later the same people were asked a few more questions, "Did you see any broken glass?" There was, in fact, no broken glass in the accident. But if you had been asked about cars smashing into each other, rather than hitting each other, you were twice as likely to say that there had been. The difference in a single word created different memories.

I have performed eyewitness experiments several times in class. It's all staged, of course. Several graduate students come in the back of the class.

One of them yells something, another throws a water balloon, a third grabs my briefcase and ducks out a side door. All three of the perpetrators are wearing colored baseball caps. The whole thing takes about 15 seconds. Immediately afterwards, students are given questionnaires to fill out. But in fact, the questionnaires are different. Half of the students are asked, "What did the guy with the green baseball cap say?" The other half asked "What did the guy with the blue baseball cap say?" There was, in fact, no blue baseball cap. But when asked about the incident a week later, lots of the kids with the second question report seeing a blue baseball cap. They all saw the same incident, but what they remember may be a factor, not of what happened, but of what they were asked later, and how. The fact that memories are so easily manipulated by later questioning has some very serious consequences. Child abuse is a real problem, but so is the prospect of false memories of child abuse. Eyewitness testimony can be all too easily be slanted by certain forms of police questioning.

So no wonder we make so many mistakes. Our perceptual system is hard wired for some contexts and not others. Our information processing comes with attention bias. We're change blind. Our memories are malleable in terms of later questions and reports. If that weren't bad enough, it's also the case that some of our patterns of reasoning come mistake prone. In the previous lecture I talked about heuristics that make us smart: fast and frugal intuitive rules of thumb that often seem to pay off. But there are also heuristics that make us dumb. Those are patterns that we use all the time and that don't pay off, habitual patterns of thought or perception that systematically mislead us.

One is called an Availability Heuristic. So here's a simple test. Are there more English words that start with the letter k or more English words that have the letter k in the third position? If you're like most people, you said there are more English words that start with the letter k. Why? It's easy to rattle off words that start with k: kite, kangaroo, kindergarten, Kool-Aid. It's easy to think of words that start with a k. Those words are available, and we follow an availability heuristic. If it's easier to think of, we tend to think it's more common. But, in fact, there are about three times as many English words that have a k in the third position: bike, ask, acknowledge. It's a lot harder to think of words in terms of a third letter rather than a first letter, but that has nothing to do with how many there are. The availability heuristic misleads us. We take how easy it is to think of something to indicate how common or probable something is, and that's where we go wrong.

In many cases we tend to think in terms of simple and vivid images. That's fine if all we're interested in is a single case. The problem comes when we generalize from those single cases as if they were genuinely representative. Sometimes our vivid and available images come from media exposure. Images in the news are there because they are newsworthy. But when those become our vivid images, our vivid images, we tend to overestimate their probability. So which should you worry about more, airplane crashes or car crashes? Clearly the latter. There are many more car crashes than plane crashes, and many more people are injured or killed in car accidents. But airplane crashes tend to be big news. Car crashes don't. Because of that, people overestimate the relative risks of air travel. What is the chance of being murdered? What is the chance of dying from stomach cancer? Murder makes the news, and is overestimated. It's about five times as likely that you'll die of stomach cancer. Shark attacks and death by lightning are vivid images. But they're not vivid because they're common, quite the opposite. When we reason about frequency or risk of events in terms of images alone, it's the availability heuristic that's the source of our mistakes.

Just as images can mislead us, our starting points can. Those starting points are called anchors. Here it's an anchoring and adjustment heuristic that can mislead us. The starting point, the anchor, can come from lots of places. It can come from something as simple as how a question is phrased. In talking about mistakes of memory, we mentioned the fact that memory can be influenced by how a question is phrased. It turns out the same holds for rational calculation. What percentage of African nations do you think are members of the United Nations? Is it more than 10%? Give me a figure. What do you think the percentage really is? People who were asked the question that way tended to give answers averaging about 25%. Other people were asked the question this way: What percentage of African nations do you think are members of the United Nations? Is it more than 65%? Give me a figure. What do you think the percentage really is? In that case, the average answer was 45%. Why? The theory is that the figure given in the question is taken as an anchor. Is it more than 10%? Yes, it must be more than that. So you adjust up from that anchor. You end up in the 25% range, perhaps. Is it more than 65%? No, that seems like a lot. People use 65% as the anchor and they adjust down. They end up in the 45% range. But of course the real probabilities have nothing to do with how the question is asked. Something like 93% of African states are members of the United Nations. The anchor-and-adjustment heuristic has led us astray.

Then there is this rationality problem: We all think we are better than average. Garrison Keillor classically began his stories with, "Welcome to Lake Wobegon, where all the women are strong, all the men are good looking, and all the children are above average." It is a joke, of course. It can't be true that all the children are above average. And yet when people are asked about their driving skill, almost everyone rates themselves above average. When asked about how attractive they are, almost everyone rates themselves above average. We also tend to be overconfident in our own abilities. We tend to be overconfident in our ability to stick to a diet, for example, or to keep with the gym plan. We tend to be overconfident in our ability to make money in the stock market, or to win at the casino. Even there we think we'll do better than average. It's become known as the Lake Wobegon effect.

According to Stefano DellaVigna, an economist at UC Berkeley, overconfidence is a general feature of human psychology. Almost everyone is overconfident. The only exception, he says, are people who are depressed. They are the ones who tend to be realists. That makes me think twice about whether overconfidence should be considered an intellectual vice after all. If the alternative is depression, we may be better off thinking of ourselves as above average. With confidence, even overconfidence, comes a life of exploring new options, trying new things, dreaming big. That sounds better to me. Maybe realism is overrated.

The take-home message? Some of the sources of mistakes, we've said, are built in. All we can do about those is know that they are there and compensate for them. Don't assume that the color things look is the way they really are. Some sources of mistakes are built into our information processing. Be aware that attention bias may blind you to other things going on. Don't bet on your ability to notice all changes around you; we all tend to be change blind. And don't treat your memory as photo perfect in all cases. What you remember is malleable to suggestion and later representation. Some of our mistakes stem from unreliable heuristics. Some of those we can deliberately avoid. We can see them coming and steer clear. In order to avoid the availability heuristic, think beyond a single image. Anchor and adjustment? Be suspicious of that anchor, wherever it comes from. Better than average? Be aware that your estimates tend in that direction and compensate for them when necessary.

In an earlier lecture we talked about simple heuristics that make us smart. So are heuristics good or bad? Both. What we need to do is to use the good ones and avoid the bad ones. Even better, be aware of the heuristic you are using and pay attention to both how it might help, and how it might lead you astray. Heuristics are rules of thumb for patterns of thought. I talked earlier about von Neumann's deliberate attempt to stock his mind with potentially useful technical formulae. But you don't have to be von Neumann, and the conceptual patterns don't have to be technical. Collecting patterns of thought across cultures can do the same thing even if the patterns are something as simple as fairy tales. Let me give you a puzzle:

A treasure hunter is going to explore a cave on a hill near a beach. He suspected there might be many paths inside the cave and was afraid he might get lost. Obviously, he did not have a map of the cave; all he had with him were some common items such as a flashlight and a bag. What could he do to make sure he did not get lost when trying to get back out of the cave?

If you want to take a minute to think about it, press pause and come back. What could he do to make sure he didn't get lost in the cave? He could take a bag full of sand from the beach and sprinkle it along his path. When that puzzle was given to American college students, about 75% gave that solution. But when the puzzle was given to Chinese students, only 25% solved it. Why? The theory is that American students had a heuristic; they had a template of how to solve this kind of story. They had the story of Hansel and Gretel. Now let me give you another puzzle. Let's see if you get the answer to this one:

In a village by a river, the chief of the region took care of its sacred stone statue. It was the chief's custom every year to go down river to the next village to collect taxes. To assess the amount of taxes, the chief would ask for the statue's weight in gold coins. His method of measuring this amount was to put the statue in a large tub at one end of a hanging balance scale and hook the other end of the scale to another large tub—this second tub was filled with gold coins—until the scale balanced the weight of the statue. During one trip to collect taxes, the chief forgot to bring his balance scale. Now he had a problem: How could he figure out how much gold to take to match the statue's weight without the benefit of a balance scale, a pulley system, or a conventional scale?

If you want, take a minute to think about that one too. That one's not so easy, right? Only 25% of American college students got a solution to this one. But 75% of Chinese college students did. That's because they had a different heuristic from a different folk tale. I'll finish with the folk tale of Chao Chong. Chinese students had grown up with this story—with this template—much as American students had grown up with Hansel and Gretel.

> Long ago in China, there lived a powerful emperor. Every year, the rulers of the surrounding countries had to give him jewelry, gold, cloth materials, and animals as presents. One day, a ruler of a southern country presented him with an elephant as a gift. The emperor was delighted to see the elephant and asked the ruler what the weight of the elephant was. The ruler was embarrassed because even the biggest scale he owned was too small to weigh the huge elephant. The emperor's youngest son, named Chao Chong, came up with an idea: You could find a boat, put the elephant in it, and mark the new water level on the boat. Then you could take the elephant out of the boat and put smaller stones into the boat until the water level reaches the same mark. Then, you could weigh those stones separately with a small scale. When you add up all the weights, you would know how heavy the elephant is.

By crossing cultures you now have two heuristics, two templates. Hansel and Gretel and the story of Chao Chong. In the next lecture I want to talk about rationality in social interaction. In particular: how to be rational in a polarized context.

Rational Discussion in a Polarized Context
Lecture 12

W e seem to find ourselves in the midst of increased polarization: polarization of argument, of opinion, and of values. What is the role of rationality in a polarized context? On a more individual level: Is it possible to have a rational discussion with people who seem to disagree with you radically, on fundamental issues? Is there a way for rationality to reduce polarization? We'll look at these questions in this lecture.

Partisan Polarization
- Before we look at the facts about **polarization** in America, we need to draw an important distinction. What we will call "partisan polarization" is polarization between political organizations, on the scale of either major political parties or smaller interest groups. What we will call "cultural polarization" is polarization within the general public on attitudes that may or may not be expressed politically.

- Partisan polarization is real. In 1980, 43 percent of Americans polled said that they thought there were important differences between the political parties. That figure is now 74 percent. In 1976, almost a third thought it didn't even make a difference who was president; that figure is now cut in half.

- Internal uniformity within each party has grown just as distance between them has. But partisan polarization isn't new. George Washington's farewell address in 1796 emphasized the danger of "factions." A year later, Thomas Jefferson complained that because of partisan polarization, "men who have been intimate all their lives cross the streets to avoid meeting."

Cultural Polarization
- Although partisan polarization has increased over the last few decades, cultural polarization—polarization in general social attitudes among the general public—may not have.

- Polarization in public attitudes is limited to a few hot-button issues. On most issues, public polarization hasn't increased between groups, regardless of what groups are being compared: the young and the old, men and women, the more and the less educated, different regions of the country, or different religious affiliations. On a number of fronts, polarization has, in fact, decreased.

- What then explains the polarization we do see? First, we'll look at some answers from political scientists and sociologists. Then, we'll explore the internal and logical dynamics of polarization, of particular interest to philosophers—the quasi-rational part of the picture.

- Political scientists concentrate on partisan polarization. They note historical changes in the parties. The Republican Party has drifted to the right; northern Democrats have become more liberal, and southern Democrats tend to be affiliated with Republicans.

- Sociologists add some observations regarding social and cultural polarization. Since the 1970s, we have sorted ourselves more residentially by income and social class. Our neighbors are increasingly like ourselves. Sociologists also note that America has a long history of fraternal and voluntary organizations that once brought people together with those who might not be exactly like them. But membership in those long-term organizations has waned, replaced with self-selecting short-term membership in groups organized for some single political purpose.

A Philosophical Analysis of Polarization

- When we think about it, it seems that a polarization of attitudes or opinions is to be expected. It may even follow something like rational lines.

- Suppose that Mr. Magoo is very much convinced of a particular opinion. Because he's convinced that the position is right, he's also convinced that any evidence offered against the position must be flawed in some way—bad research, perhaps, or an improper interpretation. In the philosophical literature, Mr. Magoo is what is called a **Kripkean dogmatist** (based on a lecture by the contemporary philosopher Saul Kripke). An extreme case, Mr. Magoo thinks his current position is right and, on that ground, is prepared to reject any information to the contrary.

- Now, imagine two people who are strongly convinced of two opposing positions: Mr. Magoo and, say, Dr. Seuss. Both are Kripkean dogmatists. Both selectively reject evidence and argument against their positions. Both are selectively receptive to evidence and argument for their positions.

- Suppose these two are given a mixed bag of new evidence and arguments, just as we get on just about every topic every day. The new information includes a handful of evidence and arguments that seem to point in one direction, but also a handful that seem to point in the other.

- One might think—one might hope—that new evidence and argument would bring the two views closer together. But if we are dealing with Kripkean dogmatists, that is not what will happen. If attitudes start out already polarized, new evidence can make them even more so.

When shown images of the earth from space, a spokesman for the Flat Earth Society said, "It's easy to see how a photograph like that could fool the untrained eye," illustrating the Kripkean dogmatist's selective rejection of evidence against his position.

- Mr. Magoo and Dr. Seuss represent an extreme case, but the same dynamics will operate even when the case isn't that extreme.

A Case Study of Polarization

- A set of classic studies in social psychology shows precisely this effect. In these studies, individuals' attitudes toward the death penalty were measured. Subjects who favored the death penalty were put in one group. Those who opposed it were put in another. Both groups were then given the same two pieces of evidence: results of one study supporting a claim of deterrence and results of another that opposed deterrence. Each group was also given further information on the two studies: experimental details, critiques in the literature, and the researchers' response to those critiques.

- The result was just as you would expect from Mr. Magoo and Dr. Seuss. All participants found the studies supporting their positions to be better conducted and less subject to criticism than the counterevidence. Both groups ended up indicating stronger commitment to their original positions than when they began.

Selection of Information

- Another dynamic in polarization that hasn't yet been strongly studied is selection of information sources.

- There was a time when America had essentially one source of television news: the nightly news on NBC, CBS, and ABC. Because of the canons of journalism these shows tried to avoid an overt political stance. Walter Cronkite was repeatedly cited as the most trusted man in America.

- Of course, we now have Fox News and MSNBC, which are clearly polarized politically. Instant access to any slanted source is immediately available on the Internet.

- Now, not only will Mr. Magoo and Dr. Seuss treat identical information differently, but they will seek out information differently. Mr. Magoo will trust information sources that reinforce his existing beliefs. Dr. Seuss will do the same on the other side. Because both will be able to find media reinforcing their existing beliefs, we can expect polarization to increase.

Warning Indicators of Polarization

- There is a good warning indicator of when two people might be falling prey to the dynamics of polarization we've been talking about. Suppose you and a friend agree on many things. On those issues, she seems as perfectly rational as you are. But her position on issue X seems radically wrong. How can she possibly think that?

- That's the warning indicator. If it's hard to believe that someone so rational in other areas could be so irrational in area X, that's a sign that it's time to lay out evidence and argument on both sides—time to recalibrate and reevaluate.

- There's no guarantee that like-mindedness will result. But strong disagreement in a small area with a person you think is generally rational is an indicator that you may be playing Mr. Magoo to her Dr. Seuss.

Rationality in a Polarized Context

- In our attempts at rationality in a polarized context, the literature on **negotiation strategies** seems to fit surprisingly well. William Lury and Roger Fisher are members of the Harvard Negotiation Project and authors of the influential *Getting to Yes*. Following are some of their tips.

- Rationality deals in cool-headed evidence and argument. A major theme in the work of Lury and Fisher is removing an issue from its ties to emotion and ego, what they call "going to the balcony." The idea is for both sides to try to view the issue as if they were *not* embedded in it. Try to see it from a distance—as if the positions were being advanced by someone entirely different.

- A correlate theme is ego-distancing. People identify with the views they hold. That's part of the problem. But of course, neither truth nor rationality cares whether a position happens to be mine or yours. If an issue can be cast as one for joint resolution, through joint effort, that's a step away from individual ego. And it's clearly a step in the right direction.

- In **positional negotiation**, people identify themselves with specific positions from the start—positions to be won or lost. In positional negotiation, you may feel forced into one of two roles. You end up either a "soft" negotiator or a "hard" negotiator. But neither of those really works. The alternative is to change the game. If the context can be changed from a contest of wills to a mutual exploration and investigation of an issue, there's a chance that the issue will be resolved.

- In leaving positional negotiation behind, Lury and Fisher urge people to look for a way to settle issues in terms of objective criteria. In the case of negotiation, that may mean an appeal to **independent standards** that allow us to look at the totality of evidence. The parties must decide in advance what kind of evidence will convince them of a particular position. If they can agree on that, they can look with new eyes at what the evidence actually is.

- Sometimes a dispute isn't about data or evidence but about background principles. In those cases, the discussion must go deeper. The parties must ask what principles are at play and what kind of principles those are.
  - For example, proponents of gun control tend to emphasize statistics regarding harm. Opponents of gun control tend to emphasize constitutional rights.

  - But if we take the principles out of the context of debate, people on both sides recognize that the Constitution says something important about rights. People on both sides recognize that we should avoid unnecessary harm. If we can get from polarized people to joint recognition that the issue is one of conflicting principles, we've made progress.

- Is rhetoric at issue in such debates and negotiations? Absolutely. We'll talk more about that subject in the next lecture.

independent standard: In negotiation or the attempt to reduce polarization, a deciding touchstone or court of appeal that is not open to manipulation and that can be agreed on by both parties in advance. In establishing a fair price for a house, for example, both parties might agree in advance to use the price that similar houses have recently sold for in the neighborhood.

Kripkean dogmatist: An individual who believes that his or her position is right and, on that ground alone, is prepared to reject any and all evidence to the contrary.

negotiation strategy: An approach to conflict resolution that may attempt to remove the "contest of wills" characteristic of positional negotiation. Among other techniques, negotiation strategies may include employing ego-distancing; talking about issues without identifying with a particular position; "going to the balcony," that is, trying to put emotions aside and view the problem from a distance; appealing to independent standards; coming to some agreement about what kinds of objective criteria could help to clarify or settle the issue; and replacing debate with collaborative research on the topic.

polarization: Radical or extreme disagreement between groups with no apparent willingness to compromise and/or with few individuals representing a middle group between the extreme positions. Polarization normally implies a wide gap between positions and increased uniformity within positions. Political polarization refers to extreme positions taken by political organizations, either major political parties or smaller interest groups; cultural polarization refers to extreme differences in attitudes of the general public that may or may not be expressed politically.

positional negotiation: In conflict resolution scenarios, an approach in which people are ego-involved or identify with their specific positions. Those involved in positional negotiation often end up in one of two roles: as the "soft" negotiator, who tries to avoid conflict by giving in and winds up feeling exploited, or as the "hard" negotiator, who is out to win at all costs and, thus, starts with an absurd extreme, allowing room to make some concessions and still hit his or her initial target.

Suggested Reading

Brownstein, *The Second Civil War*.

Fiorina (with Abrams and Pope), *Culture War?*

Fisher and Lury, *Getting to Yes*.

Lord, Ross, and Lepper, "Biased Assimilation and Attitude Polarization."

Questions to Consider

1. Many estimate that polarization in America is at a high point now. From your own experience, what periods do you remember as particularly polarized? As less polarized?

2. This lecture included some suggestions about how to reduce polarization, but it might be easier to prevent it. Can you suggest some approaches to discussion that might help avoid undue polarization from developing?

3. A general question in epistemology is as follows: It is irrational to discount new claims and new evidence simply because they conflict with what we already believe. On the other hand, we have to work from past experience: We can't examine every one of our beliefs anew at every minute. What holds for us as individuals also holds for our society and our science. Can you think of any useful rules of thumb that can guide us in deciding what beliefs are worth reexamining and when?

Exercise

This is an exercise to work on together with a friend, a friend you talk to frequently, with whom you agree on many things, but with whom you disagree on some. Together, discuss the following questions:

(a) What topics are you two most in agreement on?

(b) What topics are you most polarized on?

(c) Why those topics?

Rational Discussion in a Polarized Context
Lecture 12—Transcript

Professor Grim: We seem to find ourselves in the midst of increased polarization: polarization of argument, polarization of opinion, polarization of value. What is the role of rationality in a polarized context? What should the role of rationality be? On a more individual level, is it possible to have a rational discussion with people who seem to disagree with you so radically, on such fundamental issues? Is it possible to cut through? Is there a way for rationality to reduce polarization? That is what this lecture is about.

I want to start with the facts about polarization in America. But first it is important to make a distinction. What I will call partisan polarization is polarization between political organizations, on the scale of either major political parties or smaller interest groups. What I will call cultural polarization is polarization within the general public on attitudes that may or may not be expressed politically.

First the bad news; partisan polarization is real. In 1980, only 43% of Americans polled said they thought there were important differences between the parties. The figure is now 74%. In 1976, almost a third thought it didn't even make a difference who was president. That figure is now cut in half. Internal uniformity within each party has grown, just as distance between them has. There was a time when members of Congress often crossed party lines. Between 1969 and 1976—the Nixon and the Ford years—the rate at which Republicans voted along party lines was about 65% in both the House and the Senate. The same was true of Democrats. Between 2001 and 2004, under George W. Bush, Republicans voted with their party 90% of the time. Democrats voted with their party 85% of the time.

Partisan polarization like that isn't new. George Washington's farewell address in 1796 emphasized the danger of factions, "One of the expedients of party to acquire influence...is to misrepresent the opinions and aims of other[s]," he said. The spirit of party kindles animosity, and "agitates the community with ill-founded jealousies and false alarms." A year later, Thomas Jefferson complained that because of partisan polarization "men who have been intimate all their lives cross the streets to avoid meeting..."

Sound familiar? Jefferson observed it, but he didn't reverse it. Until the very last part of their lives, Thomas Jefferson and John Adams treated each other in precisely that way.

Partisan polarization has increased over the last few decades. But there is also some good news. Cultural polarization—polarization in general social attitudes among the general public—may not have. Morris Fiorina, an expert on the topic, says that increased polarization is primarily a political phenomenon—partisan polarization between political activists or within the political elite. But of course it is the political activists and the political elite who make the news. We have noted before that media sources may lead us to overestimate the frequency of the dangerous events that appear in the news. The impression that not just partisan politics but cultural views in general are polarizing may be a media effect as well.

Polarization in public attitudes is limited to a few hot-button issues. On most issues, public polarization hasn't increased between groups, regardless of what groups are being compared: the young and the old, men and women, the more and the less educated, different regions of the country, or different religious affiliations. On a number of fronts polarization has, in fact, decreased. Views on women's role in public life are no longer the hot-button issue they once were. Racial integration was once fought vociferously by major portions of the population, not now. Support for the death penalty has fallen, though opinions of crime have moved toward tougher enforcement, and those changes have operated in parallel across those different groups. It's polarization on a few hot-button issues that makes the news. Abortion and gay marriage are prime candidates. Views even on taxation, government, and the deficit are much less polarized than the politicians would have you think. Well over half of voters think that both spending cuts and higher taxes will be needed to reduce the deficit.

So what explains the polarization we do see? Let me give the answer the political scientists tend to give, followed by some bits of data from the sociologists. Then I'll turn to the internal and logical dynamics of polarization, of particular interest to philosophers—the quasi-rational part of the picture. Political scientists concentrate on partisan polarization, of course. They note historical changes in the parties, North and South. The Republican party has drifted strongly to the right. "Republican" and "Conservative" coincide much more closely now than in the past. In the Democratic party, Northern Democrats have become somewhat more liberal than in the 1960s. More important is the change in the South. Southern Democrats were once an important voting bloc, different, and in many respects, less liberal than their Northern counterparts. That voting bloc now affiliates itself with the Republican party.

Sociologists add some observations regarding social and cultural polarization. Since the 1970s we have sorted ourselves more residentially by income and social class. Our neighbors are increasingly like ourselves. Sociologists also note that America has a long history of fraternal and voluntary organizations: the Rotary Club, Lions, Kiwanis, Masons, Odd Fellows, the Elks. Local chapters often drew members from various occupational and class levels across a community. They brought people together with people who might not be exactly like them. But membership in those long-term organizations has waned, replaced with self-selecting short-term membership in groups organized for some single political purpose.

Let me shift from political and social observation, however, to a philosophical analysis of the dynamics within polarization. When you think about it, it seems that a polarization of attitudes or opinions is to be expected. It may even follow something like rational lines. I'll follow our earlier strategy of going to extremes, exaggerating in order to present the issue as simply as possible. Suppose that Mr. Magoo is very much convinced of a particular opinion. He's just sure that that position is right. Because he's convinced that the position is right, he's convinced that any evidence offered against the position must be flawed in some way; bad research, perhaps, or an improper interpretation. Because Magoo is convinced his position is right, he's convinced that any argument offered on the other side must somehow be a bad argument. It must rely on some trick or some logical fallacy or go wrong just because of that flawed data.

In the philosophical literature, Mr. Magoo is what is called a Kripkean dogmatist. The term comes from a lecture by the contemporary philosopher Saul Kripke. Mr. Magoo thinks a particular position is right, and that alone seems to give him reason to be suspicious of any counter evidence, any counter arguments. The position is right, so there has to be something wrong with anything that seems to go against it, even if he can't put his finger on what. Mr. Magoo is an extreme case. He thinks his current position is right and on that ground is prepared to reject any information to the contrary. Conspiracy theories seem to work very much that way.

I'm sure you're too young to remember. The Beatles were at their height in the 1960s when the rumor started that Paul McCartney was dead, killed in a car crash in 1966. The evidence? Clues from the albums. In A Day In the Life, John Lennon sings, "He blew his mind out in a car ... he hadn't noticed that the lights had changed." Aha! A reference to the accident. If you listen closely to John at the end of Strawberry Fields Forever, you can hear something that sounds like "I buried Paul." And what about the visual evidence? On Magical Mystery Tour, Paul is the only one wearing a black carnation. Aha! The Fab Four crossing the street on the cover of Abbey Road? John is dressed in white as the religious figure. Ringo is in black as the undertaker. George is in faded blue jeans; he's the gravedigger. Paul is the only one in bare feet, the corpse.

It should be easy to refute a rumor like that. Just have Paul do an interview and show everyone he's alive. This is where the strategy of the Kripkean dogmatist comes in. Not so fast. If Paul's dead, that can't be real. It seems to me I remember a Paul McCartney look-alike contest run in one of the teen magazines, and the winner was never announced. Aha! There is still a Flat Earth Society, and there have been some serious flat earthers. In 1956 a spokesman for the society was asked about satellite images from space that showed the earth as an oblate spheroid, rather than a flat disk. His response? "It's easy to see how a photograph like that could fool the untrained eye."

Here's how this offers an analysis of polarization dynamics. Imagine two people strongly convinced of two opposing positions, Mr. Magoo and Dr. Seuss, perhaps. Both are Kripkean dogmatists. Both selectively reject evidence and argument against their position. Both are selectively receptive to evidence and argument for their position. Now suppose they are given a mixed bag of new evidence and arguments, just like we get on just about every topic every day. The new information includes a handful of evidence and arguments that seem to point in one direction, but also a handful that seem to point in the other. One might think—one might hope—that new evidence and argument would bring two views closer together. But if we are dealing with Kripkean dogmatists, that's not what is going to happen. Magoo will look at the totality of evidence, reject the pieces on the other side, and accept those that favor his current position, thereby strengthening it. Dr. Seuss will do the same, rejecting exactly the opposite handful of studies and arguments, strengthening his position as well. Mr. Magoo and Dr. Seuss will become even further and more firmly polarized. There is nothing inherently polarizing about the handful of evidence and arguments they see—it may equally represent both sides. But if attitudes start out already polarized, it can make them even more so.

Mr. Magoo and Dr. Seuss are the extreme case. They are prepared to reject all contrary evidence on the basis of their current positions. But that same dynamics will operate even when the case isn't that extreme.

There are a set of classic studies in social psychology that show precisely this effect. Individuals' attitudes toward the death penalty were measured. Subjects who favored the death penalty were put in one group; those who opposed it were put in another. Both groups were then given the same two pieces of evidence. One piece of evidence supported a claim of deterrence. It was, in fact, a study that compared murder rates for the year before and after the adoption of capital punishment in 14 states. In 11 of those 14 states, murder rates were lower the year after adoption. The other piece of evidence went against deterrence. This one was a study that compared murder rates in 10 pairs of neighboring states where one state had the death penalty and the bordering state didn't. In 8 of the 10 pairs, murder rates were higher in the state that had the death penalty. Each group was also given further information on the two studies, experimental details, critiques in the literature, and the researchers' response to those critiques. So both groups studied the totality of evidence and were again asked about their opinions regarding the death penalty. The result? Just as you would expect from Mr. Magoo and Dr. Seuss. All participants found the evidence supporting their position better conducted and less subject to criticism than the counter evidence. Both groups ended up indicating stronger commitment to their position than when they began.

There's another dynamics in polarization that hasn't yet been strongly studied. My prediction is that it will get increased attention. In fact, I'm working on it in some of my own research. The issue is selection of information sources. There was a time when America had one basic television news source, the NBC, CBS, and ABC nightly news. Because of the canons of journalism, they tried to avoid an overt political stance. Walter Cronkite was repeatedly cited as the most trusted man in America. That time is past. We have Fox News and MSNBC, clearly polarized politically. Instant access to any slanted source is immediately available on the internet. Now, not only will Mr. Magoo and Dr. Seuss treat identical information differently, they will seek out information differently. Mr. Magoo will trust information sources that reinforce his existing beliefs. Dr. Seuss will do the same on the other side. Because both will be able to find media reinforcing their existing beliefs, we can expect polarization to increase.

That's what the internal dynamics of polarization might look like. And you have to admit that there seems to be something rational about that dynamics as well. Suppose you do have reason to believe a particular position. Isn't that also reason to believe that the counter position is incorrect? Isn't that reason to believe that arguments for the counter position have to be wrong? Not only do we often reason in this way, it seems we have to. We can't constantly reconsider every position on every issue from the ground up given every new piece of information that comes our way. We have to put new information through some initial plausibility filter simply in order to know what is worth paying attention to and what is not. That seems to be an essential mechanism, not just in everyday life, but in normal science. Not every wacky claim from every corner is given equal consideration. There are studies that are poorly done, trumpeting results that we have full reason to believe aren't right. Working scientists need an initial plausibility filter too. What can we base that plausibility filter on except the theories we think we already have good evidence for or the opinions we are already sure of? What differentiates our case from that of Dr. Seuss and Mr. Magoo?

The consensus in the philosophical community is that there is, indeed, something rational about the dynamics we've been talking about but not enough to justify Kripkean dogmatism, not enough to justify Mr. Magoo and Dr. Seuss. Remember that the Kripkean dogmatist rejects any new evidence and any new argument. He's sure his position is right. Nothing you can offer him to the contrary is worth looking at. What that really means, however, is that he takes whatever originally led him to his current position to trump any evidence since. That's the irrational part.

Everyone agrees that the rationality of a position depends on the total available evidence. Now if you knew there were 100 pieces of relevant evidence on a topic, would you make up your mind on the basis of just three that you happened to come across first? No. That would be irrational. But that is what the Kripkean dogmatist is doing. Of course, people decide on positions for all kinds of reasons, many of which have nothing to do with the evidence. But let's give Mr. Magoo the benefit of the doubt. Let's assume that he really did come to the position he now holds by looking at available evidence. Even so, the evidence he used to form his view originally was a long time ago. It didn't include all the relevant evidence because it didn't include the evidence that's coming in now. That means that he made up his mind on the basis of some small sample of evidence—the evidence that happened to come his way long ago. That's what's irrational.

Here, again, Magoo is an extreme case, but the lessons carry over to more realistic setups as well. Suppose it wasn't just 3% of the relevant evidence that someone used to form their opinion; suppose it was 30%. Suppose the person isn't totally impervious to new evidence from the other side; they just think they have grounds to be skeptical. That gives us a sliding scale rather than an all-or-nothing measure. Full Magoo-like irrationality comes with total imperviousness to new evidence on the basis of a position grounded in a very small or very ancient selection of the data. A position will be more rational the more data behind it—the greater proportion of the evidence that informs it. A position will be more rational the more it remains open to incoming information. If perfect rationality would demand being open to all evidence on all topics all the time, I suppose none of us can be perfectly rational. Our cognitive limitations are such that we have to do some filtering. That applies even to a social institution like normal science. We can still use the idea of an even-handed openness to new evidence as a rational ideal and a reminder to avoid becoming Kripkean dogmatists.

There is a good warning indicator of when two people might be falling prey to the dynamics of polarization we've been talking about. Suppose you and a friend agree on lots of things. On all of those, she seems as perfectly rational as you do. She handles the evidence well, she is alive to subtleties of argument, a smart cookie. Not much gets past her. I'll bet she thinks much the same of you. But her position on issue X just seems radically wrong, stupid, even. How can she possibly think that? That's the warning indicator. It is hard to believe that someone so rational in other areas could be so irrational in area X. That's a sign that it's time to lay out evidence and argument on both sides, re-calibrate and re-evaluate. There's no guarantee that like mindedness will result, but strong disagreement in a small area with a person you think is generally rational is an indicator that you may be playing Mr. Magoo to their Dr. Seuss.

How do we fix polarization? How do we try to have a rational discussion in a context that is already polarized? The literature on negotiation strategies seems to fit surprisingly well. William Ury and Robert Fisher are members of the Harvard Negotiation Project, authors of the influential *Getting to Yes*. Let me adapt some of their tips to the present case. A major theme in their work is removing an issue from its ties to emotion and ego. That applies here too. Rationality deals in cool-headed evidence and argument. To the extent that we have mere emotion on both sides, rationality doesn't have much of a chance. There is room for progress if both sides can say, "Wait a minute. I know we each feel strongly about this, but there is a real intellectual question at stake. Let's try to put the emotion aside and take a hard look at the issue itself." Ury speaks of this step back from emotional involvement as "going to the balcony." The idea is for both sides to try to view the issue as if they were not embedded in it. Try to see it from a distance, as if the positions were being advanced by someone entirely different. That's an approach to polarized discussion that could be used unilaterally, but it's going to be more effective if both sides recognize that the issue is emotionally laden and both attempt to go to the balcony.

A correlate theme is ego distancing. People identify with the views they hold. That's part of the problem. I am arguing for my position. You are arguing for yours. But of course neither truth nor rationality care whether a position happens to be mine or yours. "Look. This isn't about us." If an issue can be cast as one for joint resolution through joint effort, that's a step away from individual ego, and it's clearly a step in the right direction. Public debate is not the best context in which to drop ego identification, even if it's around a table at Thanksgiving. That's a context in which people have too much to lose. So change the context. Let's agree on some quiet discussion together on a long walk. As long as a discussion is framed in terms of winning and losing, ego involvement is inevitable. That's why debates are probably not the best ways to arrive at a clear examination of an issue. Adversarial confrontation constitutes the public part of a criminal trial, but the real decision is made by someone else, in discussion, behind closed doors—by the jury.

Professional philosophers have a major stake in logic and rationality, but I have to say that it has only been very rarely that I've heard a philosopher say, in the midst of a public discussion, "Oh, I see. Yes. I was wrong. You're right." Much more often, the change of mind is made off stage. Somehow that helps. "You know, I've been thinking about your arguments yesterday. I've got a few questions, but I think you might be right." Ury and Fisher urge people to avoid positional negotiation. In positional negotiation people identify themselves with specific positions from the start—positions to be won or lost. In positional negotiation, one seems forced into one of two roles. You end up either a soft negotiator or a hard negotiator, but neither of those really works. The soft negotiator wants to avoid conflict and so gives in. He often ends up feeling exploited, stepped upon, and bitter. The alternative is to be a hard negotiator, out to win and giving as little as possible every step of the way. In negotiation between hard negotiators, both start at absurd extremes so that even if they do have to make a few concessions they will still be within their target area. One can expect that kind of negotiation to be a lengthy and drawn-out procedure. It might well break down. Positions don't get softer. Ego involved, emotionally driven, and with an emphasis on winning, hard positions just keep getting harder.

What holds for negotiation holds for discussion in a polarized atmosphere as well. As long as the context is one of winning and losing, emotion and ego-identification, one either gives in—feeling one has sacrificed truth, principle, or rationality to buy a little peace—or positions end up hardening to the point that they threaten the relationship itself. The alternative is to change the game. If the context can be changed from a contest of wills to a mutual exploration and investigation of an issue, then there's a chance.

In leaving positional negotiation behind, Ury and Fisher urge people to look for a way to settle issues in terms of objective criteria. In the case of negotiation, that may mean an appeal to independent standards. Can't agree on the price for the house? What if we compare houses in the same area that have sold in the last six months? Can't agree on the depth of the foundation? Let's decide it on the basis of government specifications for these soil conditions. The idea is to ask in advance "What would settle the issue?" Agree on that, and then appeal to it.

The same thing can work with rational discussion. Once we have removed emotion and ego as much as possible; once we have gone to the balcony and are trying to view it as an issue for joint investigation rather than for ego-invested debate, let's try to agree on what might actually settle the issue. How would that work? The debate context is one in which opponents start with positions on, say, deterrence on the death penalty. If we can change the game, we have collaborators investigating deterrence and the death penalty. That allows us to look at the totality of evidence. But let's decide in advance. What kind of evidence should convince us that there is an effect? What kind of evidence should convince us that there isn't an effect? If we can do that, we can look with new eyes at what the evidence actually is. We can stop being Mr. Magoo and Dr. Seuss. We may actually reduce polarization.

Sometimes the dispute isn't over data or evidence, but over background principles. There discussion has to go deeper. One thing to ask is what principles are at play and what kind of principles those are. Proponents of gun control tend to emphasize statistics regarding harm. Opponents of gun control tend to emphasize Constitutional rights. But if you take them out of the context of debate, people on both sides recognize that the Constitution says something important about rights. People on both sides recognize that we should avoid unnecessary harm. If we can get from polarized people to joint recognition that the issue is one of conflicting principles, we've made progress.

There are several points I hope you'll take home from this lecture. First, some of the data on polarization. Partisan polarization is real, and the media emphasizes the hot-button controversies. When you look more generally at cultural attitudes, and consider the population as a whole, it's not true that an enormous split is growing between us. I want you to take home the image of Mr. Magoo and Dr. Seuss discounting evidence out of hand simply because it clashes with their existing positions. That dynamics is predictable. As a filtering device it may even be rational, but rationality demands that positions be sensitive to the totality of evidence. It may be time to ask whether ours are. Finally, remember some of the tips toward rational discussion taken from the negotiation literature. There is no guarantee, but if we can go to the balcony, if we can pull the plug on ego, if we can change the game from confrontational debate to joint investigation, we at least stand a chance.

Is rhetoric at issue here? Absolutely. We'll talk more about that next time.

Rhetoric versus Rationality
Lecture 13

R hetoric has acquired a bad name, as in the expressions "mere rhetoric" or "empty rhetoric." But that wasn't always the case. At one time, rhetoric was highly respected as a required skill for effective speaking and presentation of ideas. What makes a presentation effective? According to Aristotle, the character of the speaker, which he called *ethos*, resonating with the emotions of the hearer, which he called *pathos*, and the logic, or rationality, of the argument, which he called *logos*. This lecture focuses on rhetoric versus rationality.

The History of Rhetoric

- At one time, rhetoric was highly respected as a required skill for effective speaking and presentation of ideas. Plato has a dialogue on the theme, the *Gorgias*, and Aristotle wrote a major work on the topic, *Rhetoric*.

- Classical education throughout the Middle Ages was based on the **trivium**, consisting of **grammar**, logic, and rhetoric. Those were the foundations of a liberal education, the areas everyone was expected to master.

- But the dark side of rhetoric also became clear early on. In Plato's *Gorgias*, Socrates argues that rhetoric involves no real knowledge or sense of justice—that rhetoric alone, unalloyed with philosophy, is merely a form of deceptive advertising or flattery.

- In his work on the topic, Aristotle admits that rhetorical skills can be used for either good or bad. His task is to analyze persuasion. What is it that makes a presentation effective?
  - The persuasiveness of a presentation depends first of all on the character of the speaker. We are convinced by people we see as knowledgeable and wise, as effective leaders who are in the right. Aristotle calls that *ethos*.

- Second, the persuasiveness of a presentation depends on the emotions of the hearer. A message that resonates emotionally is one that is more effective. Aristotle calls that *pathos*.

- Third, persuasiveness depends on the logic of the argument. If you want to persuade, lay out the argument clearly and rationally, step by step. That is *logos*.

Rhetoric in Lincoln's Cooper Union Speech

- The political scientist Frank Myers points out that all three elements of Aristotle's *Rhetoric* are artfully managed in Lincoln's famous Cooper Union speech. The question of the day was whether the federal government had the right to control the expansion of slavery into the western territories.

- With no flowery introduction, Lincoln takes on the question in terms of the Constitution. Despite his appearance, despite his frontier accent, Lincoln uses that part of the speech to establish himself as knowledgeable and, indeed, scholarly with regard to the questions at issue. In this, he fulfills the requirements of Aristotle's quality of *ethos*.

- Another element of Aristotle's *Rhetoric* is the logic of the argument. In that aspect Lincoln's speech was accurate and well constructed, similar to a lawyer's brief. Emotion, or *pathos*, appears in the very last part of Lincoln's speech. Had he started with an emotional appeal, it is likely people would have seen him only as a ranter from the wilderness. What he needed first were character and logic.

The Dark Side of Rhetoric: Schopenhauer's Stratagems

- Our contemporary and negative sense of rhetoric is firmly in place in the work of the 19th-century philosopher Arthur Schopenhauer. Schopenhauer was the author of *The Art of Controversy*, a user's guide to rhetoric in the contemporary negative sense. He characterizes the topic as intellectual fencing used for the purpose of getting the best in a dispute. Chapter 3 of Schopenhauer's book is titled "Stratagems." It's a list of 38 rhetorical tricks. It might as well be called "How to Win an Argument by Any Means Possible."

- Some of his tricks include explicitly manipulating emotion. Make your opponent angry, Schopenhauer advises. He won't think straight. How should you do that? "By doing him repeated injustice, or practicing some kind of chicanery, and being generally insolent."

- But emotion itself isn't bad. Life and even rationality would be impossible without it. There are points, however, at which something more is required: careful weighing of alternatives, consideration of evidence, tracking of potential consequences, and, after due consideration, arriving at a decision. Appealing to emotion is intended to cross-circuit all that—pushing your buttons instead of appealing to your reason.

- **Appeal to emotion** is often by means of emotionally loaded words. Consider "Jones won the election," as opposed to "Against all odds, Jones finally triumphed over the status quo." The difference is in the emotionally loaded words.

- There are other classic fallacies that appear on Schopenhauer's list of rhetorical tricks. The stratagem known as the **straw man fallacy** appears in a number of forms. Schopenhauer recommends exaggerating the claims made on the other side. That will, of course, make them easier to refute. He recommends twisting the words used on the other side so that they mean something different than intended. Ignore your opponent's real position. Confuse the issue by imputing something different. The fallacy is called "straw man" because you are no longer arguing against the opponent's real position.

- Among Schopenhauer's recommendations is a standard debating trick: Shift the burden of proof to the opponent. One way to do this is simply to say, "This is my position. Go ahead. Prove me wrong." That relieves you of the burden of proving yourself right. It's your opponent who is on the spot. But, of course, the fact that he may not be able prove you wrong doesn't really prove that you're right.

- Schopenhauer also recommends that you lump the opponent's position into some odious category. "That's fascism," you can say, or "That's socialism." The emotionally loaded category is meant to do all the work by itself—and often does.

- If all else fails, Schopenhauer says, bluff. Or throw in so many unrelated charges that your opponent doesn't have a chance to respond to them all. Or bewilder him or her with mere bombast.

- Schopenhauer's rhetorical recommendations are ironic, of course. He didn't want to be a victim of bombast any more than we do. But look at how far rhetoric in that sense is from what Aristotle meant. Look at how far rhetoric in that sense is from the presentational dynamics of Lincoln's Cooper Union address. Unfortunately, it's not so far from what we've become accustomed to in political argument.

The Ethics of Argument
- There is another lesson here regarding the ethics of argument. To most people, rhetorical tricks are intellectually repulsive. Schopenhauer's stratagems have all the charm of a wax museum arcade of famous murderers. The best that mere rhetoric can offer is merely winning. Truth and genuine rationality demand a far higher standard than that.

- Schopenhauer says that the intellectual fencing he's talking about has nothing to do with truth. That's precisely the problem. Rational discussion, like rational argument, should be about the truth—a truth that may include the truth about people's rights, about the best available policy, or about reasonable accommodation between different interests.

- We want to work toward the truth by means of rational argument. Reasoning with others, like reasoning with ourselves, comes with a built-in goal: to work toward the truth. Persuasion alone isn't a worthy goal. That's mere rhetoric.

The Positive Side of Rhetoric

- Rationality isn't an exclusively individual undertaking. Rationality is something that we can and should practice together. Rational discussion is just that—discussion. Discussion is a social endeavor, and rationality may actually demand a social dynamic. Often, it is only when we see ideas in a context of discussion, or even civil debate, that we can see them most clearly.

- That is certainly the theory in criminal law and civil suits. The adversarial system is based on the conviction that it is from the clash of arguments, in which each side presents its best case, subject to cross-examination, that "the truth will out."

At its best, rhetoric emphasizes the social dynamics of intellectual exchange; the theory behind the adversarial system in law is that we can see the truth most clearly in the context of discussion.

- Does the truth always come out in adversarial proceedings? We all know that a case may be carried by a lawyer's rhetorical tricks—appeal to the emotion of the jury, for example—rather than by the force of evidence and argument. The consensus, however, seems to be that there is at least a better chance that the truth will out when we have arguments and evidence laid out on opposing sides, with a final verdict left to the collective deliberations of the jury of our peers.

Visualization: Graphing Logical Flow

- In a previous lecture, we used flow diagrams to graph the logical path of an argument. In that context, we were talking about an argument as a one-sided affair. We can also use visualization to graph both sides of an argument, creating a flow diagram from a simple exchange between two people about a broken washing machine.

- When a philosophical discussion becomes complex, it helps to sketch it out in advance. Include not only exchanges you have heard but possible exchanges to come, that is, graphs with branches for different possibilities.

- The analysis of a rational discussion differs from that of a single argument in that there will be more than just premises and conclusions—more than even data, warrant, and conclusions. There will be challenges raised on one side that will force qualifications on the other, and that may well evoke refinement of the argument and further backup in response.

- Are such analyses complicated? Yes, yet in normal discourse, we manage to follow the train of thought, the logical flow of argument, all the time. What graphic visualization allows is the possibility of seeing that logical flow. What we've done here is to expand our graphing tools into the social context of discourse.

Terms to Know

appeal-to-emotion fallacy: A fallacy in which positive or negative emotional tone is substituted for rational or evidential support; an argument strategy intended to cross-circuit the ability of the listener to assess whether a rational link exists between premise and conclusion by "pushing emotional buttons."

ethos: Character of the speaker; according to Aristotle, the first quality of a persuasive presentation is that the speaker must appear knowledgeable and wise.

grammar: The study of the proper structure of language in speech or writing; along with logic and rhetoric, an element of the classical medieval curriculum known as the *trivium*.

logos: Logic; according to Aristotle, the third quality of a persuasive presentation is that it will lay out the argument clearly and rationally, step by step.

pathos: Emotion; according to Aristotle, the second quality of a persuasive presentation is that it resonates emotionally with the listener.

straw man fallacy: The rhetorical strategy of exaggerating or misrepresenting the claims of the opposition in order to more easily appear to refute those claims.

trivium: The basis of a classical education throughout the Middle Ages; a three-part curriculum composed of grammar, logic, and rhetoric.

Suggested Reading

Aristotle (Roberts, trans.), *Rhetoric*.

Schopenhauer, *The Art of Controversy*.

Questions to Consider

1. Aristotle considers rhetoric as the art of persuasion, and recognizes that it can be used to either or good or bad ends. From your own experience, can you think of an example of rhetoric used for a good end? For a bad end?

2. The negative side of rhetoric: If all else fails in "intellectual fencing," Schopenhauer recommends these rhetorical tactics: (a) throw out so many unrelated charges that the opponent doesn't have time to respond to them all and (b) become personal, insulting, and rude. Have you seen examples of those tactics in debates or arguments? Do you think they worked effectively?

3. The positive side of rhetoric: Can you think of a case in which you worked out a problem by exchanging ideas with a friend—a problem that neither of you would have solved independently?

Look up Shakespeare's *Julius Caesar*, Act III, scene ii. This scene contains the whole of both Brutus's and Mark Antony's speeches. Analyze each in terms of Aristotle's categories: *ethos* (speaker's character), *pathos* (appeal to audience emotion), and *logos* (logic of the argument).

In the next political debate you see, pay attention to just the first two minutes of each speaker's presentation or to the answers to the first two questions directed to each speaker. With paper and pencil, keep track of the rhetorical pose of each speaker, specific claims made, and rhetorical tricks used.

Rhetoric versus Rationality

Lecture 13—Transcript

Professor Grim: In the last lecture I tried to offer some ideas on working toward rationality in polarized contexts. This lecture focuses on something characteristic of polarized contexts, rhetoric. In this lecture, I want to talk about rhetoric versus rationality.

Rhetoric has acquired a bad name. It's hard to say rhetoric without empty or mere in front of it. Mere rhetoric. But that wasn't always the case. Let me start with a little history. At one time, rhetoric was highly respected as a required skill for effective speaking and presentation of ideas. Plato has a dialogue on the theme, the "Gorgias," named after a prominent teacher of rhetoric. Aristotle has a major work on the topic, Aristotle's *Rhetoric*. The Roman orator Cicero was famous as a teacher of rhetoric. In her biography of Cleopatra, Stacy Schiff argues that Cleopatra's success was due far more to her skill in rhetoric than her supposed beauty. Classical education throughout the Middle Ages was based on the Trivium, consisting of grammar, logic, and rhetoric. Those were the foundations of a liberal education, the areas everyone was expected to master.

But the dark side of rhetoric also became clear early on. It's a theme in Plato's "Gorgias." There, Plato has Socrates argue that rhetoric involves no real knowledge or sense of justice, that rhetoric alone, unalloyed with philosophy, is merely a form of deceptive advertising or flattery. One has to remember that at that time the rhetoricians were in a sense philosophy's competitors. Plato's anti-rhetoric dialogue is also remarkable for its effectiveness, that is, for the effectiveness of its rhetoric. In his work on the topic, Aristotle admits that rhetorical skills can be used for either good or bad. His task is to analyze persuasion. What is it that makes a presentation effective? That's still a good question. Even contemporary attempts to answer it would do well to start with Aristotle and often do.

What makes a presentation effective? Three things. The persuasiveness of a presentation depends first of all on the character of the speaker. We are convinced by people we see as knowledgeable and wise, as effective leaders who are in the right. Aristotle calls that *ethos*. Second, the persuasiveness of a presentation depends on the emotions of the hearer. A message that resonates emotionally is one that is going to be more effective. Aristotle calls that *pathos*. Third, persuasiveness depends on the logic of the argument. Do you want to persuade? Lay out the argument clearly and rationally, step by step. That is *logos*.

When I think of rhetoric I think of Marc Antony's "Friends, Romans, countrymen, lend me your ears." But, of course, that's not a quote from Marc Antony himself. It comes from Shakespeare's Julius Caesar. Education in Elizabethan England was saturated with the classics, even elementary education. Shakespeare is reputed to have studied at the King's Free Grammar school in Stratford. No records remain, so we don't know for sure. Might Shakespeare have read Aristotle's *Rhetoric*? All we have is speculation. In support of that speculation, however, are the two prominent speeches at Caesar's funeral in Shakespeare's play. The first speech is by Brutus, one of the assassins. "Hear me for my cause, and be silent, that you may hear: believe me for mine honor, and have respect to mine honor." That looks like an attempt to establish character—Aristotle's *ethos*. The Shakespearean scholars Charles and Michelle Martindale further characterize Brutus's speech in terms of *logos*, as a "style for the intellect." If someone asks why Brutus rose against Caesar, "… this is my answer: Not that I loved Caesar less, but that I loved Rome more."

Marc Antony gives a speech in response to Brutus's, and it's very different, ironic and indirect where Brutus's is explicit and direct. Marc Antony undercuts Brutus's built-up *ethos* with a mockingly repeated phrase. Brutus says that Caesar was ambitious, and Brutus is an honorable man.

When that the poor have cried, Caesar hath wept:

Ambition should be made of sterner stuff:

Yet Brutus says that he was ambitious;

And Brutus is an honorable man.

I thrice presented him a kingly crown,

Which he did thrice refuse: was this ambition?

Yet Brutus says he was ambitious;

And Brutus is an honorable man.

As the Martindales note, the currency of Antony's speech is not logic but emotion, *pathos*.

If you have tears, prepare to shed them now… Look, in this place ran Cassius's dagger through…see what a rent the envious Casca made…through this the well-beloved Brutus stabbed…the most unkindest cut of all…

So what do you think? Is that evidence that Shakespeare knew Aristotle's *Rhetoric*?

What of Abraham Lincoln? My colleague at Stony Brook, the political scientist Frank Myers, points out that all three elements of Aristotle's *Rhetoric* are artfully managed in Lincoln's famous Cooper Union speech. Lincoln and Stephen Douglas were rival candidates for the Senate from Illinois and held debates across the state. Parts of those debates gained national attention in the newspapers. Lincoln lost the election for Senator, but shortly afterward was invited to speak at the Cooper Union in New York. Lincoln's law partner William Herndon said that no former speech-writing had cost Lincoln so much time and thought. Here is what an eye-witness said: "When Lincoln rose to speak, I was greatly disappointed. He was tall, tall—oh, how tall! and so angular and awkward that I had, for an instant, a feeling of pity for so ungainly a man." Add to that the fact that Lincoln spoke, not in the deep resonant voice we remember from so many portrayals, but in a high-pitched accent blending frontier Indiana and Kentucky. The question of the day was whether the federal governmental had the right to control the expansion of slavery into the western territories. With no flowery introduction, Lincoln takes on the question in terms of the Constitution. He gives a detailed and scholarly examination of the relevant views and related actions of each of the 39 delegates who signed it. Despite his appearance, despite his frontier accent, Lincoln uses that first part of the speech to establish himself as knowledgeable and, indeed, scholarly with regard to the questions at issue. That same eye witness who was disappointed on first impression says that once Lincoln started speaking

> His face lighted up as with an inward fire; the whole man was transfigured. I forgot his clothes, his personal appearance, and his individual peculiarities. Presently, forgetting myself, I was... cheering this wonderful man.

Lincoln has established that first element of Aristotle's *Rhetoric*, the impression of character.

The third element of Aristotle's *Rhetoric* is the logic of the argument. That, too, Lincoln worked on long and hard. His law partner Herndon says rightly that the speech was, "constructed with a view to accuracy of statement, simplicity of language, and unity of thought. In some respects like a lawyer's brief, it was logical [and] temperate in tone."

Aristotle's three modes of persuasion are character, logic, and emotion. But emotion only appears in the very last part of Lincoln's speech. Had he started with an emotional appeal, it's likely people would have seen only a ranter from the wilderness. What he needed first were character and logic. He saves emotion for his conclusion,

> Neither let us be slandered from our duty by false accusations against us, nor frightened from it by menaces of destruction to the Government nor of dungeons to ourselves. Let us have faith that right makes might, and in that faith, let us, to the end, to dare to do our duty as we understand it.

There are similar analyses of Lincoln's Gettysburg Address in terms of Aristotle's *Rhetoric*.

So why did rhetoric get a bad name? Perhaps because one element of Aristotle's three can be faked, *ethos*, the aspect of character. You will be just as persuasive if you can make yourself appear knowledgeable as you will if you really are. No need to really know what you're talking about. Just look the part. Another element of Aristotle's three is prime territory for manipulation, *pathos*. Appeal to emotion is so clearly an element beyond the rational that it appears on the list of standard fallacies. Is emotion itself wrong? No. Indeed we've seen how important it is. But when an appeal to reason is replaced by something intended merely to push our emotional buttons, it's not rational persuasion, but emotional manipulation, at work.

Our contemporary and negative sense of rhetoric is firmly in place in the work of the 19th century philosopher Arthur Schopenhauer. Schopenhauer's major work was *The World as Will and Representation*, cited as a major influence on Nietzsche, Wittgenstein, Freud, Einstein, and, unfortunately, on Adolf Hitler. Schopenhauer was also the author of *The Art of Controversy*, a user's guide to rhetoric in the contemporary negative sense. He characterizes the topic as intellectual fencing used for the purpose of getting the best of a dispute, period. He admits that it has "nothing to do with truth...thrust and parry is the whole business." Chapter three of Schopenhauer's book is titled "Stratagems." It's a list of 38 rhetorical tricks. It might as well be called "How to Win an Argument by Any Means Possible." Schopenhauer's irony is evident.

Some of tricks explicitly manipulate emotion. "Make your opponent angry," Schopenhauer says. He won't think straight. How should you do that? "By doing him repeated injustice, or practicing some kind of chicanery, and being generally insolent." If he does become angry, Schopenhauer says, you should press your case with even more zeal. Irritate him by contradiction and contention until he exaggerates his original statement. Even though what your opponent originally said was right, hone in on the exaggeration. "... when you refute this exaggerated form...you look as though you had also refuted his original statement," says Schopenhauer. If it looks like your opponent is getting the upper hand, "become personal, insulting, and rude." That's the rhetoric we're more familiar with...the kind of rhetoric we're all too familiar with.

In the last lecture we talked about how to strive for rationality in a polarized context. There a major strategy was to try to lift the dispute out of rhetoric, to remove it from Schopenhauer's intellectual fencing debate context, to go the balcony, looking at the issue at a distance removed from ego involvement and emotion. But what is one to do when that's not possible? The best defense is to know the enemy and see it coming. But by enemy I don't mean the other side. The enemy is rhetoric. One can guard against rhetoric by getting to know it, even by getting good at it. The best defense against mere rhetoric is to know its tactics well enough to call them by name and parry them when they come. Appeal to emotion is one of those tactics. Emotion itself isn't bad. Life and even rationality would be impossible without it. But there are points at which something more is required, careful weighing of alternatives, consideration of evidence, tracking potential consequences, and after due consideration, making a decision. Appeal to emotion is intended to cross circuit all that—pushing your buttons instead of appealing to your reason. Watch out for it. Appeal to emotion is often by means of emotionally loaded words. There are lots of ways to say something. Consider, "Bill came home late" as opposed to "Bill snuck into the house long after he was due." The basic facts may be the same. The portrayal is not, precisely because of the emotional language in the second case. Consider: "Jones won the election" as opposed to "Against all odds, Jones finally triumphed over the status quo." The difference is all those emotionally loaded words. So when you hear something like this, "Fanatics on the Religious Right have wormed their way into the most powerful and privileged positions of the Republican party. Now they're really going to start shoving things down our throats." Figure out what it really says when it is drained of the rhetorical phrases that are a mere appeal to emotion.

There are other classic fallacies that appear on Schopenhauer's list of rhetorical tricks. The stratagem known as the Straw Man fallacy appears in a number of forms. Schopenhauer recommends exaggerating the claims made on the other side. That will, of course, make them easier to refute. He recommends twisting the words used on the other side so that they mean something different than intended. Ignore your opponent's real position. Confuse the issue by imputing something different. The fallacy is called Straw Man because you are no longer arguing against the opponent's real position. It's as if you can't beat Billy in a fair fight, so you go home and take it out on a dummy made of straw. "Take that, Billy," you say, effectively demolishing, not the real Billy, just the straw man. We will cover other fallacies in later lectures, *ad Hominem*, appeal to Authority, and diversion. Those appear on Schopenhauer's list as well.

Among Schopenhauer's recommendations is a standard debating trick, Shift the burden of proof to the opponent. One way is simply to say "This is my position. Go ahead. Prove me wrong." That, of course, relieves you of the burden of proving yourself right. It's your opponent who is on the spot. But of course the fact that he may not be able prove you wrong doesn't really prove that you're right. Schopenhauer also recommends that you lump the opponent's position into some odious category. "That's fascism," you can say. Or "that's socialism." Don't bother to explain why the view is fascism, or even what fascism is. The emotionally-loaded category is meant to do all the work by itself, and it often does. If all else fails, Schopenhauer says, bluff, or throw so many unrelated charges that your opponent doesn't have a chance to respond to them all, or bewilder him with mere bombast, or "become personal, insulting, and rude."

Schopenhauer's rhetorical "recommendations" are ironic, of course. He didn't want to be a victim of bombast any more than we do. But look at how far rhetoric in that sense is from what Aristotle meant. Look at how far rhetoric in that sense is from the presentational dynamics of Lincoln's Cooper Union Address. Unfortunately, it's not so far from what we've become accustomed to in political argument. Rhetorical tricks may be particularly characteristic of political argument in contemporary media. What would the Cooper Union address look like today, as reported by any of the major broadcast media? There are cases in which a full speech by a candidate becomes part of the media stream, Barack Obama's speech on race in the midst of the 2008 campaign is an example. But most contemporary political media consists of sound clips extracted from some larger speech or 30-second political commercials. Those are contexts made for rhetoric. *Logos* and *ethos* don't have time to do their work in a sound bite, but *pathos* does. It only takes a split second for an appeal to emotion to do its dirty work. Rhetoric in the sense of Schopenhauer's tricks was made for the intellectual fencing characteristic of a debate in sound bites.

I have said that it is important to develop an appreciation for rhetorical tricks. I'm not advocating that you employ them to win your next argument. I urge you to learn them as a means intellectual self defense. Knowing the tricks will help you guard against being on the receiving end of those tricks. There is another lesson here regarding the ethics of argument. If you're like me, you find rhetorical tricks intellectually repulsive. Schopenhauer's list has all the charm of a wax museum arcade of famous murderers. These are tips on how to win arguments, or at least how to appear to win. These are tips for one upmanship in the fencing matches of debates. But the serious question isn't how to win arguments. The real question is when to win arguments, or even whether to win arguments. The real question is whether winning is what we're after, after all.

Schopenhauer says that the intellectual fencing he's talking about has nothing to do with truth, and that's precisely the problem. By my lights, rational discussion, like rational argument, should be about the truth, a truth that may include the truth about people's rights, about best-available policy, or about reasonable accommodation between different interests. We want to work towards the truth by means of rational argument. Reasoning with others, like reasoning with ourselves, comes with a built in goal: to work toward the truth. Persuasion alone isn't a worthy goal, that's mere rhetoric.

Let's return to rhetoric on the positive side. What study of rhetoric emphasizes is that rationality isn't an exclusively individual undertaking. Rationality is something that we can and should practice together. Rational discussion is just that, discussion. Discussion is a social endeavor, and rationality may actually demand a social dynamics. What rhetoric emphasizes—at its best, not its worst—is the social dynamics of intellectual exchange. Often it is only when we see ideas in a context of discussion, or even civil debate, that we can see them most clearly. That's certainly the theory in criminal law and in high-cost civil suits. The adversarial system is based on the conviction that it's from the clash of arguments, in which each side presents its best case, subject to cross-examination, that the truth will out. Does the truth always out in adversarial proceedings? I don't know anyone who thinks so. We all know that a case may be carried by lawyer's rhetorical tricks—appeal to emotion on the part of the jury, for example—rather than by the force of evidence and argument. But I don't know anyone who thinks we should replace that system with a single inquisitor doing all the investigating by himself, asking all the questions, and then serving as both judge and jury.

The consensus seems to be that there is, at least, a better chance that the truth will out when we have arguments and evidence laid out on opposing sides, with a final verdict left to the collective deliberations of the jury.

In lecture nine we used flow diagrams to graph the logical flow of an argument. In that context we were talking about an argument as a one-sided affair. We were trying to graph the logical flow from a set of propositions offered as reasons to the conclusion they were meant to support. Are they working independently or as a set of dependent reasons that have to work together?

In the last part of that lecture we expanded the technique to two types of reasons, data and warrant, based on the work of the twentieth-century philosopher Stephen Toulmin. The idea is that a premise may be giving you an evidential base—a piece of information or data—or maybe giving you the warrant that allows you to draw a particular conclusion from that data. Toulmin's larger project was to reintroduce, rehabilitate, and modernize the study of rhetoric as the study of argument in the sense of discourse—the exchange of ideas between two or more interlocutors. He was out to reintroduce, rehabilitate, and modernize the study of rhetoric. Let's expand our earlier visualization, then, in order to see how one can graph an argument on both sides of an issue. We'll use a simple exchange between two people.

[Video start]

Mark: I've figured out the problem with the washing machine. The belt is broken.

Tessa: What makes you think that?

Mark: You can hear the motor running, but nothing else turns.

Tessa: And that makes you think it's the belt?

Mark: Well, that would explain it. If the belt is broken, the motor can turn all it wants, but nothing else will.

Tessa: But maybe it's something is jammed in the main drum. The motor would work then, but the drum wouldn't turn even if there wasn't any problem with the belt.

Mark: I guess so. But that's pretty unlikely. Let's just say that it's probably the belt.

Tessa: Well, let's take a look.

[Video end.]

Professor Grim: You couldn't find a more normal exchange, broken appliances, a delightful break from broken politics, anyway. Here is how we might construct a flow diagram of that little exchange. Mark makes the first claim we're interested in. The belt is broken. That's really his conclusion. We don't have his reasoning yet. Tessa draws that out, "What makes you think that?" He gives the data on which his conclusion is based, "You can hear the motor running, but nothing turns." So at this point we can do a diagram that goes from Mark's data to his conclusion. Look at the two pieces of his data, by the way. "You can hear the motor running." "Nothing turns." Remember the distinction we made before. Are those working independently, or dependently? They're working dependently. Mark is presenting them as two pieces of the puzzle, which together lead him to believe that the belt is broken. Just hearing the motor running wouldn't tell him that. "Nothing turns" wouldn't be specific enough. Those are offered as pieces of evidence which together suggest the conclusion. But now Tessa asks, "And that makes you think it's the belt?" Note that she isn't asking a question about the data. She's asking a question about—maybe even challenging—the inference from the data to the conclusion. She's really asking "Do you think that data warrants the conclusion that the belt is broken?" The warrant is what's supposed to carry us from the data to the conclusion. I'll write in Tessa's question as a challenge at that point. We'll use blue for Mark's contributions, yellow indicate Tessa's. Mark answers the challenge by providing a warrant. "Well, that would explain it." He offers further backing by spelling out the explanation, "If the belt is broken, the motor can turn all it wants, but nothing else will."

At this point Tessa offers a counter-hypothesis. "Maybe it's something jammed in the main drum." Note that she doesn't deny the data Mark offered. In fact, she is going to use that data with a different warrant for the opposing conclusion. I'll change Mark's original data to a different color to indicate that both people are now using it. Tessa uses the same data toward an alternative conclusion, though a qualified conclusion. "Maybe it's something jammed in the main drum." Her warrant is an alternative explanation spelled out with further backing. "If something is jammed, the motor would work, but the drum wouldn't turn, even if there is no problem with the belt. Note that Tessa's is now a structure parallel to Mark's. She, too, uses the data as dependent premises, but her warrants is an alternative "that would explain it," with a different backing. Note that I've also marked her "maybe" as a qualification on her conclusion, "Maybe something is jammed in the main drum." Mark ends by conceding that Tessa has an alternative explanation, backing an alternative hypothesis. "I guess so." He introduces a challenge to Tessa's explanation: "That's pretty unlikely." But at the same time he adds a qualification to his own claim, "Let's just say it's probably the belt."

What we've graphed is not just an argument on one side, but the exchange of arguments on both sides. Luckily, Mark and Tessa have listened to the previous lecture on how to work rationally within a polarized context. Tessa suggests that they put the matter to an objective test. "Let's take a look." There is nothing magic about the way I've sketched the logic of the discussion. Where the pattern of an argument is really important, I urge you to try sketching it out as a graph with whatever conventions seem most useful to you. When I expect a philosophical discussion to become complex, I sometimes sketch it out in advance. I include not only exchanges I have heard but possible exchanges to come, graphs with branches for different possibilities. Perhaps he'll use this argument. Perhaps she'll make this reply. In that case my best response is probably this one. In that kind of case I'll be trying to muster the best argument that can be made on one side. But it's still Aristotle's *logos* that I'll be trying to use. I'm not plotting rhetorical tricks. I'm trying to map out the alleys we can expect to explore in search of the truth. The analysis of a rational discussion differs from that of a single argument in that there will be more than just premises and conclusions— more than even data, warrant, and conclusions. There will be challenges raised on one side, which will force qualifications on the other, and which may well evoke refinement of the argument and further backup in response. Complicated? Yes. And yet in normal discourse we manage to follow the train of thought—the logical flow of argument—all the time. What graphic visualization allows is the possibility of seeing that logical flow. What we've done here is to expand our graphing tools into the social context of discourse.

There are a number of points regarding rationality and rhetoric that I hope you'll take home from this lecture. What makes a presentation persuasive? Remember the three themes in *Aristotle's Rhetoric*, character, emotion, and reasoning—*ethos*, *pathos*, and *logos*. Remember some of the dirty tricks on Schopenhauer's list, not in order to slaughter the opposition, but so that you can defend yourself against rhetoric's dark side. Try sketching out the flow of a discussion. Develop the ability to visualize the pattern of an exchange by starting with short and simple examples. I also hope you'll remember that argument comes with an ethics. The best that mere rhetoric could offer is mere winning. Truth and genuine rationality demand a far higher standard than that.

Bogus Arguments and How to Defuse Them
Lecture 14

The first requirement for a good argument is validity—that the premises, if true, will lead us by strong logic to the conclusion. The second requirement is that the premises actually be true. A sound argument is one that is both valid and has premises that are indeed true. That is the kind of argument that will give us a conclusion we can rely on. This lecture presents bad arguments to enable you to recognize them and defuse them. The standard forms of bad arguments are called fallacies, and these appear in social and political argument all the time. In this lecture, we'll look at a number of examples.

Appeal to the Majority

- Every parent is familiar with the refrain "but everybody else is doing it." Arguments that treat majority opinion as if that alone constituted a reason for belief commit the **appeal-to-majority fallacy**.

© iStockphoto/Thinkstock.

- Here's a simple example: "A majority of the population thinks we coddle criminals. It's time we got tough on crime." How much

The argument that UFOs must be real because a significant portion of the population takes UFO sightings seriously is an example of the appeal-to-majority fallacy.

support does that premise give for that conclusion? Not much. That's the fallacy of appeal to the majority.

- There is a simple fact that defuses a number of common fallacies, including appeal to the majority. It's this: The fact that someone believes something doesn't make it so. The fact that a whole bunch of people or a majority of people believe something doesn't make it so either.

Post Hoc Ergo Propter Hoc

- ***Post hoc ergo propter hoc*** is a fallacy that just about everyone has committed. Here, two things happen in sequence, and we conclude that they must be causally connected. The name of the fallacy is a bit of Latin that means, essentially, "This came after this, so it must have been because of this."

- Many superstitions are based on this fallacy. Professional baseball outfielder John White offers a typical example: "I was jogging out to centerfield when I picked up a scrap of paper. I got some good hits that night, and I guess I decided that the paper had something to do with it. The next night I picked up a gum wrapper and had another good night. I've been picking up paper every night since."

- The thing to remember is that sometimes events are just coincidences. The fact that A happened and then B happened doesn't prove that A caused B.

Ad Hominem

- An ***ad hominem*** ("against the person") argument is one that attacks the opponent as a person, rather than his or her arguments.

- Here's an example: "The Republicans say we should shrink the deficit and cut down on government spending. But these are the same people who cranked up the deficit by getting us into two wars simultaneously. These are the same people who cranked up the deficit by passing tax breaks for the rich. These are the people who gave us the Bush years of spending more massive than any preceding administration of either party."

- What's really at issue here is the claim that we should shrink the deficit and cut down on government spending, but that claim is never addressed at all. What we get instead is an attack on the people who put it forward.

- The point is that good arguments are good arguments even in the dark; they are good arguments even if you can't see who is giving them. That's the principle violated by the fallacy of *ad hominem*.

Tu Quoque and Poisoning the Well

- A specific form of the *ad hominem* fallacy is *tu quoque*, which is Latin for "you also," or more colloquially, "You did it, too."

- Here is an example of the *tu quoque* form of *ad hominem*: "The United States says that North Korea is stockpiling nuclear weapons and is, therefore, a danger to world peace. But surely both Israel and the United States have their own stockpiles of nuclear weapons!" Does this show that North Korea is not stockpiling nuclear weapons or that such action is not a danger to world peace? No, it simply states, "You did it, too."

- A second form of *ad hominem* is the **poisoning-the-well fallacy**. Like all *ad hominems*, it involves an attack on the person rather than the argument, but it is specifically an attack on someone's motives for his or her statements.

- In 2004, when campaigning for the presidency, John Kerry published a misery index intended to show that things had gotten worse under the George W. Bush administration. The other side fired back: "John Kerry talks about the economy because he thinks it will benefit him politically." True, perhaps, but irrelevant. That's just poisoning the well.

False Alternative

- The trick of a **false alternative**, or false dilemma, is to set things up as if the opponent must choose either A or B. It's a false alternative because there may be many more options open.

- Here's an example: "You either marry Julie and have a wonderful life and family, or you live the rest of your life as a lonely bachelor. Go ahead. Make your choice."

- That's clearly a false alternative. Are those your only options? Of course not. You might live the life of a gregarious, fun-loving bachelor instead of a lonely one. You might end up marrying Julie and not having a wonderful life. You could move to Alaska, start a commune, marry someone else, and live happily ever after.

The Complex Question

- The **complex-question fallacy** takes the nefarious strategy of the false alternative a little farther.

- The classic example is a lawyer who is prosecuting someone for wife beating. The accused is put on the stand and told to answer merely yes or no to the questions put to him. The prosecutor says, "Tell me, Mr. Smith, have you stopped beating your wife yet?"

- Mr. Smith must say either yes or no. But if he says yes, the prosecutor says "Aha! You say you have stopped beating your wife. So you admit you beat her in the past. I rest my case." If he says no, the prosecutor says "What? You say you haven't stopped beating your wife? You admit you're still beating her. I rest my case."

- It's a trick question, a trap. The defendant is forced to say yes or no. Either answer is legitimate only on a certain assumption—that he has beaten his wife. He's forced into a false alternative, each side of which begs the question against him.

Hasty Generalization

- The **hasty generalization fallacy** is also known as jumping to conclusions. If you know something about a small number of things, you then jump to a conclusion about all of them.

- Here's an example: "The first person I met on Long Island was short and rude. The second person I met on Long Island was short and rude. I guess everybody on Long Island must be short and rude."

- Racist and sexist generalizations are usually hasty generalizations: "All the women I've ever known were big tippers; therefore, all women are big tippers."

Appeal to Ignorance
- Similar to the fallacy of the appeal to emotion, the **appeal to ignorance** is equally illegitimate. Here, the fallacy is to act as if our ignorance is something more than it is—as if ignorance alone gives us some kind of positive evidence.

- Here's an example: "Many ships and planes have utterly disappeared in the regions of the Bermuda Triangle, without explanation and often without a physical trace. Some mysterious force must be at work beneath the waves."

- Just as believing something doesn't make it so, sometimes ignorance is just ignorance. Ignorance isn't by itself evidence for something—a mysterious force beneath the waves, for example.

A Closer Look
- The aim of this lecture was to vaccinate you against some standard forms of bogus arguments, to inoculate you against some logical fallacies. How does that fit in with some of our earlier work? That question might be posed as a set of challenges:
  - One of the other lectures introduced the wisdom of crowds as a positive heuristic. Isn't that just the fallacy of the appeal to the majority?

  - What about the example in which students were asked which city is larger, San Antonio or San Diego? The German students made a correct inference from the fact that they had heard of San Diego and not San Antonio. Isn't that just the fallacy of an appeal to ignorance?

  - The conclusion of another lecture was that in short-range decisions, it may be rational to go with your gut. Isn't that the fallacy of hasty generalization?

- Take a closer look at each of these questions, and you'll see that there are important differences. The wisdom of crowds doesn't just recommend that you think what everybody else thinks. It's a strategy for extracting knowledge from a distribution of opinions in which some people really do know the answer. It draws a clever inference, though a fallible one, from the distribution of beliefs.

- In the San Diego/San Antonio case, the German students did better than Americans in picking the larger city precisely because they knew something about one and were largely ignorant of the other. That's a clever inference, though admittedly a risky one. It's not a simple jump from "I don't know how ancient monuments were built" to "It must have been ancient astronauts."

- There is also an important difference of context. When there is no time for careful investigation—when you have to get out of the way of an oncoming train—fast and frugal heuristics are precisely what you need. Better to go with your gut and jump to a conclusion, even if it might be wrong, than to wait on the tracks trying to calculate the velocity of an approaching locomotive.

- In the case of explicit argument, the standards for rationality are higher. There, we want more than a handy heuristic. When someone is trying to convince us of a conclusion, and when we have time to really think about it, it's important to test how tight the logic really is. Do the premises really support the conclusion? How strongly? Does the conclusion really follow? Or is this a bogus argument, built on fallacious reasoning?

Terms to Know

ad hominen: A fallacy that depends on an attack against the person making a claim instead of the claim that is being made.

appeal-to-ignorance fallacy: A fallacy in which absence of information supporting a conclusion is taken as evidence of an alternative conclusion.

This fallacy acts as if ignorance alone represents some kind of positive evidence.

appeal-to-majority fallacy: An argument that treats majority opinion as if that alone constituted evidence supporting a conclusion or gave a reason for belief. This fallacy ignores the fact that people, even large numbers of people, are fallible.

complex-question fallacy: A "trick question" presenting a false dilemma, or forced-choice alternative, presented in such a way that any answer is incriminating. For example: "Answer yes or no: Have you stopped beating your wife?" If you say yes, you have essentially admitted that at one time, you did beat your wife; if you say no, you have admitted that you are still beating her.

false alternative: A fallacy in which a problem is presented as an either/or choice between two alternatives when, in fact, those are not the only options. Also called a "false dilemma."

hasty generalization fallacy: Also known as jumping to conclusions. This fallacy occurs when one jumps to a conclusion about "all" things from what is known in a small number of individual cases. Racism and sexism often take the form of hasty generalizations.

poisoning-the-well fallacy: A fallacy that depends on an attack against a person's motives for saying something rather than a refutation of the claims being made; a subtype of *ad hominem*.

post hoc ergo propter hoc: "After it, therefore because of it"; a fallacy based on the claim that because something followed another thing, it must have been because of that other thing. This fallacy overlooks the possibility of coincidental occurrence. Abbreviated as *post hoc*.

tu quoque: A fallacy in reasoning that tries to defuse an argument by claiming, "You did it, too."

Suggested Reading

Salmon, "Index of Fallacies."

White, *Crimes against Logic*.

Questions to Consider

1. Here is a standard form for some magic tricks: The magician seems to make a coin "travel" from point A to point B. The trick is actually done by "disappearing" one coin at point A and producing a different but similar coin at point B. In order to create the illusion of "transfer," the magician relies on the audience to commit a logical fallacy. What fallacy is that? (See "Answers" section at the end of this guidebook.)

2. Sometimes fallacious forms of argument are just a short distance from practical conceptual strategies. What is the difference between these?
   (a) We have to choose between the options available. How does that differ from false alternative?

   (b) We need to work from past experience. How does that differ from hasty generalization?

   (c) We need to consider the plausibility of an information source. How does that differ from *ad hominem*?

   (d) We often decide policy by putting the question to a vote. How does that differ from appeal to the majority?

Exercise

I once tried to give a midterm by showing only the commercials in a half-hour of commercial television and having students write down all the fallacies they saw committed. This was a good plan, but it didn't work because there were just too many fallacies.

The next time you watch a half-hour of television, pay attention to the commercials. Within three months of an election, give yourself 10 points for each instance of the following fallacies you find:

- Appeal to the majority

- Hasty generalization

- *Post hoc ergo propter hoc*

- False alternative

- Appeal to ignorance

- *Ad hominem* (more common in political than in commercial advertising)

- Appeal to dubious authority (coming in a later lecture)

If you're trying this exercise outside of an election year, give yourself 15 points for each instance. It should be easy to get 100 points in a half-hour of television; 200 points is a good, solid job; 300 points is outstanding; and beyond that, give yourself credit for having a touch of genius.

Bogus Arguments and How to Defuse Them
Lecture 14—Transcript

Professor Grim: Let me remind you of the requirements for a good argument. The first requirement is validity, a tight logical connection between premises and conclusion. In order to be valid, the reasons given for the conclusion must really support that conclusion. The conclusion must really follow from the premises. At its strongest, validity is deductive validity, Aristotle style. There the connection is air tight. Given the truth of the premises, the truth of the conclusion is absolutely guaranteed. More broadly, validity demands a strong logical connection between premises and conclusion, even if not absolutely guaranteed. Do the reasons give us convincing evidence for the conclusion? Do they give solid reason to believe it? The first requirement is validity, that the premises, if true, will lead us by strong logic to the conclusion. The second requirement for a good argument is that the premises actually be true. A sound argument is one that is both valid and has premises which are indeed true. That's the kind of argument that will give us a conclusion we can rely on. That's a good argument. In this lecture I want to go over to the dark side. In this lecture I want to give you some really bad arguments. That sounds terrible, doesn't it? Of course I'm not going to try to persuade you with bad arguments; I'm going to try to vaccinate you against them.

Bad arguments, like flu germs, are all around us, but we can prepare ourselves with a flu shot that strengthens the immune system against expected types. With bad arguments, we can prepare ourselves by learning to recognize the most common forms, the usual suspects. If you can see them coming, you can more easily detect and defuse them. There are bad argument types that reappear over and over again. Those standard forms of bad arguments are called fallacies. What I want to give you is a tour of some standard fallacies. The idea is, once again, to rely on your pattern-recognition abilities. I want you to be able to say, ah, I recognize that pattern of argument, one of the usual suspects. It even has a name; that's an *ad hominem* argument.

We've already talked about some standard forms of bad argument. Appeal to emotion and the straw man fallacy appeared in Schopenhauer's list of rhetorical tricks. Here we'll cover some more. In each case I'll start out

273

with simple examples. As I go along I'll occasionally stop to review by asking you to name that fallacy. How many kinds of bogus arguments are there? All too many. I'm not even sure there's a limit. Were I to try to give you a complete list of all the ways an argument can go bad, I'll bet some clever advertising executive could come up with a new one. A first warning, then, is that I can't give you an exhaustive list. We'll just hit some of the most common offenders. A second warning is this: When people start to recognize bogus arguments, they often want a single answer to the question, "What's wrong with this argument?" But an argument can go wrong lots of different ways at once. There's no guarantee that only one fallacy will be committed. I'm going to be using lots of examples. Fallacies appear in social and political argument all the time, and I'll be using a number of social and political examples. Let me make it clear that those are only examples. I'm not pushing any particular agenda on any of those issues. Bad arguments appear on just about every side of every issue. And of course, the fact that someone supports a conclusion with a bad argument doesn't mean that there might not be a better argument that supports the same conclusion.

First stop on the tour of bad arguments, every parent is familiar with the refrain, "but everybody else is doing it." Every parent is also familiar with the appropriate reply, "That's no reason why you should." The fact that everybody else is doing it is no reason why you should, and the fact that everybody else believes something is no reason why you should. Arguments which treat majority opinion as if that alone constituted a reason for belief commit the fallacy of appeal to the majority. Here's a simple example.

[Video start.]

Woman: A majority of the population thinks we coddle criminals. It's time we got tough on crime.

[Video end.]

Professor Grim: The premise? A majority of the population thinks we coddle criminals. The conclusion? It's time we got tough on crime. Just how much support does that premise give for that conclusion? Not much, that's the fallacy of appeal to the majority. Sometimes the appeal isn't even to a majority.

[Video start.]

Man: A significant portion of the population takes UFO sightings seriously. UFOs are real.

[Video end.]

Professor Grim: Fallacious for the same reason, right? There is a simple fact that defuses a number of common fallacies, including appeal to the majority. It's this. The fact that someone believes something doesn't make it so. We're all fallible. Anybody could be wrong. I don't care who you are, the President of the United States, or the head of NASA, or the Chair of the Physics Department. Nobody is exempt from the possibility of error. It's never the case that something is true just because you believe it's true. You aren't that important to the universe. It's not all about you. You hope that what you believe is true, but the truth doesn't depend on what you happen to believe.

The fact that someone believes something doesn't make it so. The fact that a whole bunch of people believe something doesn't make it so. The fact that a majority of people believe something doesn't make it so. So maybe it's true that we coddle criminals. Maybe it's true that there are UFOs, but in order for an argument to give us a solid reason to believe those things it had better offer something stronger than the fact that some people, lots of people, maybe even a majority of people believe those things.

You may not be tempted by appeal to the majority. It's too obviously fallacious. Here's a fallacy that just about everyone has committed and has committed on themselves. Two things happen in sequence and we conclude that they must be causally connected. The name of the fallacy is a bit of Latin, *post hoc ergo propter hoc*. What it means is this came after this, so it must have been because of this. *Post hoc ergo propter hoc*, or *post hoc*, for short. Here's how we fall victim. I'm waiting for my friends to arrive. They said they were coming over this afternoon, but I've been waiting and waiting. Finally I decide to call them. I pick up the phone and they drive in the driveway. Good thing I picked up that phone. Brains seem to be wired to make connections, even connections that aren't really there. Much of what is termed superstitious behavior is grounded in *post hoc ergo propter hoc.* Professional baseball is a prime example. Outfielder John White gave the typical story: "I was jogging out to centerfield … when I picked up a scrap of paper. I got some good hits that night, and I guess I decided that the paper had something to do with it. The next night I picked up a gum wrapper and had another good night … I've been picking up paper every night since."

Wade Boggs eats chicken before every game because he thought he did well as a rookie after eating chicken. Clint Barmes's ankle was taped after he sprained it. Because he started hitting well, he insisted it be taped up long after the sprain had healed. Elements of ritual build up. As Dennis Grossini, a pitcher on the Detroit farm team, said, "You can't really tell what's most important, so it all becomes important. I'd be afraid to change anything. As long as I'm winning, I do everything the same." So Turk Wendell chews four pieces of black licorice while pitching, insists on throwing the rosin bag down hard, draws three crosses in the dirt. Mike Hargrove was a first baseman for the Cleveland Indians. He built up such an elaborate routine that he was known as the human rain delay, all because of *post hoc ergo propter hoc.* Sometimes things are just coincidences. The fact that A happened and then B happened doesn't show that A caused B.

Let me give you third bad argument type with another Latin name, *ad hominem.* An *ad hominem* argument is one that attacks the opponent himself as a person, rather than his position. *Ad hominem* means against the person. Here it is in a simple form. You're in a debate. Your opponent has argued brilliantly that racial profiling is both unconstitutional and morally wrong, that it violates inherent human rights. It's your turn, and you say, "Where'd you get that shirt, nutmeg? Have you noticed that your socks don't match? Where'd you get that argument, off the back of a candy wrapper? Your mother wears army boots."

Of course, *ad hominem* is often a bit more subtle. Here's another example.

[Video start.]

Woman: The Republicans say we should shrink the deficit and cut down on government spending, but these are the same people who cranked up the deficit by getting us into two wars simultaneously. These are the same people who cranked up the deficit by giving tax breaks to the rich. These are the same people who gave us the Bush years of massive spending, which was more than any other administration from either party.

[Video end.]

Professor Grim: What's really at issue here is the claim that we should shrink the deficit and cut down on government spending, but in fact, that claim is never addressed at all. What we get, instead, is an attack on the people who put it forward; that's *ad hominem*. An attack on the person rather than the claim. With truth and rationality on the line, what's important is the quality of the argument itself. If the argument is solid, it doesn't matter where it came from. Put it this way. Good arguments are good arguments in the dark. They are good arguments even if you can't see who is giving them. That's the principle violated by the fallacy of *ad hominem*.

Ad hominem applies to any case in which the attack is on the person rather than the argument. There are two specific forms of *ad hominem* that reappear again and again, particularly in election years. The first is *tu quoque*, Latin for "you also," or more colloquially, "you did it too." The last example I gave you was a *tu quoque* form of *ad hominem*.

[Video start.]

Woman: The Republicans say we should shrink the deficit and cut down on government spending, but these are the same people who cranked up the deficit by getting us into two wars simultaneously.

[Video end.]

Professor Grim: That's *tu quoque*. Does it show that they're wrong about the deficit and government spending? No. The most it shows is that Republicans have also been part of the problem. Here's another *tu quoque*.

[Video start.]

Woman: The U.S. says that North Korea is stockpiling nuclear weapons, and is, therefore, a threat to world peace. But surely both Israel and the U.S. have their own stockpiles of nuclear weapons.

[Video end.]

Professor Grim: Does this show that North Korea is not stockpiling nuclear weapons, or that it is not a danger to world peace? No. It's just a "you did it too."

A second form of *ad hominem* is poisoning the well. Like all *ad hominems*, it involves an attack on the person rather than the argument, but specifically an attack on someone's motives for saying what they do. "They're just saying that because" … "you can't trust water from that well" … That's poisoning the well. Those who opposed the invasion of Iraq were quick to speculate regarding motives. Bush wanted to finish his father's work or show his father he was stronger. Dick Cheney's Halliburton just wanted bigger contracts, or they were all out for the oil. If the invasion of Iraq was a bad idea, it was a bad idea independently of what motivated it. The same is true if it was a good idea. Talk of motivations didn't touch the real issue. In 2004, when campaigning for the presidency, John Kerry published a misery index intended to show that things had gotten worse under the Bush administration. The other side fired back, John Kerry talks about the economy because he thinks "it will benefit him politically." True, perhaps, but irrelevant. That's just poisoning the well, *ad hominem*

Are you ready? It's time to play Name that Fallacy. I want you to listen to these three arguments. In each case, I want you to tell me what fallacy is being committed. Here's the first one.

[Video start.]

Woman: As you know, most European countries have abolished the death penalty, with Turkey being the only country in recent years to carry out the death sentence. So it's clear that the death penalty is immoral.

[Video end.]

Professor Grim: Here's argument number two.

[Video start.]

Woman: All you hear in the current campaign is we need someone in Washington who is not a professional politician. But the people making that claim are lobbyists and politicians.. So much for that claim.

[Video end.]

Professor Grim: Here's argument number three.

[Video start.]

Woman: In 2011 Illinois abolished the death penalty. Did murders increase after abolition? No, they actually continued to decline. Why can't people understand that the death penalty doesn't prevent murders, it actually causes them?

[Video end.]

Did you get those? The first one was an appeal to the majority. The premise was that most European countries have abolished the death penalty. Here the majority was countries, but the spirit is exactly the same. The conclusion was that the death penalty is immoral. But believing it's so doesn't make it so, even when we're talking about the beliefs of a bunch of countries. That's the fallacy of appeal to the majority. The second one? *Ad hominem.* The claim at issue is that we need a Washington outsider. A good argument would give us reasons to believe that's not true. But that's not what we get. What we get instead is an attack on the people making that claim—that they themselves are Washington insiders. The attack is on the people making the claim, not the claim itself, *ad hominem.* Can you tell me the specific kind? That's right, *tu quoque*, you too. And the last one? *Post hoc ergo propter hoc.* What the premise gives us is two data points. Illinois abolished the death penalty in 2011. Murders continued to fall after abolition. That's all we've got. The conclusion drawn is that there was a causal connection between those data points. *Post hoc ergo propter hoc.*

Roll up your sleeve. I want to inoculate you against another fallacy, the fallacy of false alternative, also called false dilemma. The trick of a false alternative, or false dilemma, is to set things up as if the opponent has to choose either A or B. It's a false alternative because it may not be true at all. There may be lots more options open.

[Video start.]

Mom: Look, you either marry Julie and have a wonderful life and family, or you spend the rest of your life as a lonely bachelor. Go ahead. Make up your mind. Julie's waiting.

[Video end.]

Professor Grim: That's clearly a false alternative. Are those your only options? Of course not. A lonely bachelor? Maybe you'll live your life as a gregarious, fun-loving bachelor instead. You might end up marrying Julie and not having a wonderful life. You could move to Alaska, start a commune, marry someone else, and live happily ever after. Sorry, Mom. Bogus argument. That's a false alternative.

There's a famous case of false alternative used by George W. Bush in a State of the Union address: "Either you're with us, or you're with the terrorists." But as I said in the beginning, neither side has a lock-down on fallacies. Barack Obama used false alternatives in a State of the Union address as well, not once but twice. "Either we ask the wealthiest Americans to pay their fair share in taxes, or we're going to have to ask seniors to pay more for Medicare. We can't afford to do both." "Either we gut education and medical research, or we've got to reform the tax code so that the most profitable corporations have to give up tax loopholes that other companies don't get. We can't afford to do both." Options in both cases are far more complex than that. It's not just an either, or. Trying to make it look that way is the fallacy of false alternative.

Here's a fallacy that takes the nefarious strategy of the false alternative a little farther. It's called the complex question. The classic example is a lawyer who is prosecuting someone for wife beating. The accused is put on the stand and told to answer merely yes or no to the questions put to him. The prosecutor says, "Tell me, Mr. Smith, have you stopped beating your wife yet? He's got to say yes or no, but if he says yes, the prosecutor says, "Aha! You say you have stopped beating your wife, so you admit you beat her in the past. I rest my case." If he says no, the prosecutor says, "What? You say you haven't stopped beating your wife? You admit you're still beating her. I rest my case." It's a trick question; it's a trap. The defendant is forced to say yes or no. Either answer is legitimate only on a certain assumption, that he has beaten his wife. He's forced into a false alternative, each side of which begs the question against him.

Hasty generalization is a fallacy of jumping to a conclusion. You know something about a small number of things. You jump to a conclusion about all of them.

[Video start.]

Man: The first person I met on Long Island was short and rude. The second person I met on Long Island was short and rude. I guess everybody on Long Island must be short and rude."

[Video end.]

Professor Grim: Racist and sexist generalizations are one of the worst kinds of generalizations, and those are quite generally hasty generalizations.

[Video start.]

Woman: Did you see what that Polish guy did? Polish people must be stupid.

Man: Every Italian guy that's come in this week bought shoe polish. Italians just want to buy shoe polish.

Tall Man: Every woman I've ever dated was a big tipper. Women are just big tippers.

[Video end.]

Professor Grim: Hasty generalization's wrong for all kinds of reasons. In other lectures we've talked about the fallacy of appeal to emotion. The purpose of argument is to give solid reasons for believing a conclusion. An appeal to emotion is intended to short-circuit rationality, to manipulate a hearer by pressing emotional buttons instead. Appeal to ignorance is equally illegitimate. Here the fallacy is to act as if our ignorance is something more than it is, as if ignorance alone gives us some kind of positive evidence. Here's an example.

[Video start.]

Man: Many ships and planes have utterly disappeared in the regions of the Bermuda Triangle often without explanation, often without a physical trace. There must be some kind of mysterious force at work beneath the waves.

[Video end.]

Professor Grim: Earlier I emphasized a simple fact. Believing it is so doesn't make it so. Here's another simple fact. Ignorance is just ignorance. If we don't know something, we just don't know it. Ignorance isn't, by itself, evidence for something, a mysterious force beneath the waves, for example. Evidence for a mysterious force would require real evidence, not just ignorance, not just lack of evidence. Here's another example.

[Video start.]

Man: How do we know that Oswald acted alone? How do we know this wasn't some massive conspiracy, and that our own government hasn't simply covered it up? We haven't yet heard the real truth about the Kennedy assassination.

[Video end.]

Professor Grim: You'll notice that's driven entirely by nothing more than questions. The argument works entirely from ignorance and yet treats "how do we know?" as if it offered a definite conclusion. "We haven't yet heard the truth about the Kennedy assassination." Could there be grounds to think that? Certainly, but a bunch of questions aren't grounds. They're just questions.

Are you ready for another round? It's time to play Name that Fallacy. This time it will be cumulative. Watch for all the fallacies we've mentioned. I'll give you three arguments. You tell me the fallacies.

[Video start.]

Woman: First they tell us that hormone replacement therapy is good for the health of women after menopause. Comes to find out that hormone replacement therapy actually leads to some serious health risks. First they tell us that a certain medication is good for avoiding high blood pressure and heart disease. Then they say that that same medication is linked to diabetes, which is itself linked to heart disease. I guess you really just can't trust medical research.

[Video end.]

Professor Grim: Here's a second argument.

[Video start.]

Man: No one knows exactly how the ancient pyramids were built. Why did the ancient South Americans, separated by thousands of miles, build pyramids just as the ancient Egyptians did? How did primitive peoples erect the mammoth pillars of Stonehenge, and how were the stones of Machu Picchu fitted so closely together that even a knife cannot slip between them? The answer is obviously the influence of ancient astronauts.

[Video end.]

Professor Grim: Argument number three:

[Video start.]

Woman: It's time to use the three strikes law to get tough on crime. Anyone who gets convicted of three felonies should go to jail for life. Look, we've got to make a choice. We either slam the door and throw away the key, or we're just coddling these people, encouraging recidivism, and using our prison system to educate career criminals. In Washington State the law passed with a majority. In California it was 72% of the vote. It's time to use three strikes to get tough on crime.

[Video end.]

Professor Grim: Did you catch the fallacies? The first one was a hasty generalization. We have two cases in which health recommendations have been reversed. That's hardly grounds for the conclusion that we can't trust any medical research—hasty generalization. No one knows exactly how the pyramids were built. Exactly. No one knows. From that ignorance it doesn't follow that ancient astronauts were responsible. Nor, of course, that they weren't. The third argument was a tricky one because it involved two fallacies. "Look. We've got to make a choice, either we coddle criminals or go for three strikes." That's a false alternative. California passed it with 72% voter approval; that's appeal to the majority. As I said at the beginning, there's no guarantee that you won't find two or more fallacies being committed at the same time.

My aim in this lecture was to vaccinate you against some standard forms of bogus arguments; to inoculate you against some fallacies. I want to finish with a question as to how this fits in with some of our earlier work. That question might be posed as a set of challenges. Wait a minute. One of the other lectures introduced wisdom of crowds as a positive heuristic. Isn't that just appeal to the majority? What about the example of which is larger, San Antonio or San Diego? The German students made a correct inference from the fact that they had heard of San Diego, and not of San Antonio. Isn't that just appeal to ignorance? The conclusion of another lecture was that in short-range decisions it may be rational to go with your gut. Isn't that just an instance of hasty generalization?

Take a closer look and you'll see that there are important differences. Wisdom of crowds doesn't just recommend that you think what everybody else thinks. It's a strategy for extracting knowledge from a distribution of opinions where some people really do know the answer, like the studio audience of *Who Wants to Be a Millionaire?* It doesn't just copy other people's beliefs like appeal to the majority does. It draws a clever inference, though a fallible one, from the distribution of beliefs. In the San Diego/San Antonio case the German students did better than Americans in picking the larger city precisely because they knew something about one and were largely ignorant of the other. Why have I heard of San Diego and I'm not so sure about San Antonio? Maybe because it's larger. That's a clever inference, though admittedly a risky one. It's not a simple jump from "I don't know how ancient monuments were built" to "It must have been Ancient Astronauts." There is also an important difference of context. When there's no time for careful investigation, when you have to get out of the way of an oncoming train, fast and frugal heuristics are precisely what you need. Better to go with your gut and jump to a conclusion, even if it might be wrong, than to wait on the tracks trying to calculate the velocity of an approaching locomotive.

In the case of explicit argument the standards for rationality are higher. There we want more than a handy heuristic. When someone is trying to convince us of a conclusion, and when we do have time to really think about it, it's important to test how tight the logic really is. Do the premises really support the conclusion? How strongly? Does the conclusion really follow? Or is this a bogus argument built with fallacious reasoning? We'll do more with that next time. It will be your turn to evaluate arguments in the Great Debate.

The Great Debate
Lecture 15

In this lecture, we'll analyze a "great debate" between two speakers, Mr. McFirst and Ms. O'Second, on the topic of democracy. The purpose of all debates is to mold opinion. As Lincoln said in the first of a series of debates with his rival, Stephen Douglas, "Public sentiment is everything. With public sentiment, nothing can fail; without it, nothing can succeed." The debate we'll look at in this lecture won't rise to the level of Lincoln and Douglas, but you can use it to hone your skills in argument analysis.

McFirst Opens

- McFirst opens the debate by stating that democracy is the "wave of the future." As evidence, he cites the Arab Spring, as well as some data from Freedom House, an organization that conducts research on the topic of freedom around the world.

- McFirst's opening argument has two conclusions: (1) "What we have witnessed is the increasing democratization of the world" and (2) "Democracy is the wave of the future."

- His support for the first conclusion includes some vague images (dictatorships are falling), in addition to the data from Freedom House. His argument isn't bad, at least if we can trust the data and the source.

- In support of the second conclusion, McFirst says that if we project the Freedom House numbers into the future, we will have worldwide democracy. But we don't really have any further reason to believe that. In his emotional language, McFirst also makes use of the fallacy of appeal to emotion.

- In considering the statements of both speakers, we should ask ourselves three questions: (1) What conclusion or conclusions are at issue? (2) What argument or arguments are given? How does the reasoning go? How good are those arguments? (3) Have any fallacies been committed?

O'Second Responds

- Ms. O'Second responds to McFirst's conclusions by noting that not all players in the events of the Arab Spring were advocates of democracy and questioning the objectivity of Freedom House. She also states that there is no reason to believe that the data from Freedom House can be extended into the future to predict worldwide democracy.

- What is O'Second's conclusion? That her opponent is wrong, of course. Which of McFirst's two claims was O'Second attacking? The answer is both. Part of her response was directed at his claim that we've seen democracy increasing. The other part of her response was directed at the claim that democracy is the wave of the future. But she used different arguments in those two parts.

- In addressing the claim that democracy is increasing, O'Second had two arguments. Her first is that we can't know what we have seen until we see how it eventually pans out. But that argument doesn't really prove that McFirst is wrong. It attacks not so much the claim as McFirst's confidence in it.

- In her second argument, O'Second attempts to knock down the one data-driven argument McFirst gives by impugning the source Freedom House. But that argument doesn't show that the data are wrong. Like the first argument, its goal is to raise doubts.

- What about O'Second's challenge to McFirst's claim that democracy is the wave of the future? At that point, she uses an argument from analogy: the multiple-blade razor analogy. But the analogy doesn't really work here. At best, it simply shows that we can't always project current trends into the future. It doesn't give us any particular reason to believe that this is one of those times.

- O'Second uses an *ad hominem* fallacy, poisoning the well: "It is surely significant that Freedom House is located in Washington, DC." She adds to that fallacy with an appeal to ignorance. She doesn't really give us any solid reason to think Freedom House should be discredited.

McFirst Uses Schopenhauer's Strategems

- O'Second challenged McFirst's reasons for believing that democracy is on the rise, but instead of addressing her challenge, McFirst, in his response, charges O'Second with arguing that democracy is not a good thing. In other words, he attacks a position that she never held—a straw man.

- At the same time, McFirst gives himself an easier position to argue for—not that democracy is historically on the rise or that we can predict it will spread but simply that it's a good thing. And if we're looking for fallacies, McFirst's quote from Churchill might count as an emotion-loaded **appeal to authority**.

- Given a chance to improve his argument, McFirst focuses on the real points of disagreement. He backs up the data claim for his first conclusion—that democracy is increasing worldwide—countering O'Second's appeal to ignorance with real information.

- He also seems to have backpedaled on his second conclusion a bit. He replaces his initial confident claim that the trend will continue with the more modest claim that there's no reason to think it won't.

O'Second Casts Doubt

- O'Second responds by noting that the American democracy arose from a particular historical background and set of precedents. Without that history, she says, democracy may not come as easily to other societies or it may come in a different form.

- This response by O'Second may be a little more subtle and a little harder to analyze. She begins with an Aristotelian syllogism:

In analyzing the statements of any speaker in a debate, ask yourself: (1) What conclusions are at issue? (2) What arguments are given? (3) Have any fallacies been committed?

> "Democracy...real democracy...is important. Important things take real work. So democracy is going to take real work."

- With that reasoning, she gives her initial challenge a different spin than we saw before. She argued earlier that O'First's projection into the future was unjustified. Here, she says that we shouldn't assume the inevitability of democracy because democratization will take real work.

- O'Second then uses a brief sketch of the American Revolution as the backdrop for another argument.
    - She starts with the claim that our democracy would not have the form it does without particular aspects of our history, such as the Protestant Reformation or our existence as colonies.

    - From that, she draws the inference that we should, therefore, not expect other paths to democracy in other cultures to look like ours. We should not expect the outcome to be the same.

- How good is that reasoning? Although it's suggestive, it's not solid proof. She doesn't prove that our institutions demanded those particular historical precedents, nor does she prove that other cultures with other histories couldn't reach the same outcome.

- If we look at her language carefully, she seems to recognize that this line of thought is merely suggestive. For example, she doesn't say that democracy is bound to come in different forms; she says that it "may."

- What O'Second has given us, then, are some skeptical doubts and some suggestive counter-suggestions. In urging us not to immediately accept her opponent's stirring forecast of a predictable rise in democratization worldwide, however, that may be enough.

Term to Know

appeal-to-authority fallacy: A fallacy in which the opinion of some prominent person is substituted for rational or evidential support; often used in advertising by linking a product to a celebrity "expert" rather than providing rational or evidential support for a claim about the product.

Suggested Reading

Fineman, *The Thirteen American Arguments*.

The Lincoln-Douglas Debates.

Questions to Consider

1. Who do you think won the debate: Mr. McFirst or Ms. O'Second? If that questions is too simple, explain why.

2. Consider how you think through an issue by yourself, and consider how an issue is addressed in a debate. In what way are those the same? In what ways do they differ?

3. Our adversarial system of justice has each side present its best case, with the opportunity to cross-examine the other side. The decision is ultimately made by some neutral party—a judge or jury. What are the pros of such a system with regard to finding the truth? What are the cons?

Exercises

You've undoubtedly watched a presidential debate live. Look for a video of a recent debate online (http://elections.nytimes.com [select the Video tab, then search for "Presidential Debates"] or search on YouTube). Watch the video, pausing after each contribution to analyze its strengths and weaknesses. Ask the same questions that we asked in this lecture:

(a) What precisely is the position?

(b) What precisely is the argument given?

(c) How good is that argument?

(d) Did the speaker commit any fallacies?

The Great Debate

Lecture 15—Transcript

Professor Grim: Welcome to the Great Debate. This isn't the Lincoln-Douglas debate, though that's not a bad precedent. As noted in an earlier lecture, Lincoln and Stephen Douglas were rival candidates for the Senate from Illinois. In August through October of 1858 they held seven face-to-face debates in towns across the state. Papers across the country printed those debates in full. The purpose of all debates is to mold opinion. Lincoln knew that. In the very first debate he says:

> Public sentiment is everything. With public sentiment, nothing can fail; without it, nothing can succeed. ... [He] who molds public sentiment, goes deeper than he who enacts statutes or pronounces decisions. He makes statutes and decisions possible or impossible...

The debate you're about to hear won't rise to the level of Lincoln and Douglas, but I want you to use it to hone your skills in argument analysis putting to work some of what we've covered. I'll interrupt at various points to ask you to analyze what you've heard. I'm particularly interested in your applying what we've talked about in evaluating parts of the arguments. There is nothing that says this has to be done fast. If you want to pause the debate at any point, to repeat a passage, or to give yourself some time to think, go right ahead. High school and college debate teams use a format named after the Lincoln-Douglas debates. An initial statement by side A is followed by a brief questioning by side B. B's full initial statement is then followed by a brief questioning by A. Each side is then given a chance for rebuttal with an interesting structure, a short rebuttal for A, a longer rebuttal for B, and a short final word for A. I know, it's complicated, and our debate won't follow that strict form. In fact, the Lincoln-Douglas debates didn't either, although they did allow equal time for each side in an A-B-A structure. Douglas spoke first in four of the debates. Lincoln spoke first in three.

And now for our great debate. As you listen, I want you to pay attention to the flow of argument. What is the speaker's intended conclusion? What are the reasons given in support of that conclusion? And how, precisely, are they supposed to support the conclusion? I'll also ask you whether the reasons given really do support the conclusion. Is the argument valid? In another lecture we've talked about standard forms of bogus arguments—fallacies. I want you to keep a sharp eye open for those. Without further ado, the first debate. The topic: Democracy.

[Video start.]

McFirst: It is becoming clearer every day that democracy is the wave of the future. What we have witnessed is the increasing democratization of the world. The Arab Spring is only one of the more recent signs. Dictatorships are falling. Monarchies and hereditary aristocracies and oligarchies are tumbling. There is a scent of freedom in the air. Power to the people! That's the stirring cry of human progress. That's the sound of democracy on the march.

But don't take my word for it. Here is some hard data from Freedom House, an international organization that conducts research on the topic. Freedom House categorizes countries as either free, partly free, or not free. If you look at their data, you'll see that the percentage of free countries in the world has risen approximately 20% over the last 35 years. We can project that trend into the future. By 2045 over 60 percent of all countries will be free. By the turn of the next century, the transition will be complete. Almost every country in the world will be free. We won't have long to wait. Democracy is the wave of the future.

[Video end.]

Professor Grim: I know we haven't heard from the other side yet, but let's stop things right there. Let's look at the flow of that first argument. I have three questions for you as an observer and judge in the debate. Question 1, what is McFirst's conclusion? Question two, what is the argument he gives for that conclusion? If there are multiple arguments, what are they? How precisely does his reasoning go? And question three, did you notice any fallacies going by? If you want to pause and review, go right ahead.

So start with question number 1. What is the conclusion of the argument? Here's a good candidate; he said "What we have witnessed is the increasing democratization of the world." Interestingly, there's also another good candidate for the conclusion: "Democracy is the wave of the future." In what you just heard, McFirst tended to run those two together. But they're really separate propositions. One proposition is that more and more countries have become democratic. That's a claim about what we are seeing now. The other is the claim that democracy is the wave of the future, that such a trend will continue into the future. The first proposition is the one he represented with his initial chart. The second proposition is the one he put forward by talking of extending the line.

So let's examine the first conclusion: "We are witnessing the increasing democratization of the world." What support does he give for that conclusion? What reasons does he give for us to believe it? At the beginning, he just gave us a few vague images: dictatorships are falling. monarchies and hereditary aristocracies are tumbling. But then he gave us what he said is hard data. According to Freedom House, free countries—we assume he means democracies—have increased by 20 percent over the last 35 years. Well, that's not a bad argument, at least if we can trust the data. If we can trust the source, that does make it look like we've seen an increase in democratization. Good for you, McFirst. That looks like a pretty solid argument for your first conclusion, that we are currently witnessing increased democratization. But that was only one conclusion offered in the argument. The other was "Democracy is the wave of the future." He represented that by extending the lines into the future. Do we have any additional argument that the observed trend will continue? He says we can project that into the future. If we do, we get his optimistic estimates for worldwide democracy in 2045 and the turn of the next century. Do we really have any further reason to believe that? It doesn't look like it.

What about fallacies? Did McFirst use any fallacious sleights of hand? If you said appeal to emotion, I think you're right. It's evident in the emotional language he used: "... a scent of freedom in the air, power to the people, the stirring cry of human decency, the sound of democracy on the march." Those just amount to emotional cheerleading for democracy. We may agree that democracy is a wonderful thing, but that doesn't really advance McFirst's conclusion that it's becoming more widespread. Those phrases are just designed to push your emotional buttons.

That's just our first bit of debate, and I've already talked longer than Mr. McFirst has. Let's hear what Ms. O'Second has to say in response. In a minute, I'll again be asking you question number one, what conclusion or conclusions are at stake? Question two, what argument or arguments are given? How does the reasoning go? How good are those arguments? And question three, have any fallacies been committed?

[Video start.]

O'Second: I'll give you one thing, Mr. McFirst. We would certainly all like to think that democracy is the wave of the future. But wanting doesn't make it so. I'm afraid I have to challenge your oh so rosy and optimistic conclusions.

The Arab Spring stirred great hopes in freedom-loving people, but that's because we chose to interpret the Arab Spring a very particular way. We viewed it as something like an American Revolution in the Middle East. Down with the dictator! Up with human rights and constitutional democracy!

I invite you to look at that event more carefully. There were many players in the events of the Arab Spring, and they weren't all advocates of democracy, at least not democracy in the sense we think of it. Whether what we've seen is really a triumph for democracy is something we can't know yet. It is just as likely that the result of those events will be not democracy but theocracy— governmental control by a religious elite. It is just as likely that the result will not be human rights for everybody, but expanded rights for a few, at the cost of the rights of others. It is also very possible that governments set up as a result of those events will be flimsy and fragile, ripe for take-over by a military junta or, I'm sorry to say, by another dictator.

Your one piece of real data, McFirst, is based on a report card published by Freedom House. Much as it claims to be an international organization unaffiliated with any government, surely it's significant that Freedom House is based in Washington DC. How do we know it's not just a front for the United States government? Or for the CIA?

Let me finally take issue with your future projections. If we do accept the Freedom House data, there is no particular reason to think we can extend your blue line into the future. Let me give you an analogy. In 1971, Gillette introduced the twin razor blade. In 1998, they introduced the first triple-blade razor, the Mach 3. Four-blade razors appeared between 1998 and 2005, when a 5-blade razor was introduced. Extend that history in a line like my opponent wants to do and we'll be seeing hundreds of blades in our razors by 2050. Ridiculous. But that's just like your future projections for democracy. Democracy worldwide? You wish.

[Video end.]

Professor Grim: Now things are getting a little heated. Let's go to our questions. What were the conclusions? What is the flow of argument—the pattern of reasoning offered? And, is her argument valid? Just how good is it? And did you notice any fallacies going by? Press rewind if you want to hear the argument again. Press pause if you want to think about it a minute. Then come back.

What is Ms. O'Second's conclusion? That her opponent is wrong, of course. We said that McFirst really had two claims, one, that democracy is increasing in the world, and two, that it is the wave of the future. Which of those was Ms. O'Second attacking? If you said both, you're right. Part of her response was directed at his claim that we've seen democracy increasing. The other part of her response was directed at the claim that democracy is the wave of the future. But she used importantly different arguments in those two parts. Try to reconstruct in your mind her argument against the claim that democracy is increasing. Now try to reconstruct what her argument was against the future projection of democracy as the wave of the future. What precisely were those arguments? Take the claim that democracy is increasing. Here Ms. O'Second really had two arguments. The Arab Spring had been mentioned as a recent sign of increased democratization. Ms. O'Second attempts to throw skeptical cold water on that example. Her argument is that we can't know what we have seen until we see how it eventually pans out. She says "Whether what we've seen is really a triumph for democracy is something we can't know yet." Maybe it will end up in a theocracy, she suggests, in a military junta, or in another dictator. That's a fairly direct challenge to McFirst's claim.

Maybe we aren't really seeing an increase in democratization. Maybe we're reading more into it than we really have reason to hope for. How good is that argument? You'll note that it doesn't really prove that McFirst is wrong. It doesn't show that the Arab Spring will result in some non-democratic future. It attacks not so much the claim as McFirst's confidence in it. Maybe we can't yet tell. That's one argument against the democracy-is-increasing claim, an attack on its certainty.

In a second argument, Ms. O'Second attempts to knock down the one data-driven argument that McFirst gives. McFirst relied on the Freedom-House data. Ms. O'Second attempts to undercut his argument by impugning that source, remember? "… surely it's significant that Freedom House is based in Washington D.C. How do we know it's not just a front for the United States government or for the CIA?" That argument doesn't show that the data is wrong. Like the first argument, its goal is to raise doubts. O'Second's arguments, then, are skeptical arguments intended to raise doubts about the evaluations we tend to make of the Arab Spring; intended to raise doubts about the reliability of Freedom House data.

What about her challenge to McFirst's second claim that democracy is the wave of the future? She attacks that claim even if the first goes through. She says, "even if we do accept the Freedom House data, there is no particular reason to think we can extend your blue line into the future." At that point she uses an argument from analogy, the multiple-blade razor analogy. Twin-blade razors in 1971, three blades in 1998, four blades by 2005. Extend a linear projection of that history, she says, and we'll see razors with hundreds of blades by 2050. "Ridiculous," she says, "just like linear projections for future democratization."

How strong an argument by analogy is depends on how strong the analogy is. What's your take on that in this case? I think this particular analogy is probably not particularly good here. At best it just shows that sometimes we can't project current trends into the future. That doesn't give us any particular reason to believe that this is one of those times. All in all, then, how strong do you think O'Second's response was? If she was out to refute McFirst, I'm afraid she didn't fully succeed. She certainly didn't show his claims to be false. If she was out to raise a few doubts about his confidence, or ours, well, maybe she did succeed at that.

What about fallacies? You probably saw at least one. I think there are actually a couple in here, entwined. Both have to do with her attack on Freedom House. McFirst used data from Freedom House. O'Second seems to think we shouldn't accept or rely on that data. Why not? She says, "it is surely significant that Freedom House is located in Washington D.C." That sounds like the ad-hominem fallacy. She's implying some kind of underhanded or ulterior motive, an implication that Freedom House is just pushing U.S. policy, perhaps. She's poisoning the well. She adds to that fallacy with an appeal to ignorance. She doesn't really give us any solid reason to think Freedom House should be discredited. She relies, instead, on leading questions: How do we know it's not just a front for the United States government? How do we know it's not just a front for the CIA? Appeal to ignorance.

It's Mr. McFirst's turn again. Let's see how he responds. Here, again, I want you to try to identify major conclusions. I want you to identify and evaluate patterns of reasoning, and keep an eye out for fallacies.

[Video start.]

McFirst: Well, I guess even democracy has some enemies. It's hard to believe, in this day and age, that anyone could think that a theocracy or a benevolent dictatorship was preferable to our tradition of individual rights and majority rule, but that seems to be what my opponent is arguing for. I'm the first to admit that the majority isn't always right—that would be the fallacy of appeal to the majority, after all. Allow me to quote a great statesman, Winston Churchill:

> Many forms of Government have been tried, and will be tried in this world of sin and woe. No one pretends that democracy is perfect or all-wise. Indeed, it has been said that democracy is the worst form of government except all those other forms that have been tried.

When the impact of a policy is spread across the entire population, surely the whole population has a right to decide on that policy. It's not a single individual that has that right. It's not a hereditary aristocracy. It's not a chosen few. It's the people who have the right to decide policy that will govern the people. That's democracy. That's the form of government that is on the rise, however much my opponent may dislike it.

[Video end.]

Professor Grim: Okay, I'm going to stop the debate right there. That was a really poor response Mr. McFirst. You disappoint me. You're using rhetorical tricks right off of Schopenhauer's list. The basic problem is this: McFirst isn't arguing against O'Second's real position. Rather than taking on the position that is really on the floor, he has picked an easier position to argue against, and by the same token an easier position to argue for. What McFirst just gave us was almost entirely a Straw Man argument. At the beginning, you remember, McFirst was arguing that worldwide adoption of democracy was on the rise and would continue. Ms. O'Second challenged his reasons for believing that. This time around McFirst has charged her with arguing that democracy is not a good thing, that theocracy or benevolent dictatorship would be better. Press replay and you'll find that she never said anything like that. He's attacking a position she never held. He is attacking a Straw Man or in this case, maybe, a straw woman.

By the same token, McFirst is giving himself an easier position to argue for. Not that democracy is historically on the rise, that we can predict that it will spread, but just that it's a good thing, or, as Churchill said, the worst except for all those other forms of government. If we're looking for fallacies, that quote from Churchill might also count as an emotion-loaded appeal to authority. You're using rhetoric rather than appealing to rationality, McFirst. Let's go back and see if you can present a better argument.

[Video start.]

McFirst: OK. Let me try it again. Ms. O'Second challenged my first conclusion by casting aspersions on the data from Freedom House. Most of her attack was a fallacious appeal to ignorance.

Let me address that directly by reducing some of the ignorance. Freedom House is not an agency of any government. I admit that American foreign policy is a focus and that the organization is based in the United States. Why shouldn't it be? Founded in 1941, it is an advocacy group for political liberty and human rights worldwide. In the 1940s, its target was fascism. In the 1950s, foreign communism was its target, though it took a strong stand against McCarthyism at home.

The data I offered comes from an annual survey called Freedom in the World, published each year since 1972. It uses a strategy devised by leading social scientists, based on the United Nation's Universal Declaration of Human Rights, in order to objectively rate every country in the world on indicators for both political rights and civil liberties. Political indicators include electoral process, political pluralism, and lack of corruption. Indicators for civil liberties include freedom of expression and belief, rights of free association, and the rule of law. That's the data that says that the proportion of free countries in the world has risen 20 percent in the last 35 years. That's the data that says we're witnessing an increase in democracy.

As to my second claim, that we can extend that data line into the future, I guess I have to admit that one can't always project current trends directly. But you'll note that Ms. O'Second didn't suggest any forces that might buck that trend. She didn't give any reason to think that the forces that have been driving democratic progress over the last 35 years won't continue. That's my point.

[Video end.]

Professor Grim: That's a lot better Mr. McFirst. Now you're focusing on the real points of disagreement, not some made up for rhetorical effect. You've backed up your data claim for your first conclusion that we have been witnessing increasing democratization worldwide. We might still want to look at the details, but you have certainly countered the appeal to ignorance with real information. I noticed that you've also backpedalled on your second conclusion a bit. Now you have replaced your initial confident claim that the trend will continue with the more modest claim that there's no reason to think that it won't. I don't take that as a mark of rhetorical weakness. I take that as a mark of rational strength. Good for you, Mr. McFirst. Let's see how Ms. O'Second responds.

[Video start.]

O'Second: I'm glad that my opponent has seen the light. I am as much an advocate of democracy as he is.

Democracy—real democracy—is important. Important things take real work, so democracy is going to take real work. That's why we shouldn't fool ourselves into thinking that it's easier or more inevitable than it in fact is. Is democracy the wave of the future? That won't be true, I think, unless we work to make it so.

We have been raised in a tradition of individual rights that goes back to the American Revolution, including rights to religious freedom. Because of our background, it is hard for us to think of our form of democracy as anything other than the natural form of self-governance.

But you have to remember that we got here only by a very specific historical path. Our history was one of Protestant Reformation long before the American Revolution, and of colonies physically separated from the mother country.

It's not clear that our democracy would have formed without those precedents. It's not clear that it's exportable without such a history. Consider the French Revolution. That too announced the "rights of man." But it had a very different outcome—an outcome that led directly to Emperor Napoleon.

Other societies don't have the same historical precedents that we did. Democracy—with religious freedom a major component—may not come so easily to them. And even when democracy does come in other societies, it may come in a different form. Other societies may become more democratic without adopting a form that ultimately looks very much like our system at all. That's my point.

[Video end.]

Professor Grim: This may be a little more subtle and a little harder. We want to know: What is the conclusion? What is the intended flow of argument? How solid is the reasoning? And are there any fallacies involved? Did you notice that she began with an Aristotelian syllogism? Right here: "Democracy—real democracy—is important. Important things take real work. So democracy is going to take real work." Aristotle would be pleased. With that reasoning, she gives her initial challenge a different spin than we saw before. She argued earlier that O'First's projection into the future was unjustified. Here she says that we shouldn't assume the inevitability of democracy because democratization is going to take real work.

What happens in the rest of the argument? O'Second uses a thumbnail sketch of the American Revolution as the backdrop for another little argument. It starts from the claim is that our democracy would not have the form it does without particular aspects of its historical background, the Protestant Reformation, for example, and our existence as colonies. From that she draws the inference that we should, therefore, not expect other paths to democracy in other cultures to look like ours. We should not expect the outcome to be the same. How good is that reasoning? Interesting, but I think we have to say that it's suggestive, rather than a solid proof. She doesn't prove that our institutions demanded those particular historical precedents. Nor does she prove that other cultures with other histories couldn't reach the same outcome. If you look at her language carefully, she seems to recognize that this line of thought is merely suggestive. She doesn't really say that our democracy wouldn't have formed without the historical precedents she names. She says it's not clear that it would have. She doesn't say democracy is bound to come in different forms. She said it may. What O'Second has given us in the whole course of the debate, then, are some skeptical doubts and some suggestive counter-suggestions. In urging us not to immediately accept her opponent's stirring forecast of a predictable rise in democratization worldwide, however, that may be enough.

An interesting debate, not up to the standards of Lincoln and Douglas, perhaps, but a good exercise for analysis. What happened in that other debate? A major topic there was the expansion of slavery into the territories and, by implication, slavery in general. Douglas was clearly on one side. He said,

> I believe this Government ... was made by white men for the benefit of white men and their posterity for ever, and I am in favor of confining citizenship to white men, men of European birth and descent, instead of conferring it upon Negroes, Indians, and other inferior races.

There were shouts of encouragement from the audience.

Lincoln's position? "There is no reason in the world why the Negro is not entitled to all the natural rights enumerated in the Declaration of Independence, the right to life, liberty, and the pursuit of happiness ..." But at other stops on the debate trail he says things that are far less inspiring. He says,

> I am not, nor ever have been, in favor of bringing about in any way the social and political equality of the white and black races ... [nor of making] voters or jurors of Negroes, nor of qualifying them to hold office, nor to intermarry with white people.

Douglas accused Lincoln of contradicting himself. Lincoln accused Douglas of twisting language so as to make a horse chestnut into a chestnut horse. Ah, debates.

Next time we'll look at another place that rhetoric and fallacies appear: the realm of advertising.

Outwitting the Advertiser
Lecture 16

A dvertising is ubiquitous. The average American sees 2 million television commercials in a lifetime, to say nothing of billboards, newspapers, magazines, and much more. The advertiser's job is to get you to buy something or to vote for someone, not to encourage you to reason logically. Although philosophers sometimes analyze advertising in terms of logical fallacies, it is often more effective to think of advertising in terms of psychological techniques. Those techniques fall into two categories: The first relies not on information or content but on emotion, and the second does rely on content and does use information, but that information is used in misleadingly persuasive ways.

Attractiveness Attracts

- The people in television commercials are a strange breed. They're far more attractive than the average person next door. Ads use good-looking people because attractiveness attracts. Regardless of what product is being sold or its quality, an attractive person will be used to sell it. The **"attractiveness attracts" advertising strategy** is used to appeal to your nonrational psychology rather than your logical rationality.

- To process advertising rationally, we have to be aware of, and compensate for, the fact that attractiveness attracts. In judging whether to buy what a commercial wants us to buy or to vote how a commercial wants us to vote, we need to filter out that aspect of persuasion that relies merely on an irrelevant attractiveness factor.

- Of course, it's not just the people in advertisements who are at their attractive best. The settings are also pristine. People drink beer on gorgeous beaches that have no crowds or speak to you from clean, spacious kitchens. Car ads feature single cars gliding through isolated and beautiful surroundings. And next time you go to a fast food restaurant, look at the image pictured on the overhead menu.

Hold up what you purchase and see how representative that image is of what you just bought.

- Precisely because attractiveness attracts is a standard strategy, some advertising campaigns catch our attention by deliberately flaunting the strategy. For example, Frank Perdue was a skinny, bald man who reminded people of the chickens he sold. His very unattractiveness was used to convey sincerity and concern for quality.

- A related form of emotional manipulation is the **appeal to prestige**. Because an association with Europe can carry that prestige, in some products, we see **foreign branding**.

- Attractiveness attracts works not only for products but also for politics. Candidates for office are often filmed with dramatic lighting, positioned behind a prestigious podium and with an American flag in the background.

Appeal to Emotion

- The appeal to emotion is a classic fallacy. One of the places it shows up is in political advertising. There, the emotions most often exploited are pride and fear. A political commercial might show an American flag rippling in the breeze or pan across amber waves of grain with "America the Beautiful" playing on the soundtrack. Whoever the candidate and whatever the message, remember that what you're seeing is a fallacious appeal to emotion.

- The other emotion exploited in political advertising is fear. A classic example is the television commercial used by Lyndon Johnson against Barry Goldwater in 1964. It opens with a little girl happily counting petals on a daisy. The count is then taken over by a male voiceover and becomes a countdown, culminating in the flash of a nuclear explosion. The implication was that Goldwater was a reckless man ready to push the nuclear trigger. This is an example of the fallacy of false alternative, but it's also clearly an appeal to emotion: fear.

Better Thinking: Detecting Fallacies in Advertising

- Here is a simple philosophical tool for detecting the kinds of fallacies we've been talking about.

- Choose any advertisement. Read it, or listen to it, or watch it with care. Then sit down with a piece of paper and list the reasons given for buying the product; next, list the reasons given for buying this product rather than a competitor's. A rational decision is one based on reasons.

- If the advertisement is relying merely on an attractive image, it may turn out that you can find no reasons to list at all. Sometimes you can list reasons, but once examined, it becomes clear that they aren't very good reasons.

Appeal to Dubious Authority

- In the appeal to dubious authority, an instantly recognizable celebrity touts the virtues of a product, recommends it, or says that he or she uses it. What makes it an appeal to dubious authority is that you're given no reason to believe that the celebrity knows what he or she is talking about.

- For example, a professional athlete may know a good deal about sports but may not know much about cars, watches, or brokerage companies.

Laws against False Advertising

- Most of the advertisements we've talked about so far rely on psychological manipulation rather than on claims or statements of fact. Some commercials do make claims, however, either explicitly or by implication.

- There are laws against false advertising. The core piece of federal legislation in this regard is the Lanham Act, which prohibits "any commercial advertising or promotion [that] misrepresents the nature, characteristics, qualities, or geographic origin of goods, services or commercial activities." The Federal Trade Commission is empowered to enforce regulations against "unfair and deceptive practices in commerce."

- The law also allows consumers and rival companies to pursue civil action. The suit must prove a false statement of fact that could be expected to deceive, that it was likely to affect purchasing decisions, and that the plaintiff has suffered injury of some kind. Short of a class-action case, that kind of litigation is a tall and expensive order for any consumer. And, of course, these legal restrictions extend only to commercial advertising. Political advertising isn't included.

Not the Whole Truth
- One way of misleading people is to tell them the truth but to be careful not to tell them the whole truth.

- Consider just a few of the phrases that tend to show up in advertising: (1) "none proven more effective" (true even if all competing brands have been shown to be equally effective); (2) "faster acting than the leading brand" (it is not obvious what the leading brand is, and it could be slower acting than any of the other contenders); and (3) "recommended by doctors" (it would be hard to find a pharmaceutical that wasn't recommended by some doctors, but any implication of general medical consensus is a false implication).

The number of 30-second television commercials seen by the average American child in a year has been estimated at 20,000.

Say the Magic Word

- Here's a final category of advertising tricks: the magic word. There are a whole set of magic words that reappear in advertising. Some may not mean what you think they mean. Some may not mean much of anything at all. "As seen on TV" is a perfect example.

- Consider, for instance, foods labeled "low fat," "reduced fat," "light," "98% fat free," "low calorie," "reduced calorie," "lower sodium," or "lite." All of those labels sound great, but what do the words mean?
    - When you check the Food and Drug Administration guidelines, some of those terms carry absolute standards. A main dish counts as "low fat" if it contains less than 3 percent fat and not more than 30 percent of its calories come from fat.

    - But some of those terms are relative. A potato chip can boast of "reduced sodium" if it has at least 25 percent less sodium than a standard potato chip. The standard is higher for "light": A can of beer, for example, must have 50 percent fewer calories from fat than the standard.

- Such standards often have no legal "teeth"; they are merely non-binding industry recommendations.

- There is also another problem here. If a product is labeled "low fat," that means it contains a certain percentage of fat. It doesn't mean it won't make you fat. A product can be labeled "low fat" despite the fact that it is packed with calories, sugar, carbohydrates, and salt. The label that says "low salt" may speak the truth, but what it labels may still be unhealthful in other ways.

Test Your Skills

- As you look at advertising and try to identify what may be wrong or suspicious about ads you see, ask yourself these questions: What psychological buttons are being pushed? What is the precise claim that is being made? What relevant information may have been left out?

- Consider an imaginary ad urging voters to reject a gun control measure. Gun control is a difficult issue, worthy of rational discussion, but this ad merely stimulates fear.

- Another imaginary ad, for an energy drink, appeals to dubious authority by using a woman in a lab coat as the spokesperson and referring to what physicians say about fatigue and energy.

- In order to outwit the advertiser, we have to see through the attractiveness attracts appeal; the appeals to prestige, emotion, and dubious authority; misleading half-truths; and magic words. In other words, we have to learn to see through the spin. In the next lecture, we'll talk about spin again, not in advertising but in putting a spin on statistics.

Terms to Know

appeal-to-prestige advertising strategy: Linking a product to status symbols in order to enhance the product's desirability. Foreign branding is a specific form of the appeal-to-prestige strategy.

"attractiveness attracts" advertising strategy: Enhancing the desirability of a product by using spokespeople who are more attractive than average, by showing the product in beautiful settings, or by doctoring the product's image or the product itself so that it is more photogenic. The flip side, "unattractiveness detracts," is common in political campaign literature, which often seeks unflattering images of the opponent.

foreign branding: In advertising, associating a product with a foreign country in order to increase its desirability; a particular type of prestige advertising.

Suggested Reading

Packard, *The Hidden Persuaders*.

University of Illinois Advertising Exhibit.

1. You can probably think of an advertisement that influenced you to purchase the product. What was it about the advertisement that influenced you? Were you influenced on the basis of information and argument or something else?

2. Think of an advertisement that you think definitely didn't influence you to purchase the product—that perhaps convinced you not to. What was it about the advertisement that had that effect?

3. What legal limits should be put on advertising? Should advertising be unregulated? Should there be a restriction on "truth in advertising," penalizing anything other than the literal truth? Should there be restrictions beyond that, on implication or suggestion? How far would be going too far?

Exercises

Next time you are in a fast food restaurant, look first at the image of the food portrayed on the menu board. When your food comes, take it out of its wrapping and hold it up for comparison. How do the two differ?

The Clio awards are the Academy Awards of advertising. You can find samples and Clio winners at http://www.clioawards.com/. Take a look at last year's winners—choose the radio ads, perhaps, and focus on ads you are not already familiar with. Construct your own ratings by writing down answers to each of these questions:

(a) How entertaining is the clip?

(b) How much information does it give you about the product?

(c) The reasons given for buying the product were:

(d) Does the advertisement offer reasons to believe that this product is superior to the competition?

(e) Does the advertisement commit any fallacies?

Outwitting the Advertiser
Lecture 16—Transcript

Professor Grim: If there is one thing we are all exposed to, it's advertisements. The number of 30-second television commercials seen by the average American child in a year has been estimated at 20,000. At that rate, the average American sees two million television commercials in a lifetime. Add billboards, newspapers, magazines, mailed flyers, phone solicitations, online advertising and in-store signage. Advertising is ubiquitous. The advertiser's job is to get you to buy something or to vote for someone. Getting you to reason logically isn't part of the job description. If they can sell you something without having you think logically, or by making you think illogically, or not think at all, that's all in a day's work.

Philosophers often analyze advertising in terms of logical fallacies. You know a number of those already, *ad hominem*, false alternative, *post hoc ergo propter hoc*, appeal to ignorance, hasty generalization. In this lecture I'll take a different tack. When it comes to advertising, it is often more effective to think in terms of psychological categories. Advertisers use every psychological strategy available. Outwitting the advertiser demands a philosophically critical attitude attuned both to logic and to a range of psychological techniques. I'm going to divide those psychological techniques into two rough categories. The first is a category that relies, not on information, not on content, but merely on pushing your buttons. The second category does rely on content and does use information, but that information is used in misleadingly persuasive ways.

Let me start with a particular kind of button pushing that I'll call attractiveness attracts. It has been estimated that 25 percent of men begin balding by age 30, two thirds by age 60. All in all, 35 million American men are losing their hair. That's about half. Outside of commercials for baldness remedies, how many bald men do you see as spokesmen on television commercials? The Gallup-Healthways Well-Being Index indicates that 63 percent of Americans are either overweight or obese. Outside of weight-loss commercials, how many overweight and obese people do you see as spokespeople on television commercials? That's right. The people in television commercials are a strange breed, a different species. They're far more attractive than the average person-next-door. Why do ads use good-looking people? Because attractiveness attracts. Regardless of the quality of what is being sold, regardless of even what product is being sold, you can bet that an attractive person will be used to sell it. It is your non-rational psychology, rather than your logical rationality, that is being appealed to.

Attractiveness is ubiquitous in advertising, and I don't think we can expect that to change. I am not arguing that advertisers should be forced to pick spokespeople at random—overweight, bald and all, of course not. The point is just that to process advertising rationally we have to be aware of, and have to compensate for, the fact that attractiveness attracts. In judging whether to buy what a commercial wants us to buy or to vote how a commercial wants us to vote, we need to filter out that aspect of persuasion that relies merely on an irrelevant attractiveness factor. But of course it's not just the people in advertisements that are at their attractive best. The settings are also pristine. Beer is shown on gorgeous beaches that somehow have no crowds. People in ads speak to you from kitchens that are as clean and spacious as stage sets. That's because they are stage sets. So commercials feature unrealistically attractive people in unrealistically attractive locations. We shouldn't be surprised that they picture the product in the most attractive way as well.

Car ads feature single cars gliding through isolated and beautiful surroundings. They're photographed from helicopters—without the sound of helicopters—as they smoothly curve mountain roads. Or as they execute a choreographed pirouette on what looks like an enormous air strip. Never another car in sight. Who wants to buy a car that gets stuck in traffic jams? Guess what, all cars get stuck in traffic jams. I've seen wonderful displays in which people put images from advertisements for fast food next to the most attractive photographs they were able to make of the same item as actually purchased. Advertising images of tidy well-stuffed tacos next to the peculiarly flat and greasy or partially filled shells you have just paid for. Advertising images of tall super burgers with crisp lettuce, red tomatoes and attractively melted cheese next to the real sandwich, a little squished and lopsided, the one that actually comes over the counter. Those tomatoes just don't seem to look the same. I invite you to duplicate the experiment yourself next time you go to a fast-food restaurant. Look at the image pictured on the overhead menu. Hold up what you purchase and see how representative that image is of what you just bought.

In 1970 the Federal Trade Commission settled a complaint against Campbell Soup who apparently put marbles in the bottom of the bowl to push the vegetables to the top. Each year fast-food and casual dining chains spend more than four billion dollars on television advertising, and it's on the rise. It's not surprising that there is a small industry of specialists working to make food look good in the ads. But what you see may not always be what you think you see. Ice cream melts under the photographic lights, so lard has been used instead. The ice in cold drinks may actually be acrylic. Glue has been used to keep spaghetti on the forks, off-camera hypodermics to make the sauce drip, machines under the table to give that appealing waft of steam. All because attractiveness attracts. So here's a thought experiment for you: What would an ad look like that didn't rely on an attractiveness factor? A car ad would show a photograph of the car in a normal environment, under normal lighting, with perhaps facts on gas mileage and repair record. A food ad would show a photograph of the hamburger precisely as delivered. Neither of those sounds very attractive. But that's just the point: what you're imagining right now is, in fact, precisely what you are buying. All of the attractiveness you see in commercials above and beyond that isn't appealing to your logic or your rationality. It's just manipulating you psychologically.

What I have been emphasizing is the ubiquitous use of attractiveness in advertising. But let me add an exception-that-proves-the-rule point as well. Precisely because attractiveness attracts is the standard strategy, some advertising campaigns catch your attention by deliberately flaunting that strategy. Frank Perdue was a bald and scrawny man who reminded people of the chickens he was selling. His very unattractiveness was used to convey sincerity and quality concern, and it worked. Dave Thomas, the man who founded Wendy's, was bashful, soft-spoken, and on the plump side. The fact that he broke the stereotype for attractive spokespeople lent an air of homey honesty to Wendy's commercials. That, I guess, is reverse psychology in terms of attractiveness in advertising. But it is still mere image. The hamburger pictured in the ad may still have only a remote resemblance to what they hand you over the counter.

Another form of button-pushing is appeal to prestige. European associations tend to carry that prestige—Perrier water and Grey Poupon mustard from France, for example, or Häagen-Dazs ice cream from—where exactly is Häagen-Dazs from? Denmark, maybe? Someplace in Scandinavia? Actually, Häagen-Dazs was founded in the Bronx in 1960 by Polish immigrants Reuben and Rose Mattus. The name Häagen-Dazs was made up to look European to American eyes. Neither a double a with an initial umlaut nor the combination "zs" at the end are part of any Scandinavian language. Dutch has double a, but no umlaut. I'm told that the only European language with a "zs" is Hungarian. This is called foreign branding in the marketing industry. Häagen-Dazs is an entirely fictional piece of foreign branding. Indeed, early packaging also included an outline map of Denmark on the label, despite lack of any real association with Denmark. A button-pushing fallacious appeal to prestige.

Fake foreign branding actually ended up in a legal case. There was another foreign-sounding ice cream called Frusen Glädjé. If you leave the accent off that final e, that name actually means something. Frusen Glädjé means "frozen delight" in Swedish, but despite the fact that the package carried three lines of Swedish, Frusen Glädjé was never sold in Sweden. It was produced, packed by an American company, and distributed entirely within the United States. In 1980, the non-Danish Häagen-Dazs sued the non-Swedish Frusen Glädjé for stealing their trade dress—the general look of a product, which is protected under American law. They lost. The court said there was little chance of confusion between the two. Häagen-Dazs had also claimed that Frusen Glädjé was illegitimately labeling their product as if it were Swedish. The court noted that Häagen-Dazs had equally dirty hands.

Attractiveness attracts works, not only for products, but for politics. In filming a political ad, you want your candidate to look presidential. You want your candidate seen with dramatic lighting, a prestigious podium in front, and an American flag in the background. But, of course, your candidate is running against someone else. The flip side of attractiveness attracts is that unattractiveness detracts. Political advertisements—television, internet, newspaper and mass-distribution flyers—standardly seek out the least attractive image of their opponent that they can find. If you can find an image of your opponent looking annoyed, or bewildered, or shaking hands with Hitler, you'll use it. You notice that I just switched to speaking of you as the advertiser. And that's not a bad strategy for outwitting the advertiser. You can imagine what you would do if you were just out to push buttons regardless of content. Because you can imagine what you would do on the sending end, you'll know it when you see it on the receiving end.

Appeal to emotion is a classic fallacy. One of the places it shows up is in political advertising. There the emotions most often exploited are political pride and fear. If you're filming a political commercial, it would be great to have an American flag rippling in the breeze. If you pan across amber waves of grain with "America the Beautiful" rising on the sound track, all the better. You note that I haven't said anything about who your candidate is or what your candidate stands for. I've just told you to use emotionally charged patriotic images. Next time you see those, whoever the candidate and whatever the message, remember that what you're seeing is a fallacious appeal to emotion.

The other emotion exploited in political advertising is fear. A classic example is the television commercial used by Lyndon Johnson against Barry Goldwater in 1964.

[Video start.]

Professor Grim voiceover: It opens with a little girl happily counting petals on a daisy.

Girl: One, two, three, four, five, seven, six, six, eight, nine ...

Professor Grim voiceover: The count then starts in the other direction by a male voiceover.

Male voiceover: Ten, nine, eight, seven, six, five, four, three, two one ...

Professor Grim voiceover: ... culminating in the flash of a nuclear explosion. Johnson then comes on in voiceover.

Lyndon Johnson voiceover: These are the stakes: To make a world in which all of God's children can live, or to go into the dark. We must either love each other, or we must die.

Male voiceover: Vote for President Johnson on November 3rd. The stakes are too high for you to stay home.

[Video end.]

Professor Grim: The implication is that Johnson's opponent—Barry Goldwater of Arizona—is reckless man ready to pull the nuclear trigger. We've talked about the fallacy of false alternative, and this is another example. But it's also clearly an appeal to emotion, and that emotion is fear.

Fear of what will happen if you elect the other guy has probably been used in every political campaign since George Washington. It's a good bet it will be used in the next one. The fear might be of a terrorist attack, of over-competition by China, of what will happen if we don't have mandated health insurance, of what will happen if we do, of future generations burdened by our debt, or a future without social security. All of those represent issues that should be rationally considered in a political decision, that should be addressed with relevant evidence and careful argument in political debate. The problem is that they almost never are. They're used as hair triggers for fear instead.

One way of sensitizing oneself to fallacious advertising is to learn the standard categories. Appeal to emotion is one of those, as is appeal to prestige. But you don't have to know the names. Here is a simple philosophical tool for detecting the kinds of fallacies we've been talking about. Take any advertisement. Read it, or listen to it, or watch it with care. And then sit down with a piece of paper and write at the top, the reasons given for buying the product were. List them. One, two, three. Also write, the reasons given for buying this product rather than a competitor were. List those one by one. A rational decision is one based on reasons. Rational means reasonable, and reasonable demands reasons. If you're going to do one thing rather than another, or buy one brand rather than another, or vote for one candidate rather than another, any rational decision is going to be based on reasons. So sit down with any advertisement and try to cash it out in terms of reasons. The reasons given for buying the product were... If the advertisement is relying merely on an attractive image, it may turn out you can find no reasons to list at all. Sometimes you can list reasons, but once examined it becomes clear that they aren't very good reasons. Like, it sounds Swedish. Once you try to cash out standard commercials in terms of reasons, it becomes pretty clear how far from rational their appeal often is.

Here's another fallacy common in advertising, the appeal to dubious authority. An instantly recognizable celebrity touts the virtues of the product, recommends it, or says that they use it. What makes it an appeal to dubious authority is that you're given no reason to believe that the celebrity knows what they're talking about. Here are some examples. Tiger Woods did a series of commercials touting Buicks. Tiger's a great golfer. He's had some serious marital problems, but we've got no reason to believe he knows much about cars. Johnny Depp does Mont Blanc watches. A good-looking guy. Great fun as an actor. But I'm not sure he's an authority on watches.

Most of the advertisements we've talked about so far rely on psychological manipulation rather than on claims or statements of fact. Some commercials do make claims, however, either explicitly or by implication. Aren't there laws against false advertising? Indeed there are. The core piece of Federal legislation is the Lanham Act, which prohibits "any commercial advertising or promotion [that] misrepresents the nature, characteristics, qualities, or geographic origin of goods, services or commercial activities." The Federal Trade Commission is empowered to enforce regulations against "unfair and deceptive practices in commerce." The FTC has acted against false and misleading advertising, against the marbles in the bottom of the bowl, for example, and against an ad campaign by the California Milk Producers Advisory Board that touted "Every body needs milk." Of course that's not true. Milk is the last thing that the lactose intolerant need. The FTC acted against Amoco oil for claiming that Crystal Clear Amoco Ultimate was better for your car and better for the environment because it was clear. Amoco had no evidence to back up those claims, but the result of such action rarely goes beyond an order to pull the ads or to demand corrective advertising. No advertising executive is doing hard time for false advertising.

The law also allows consumers and rival companies to pursue civil action. The suit must prove a false statement of fact that could be expected to deceive, must prove that it was likely to affect purchasing decisions, and that the plaintiff suffered injury of some kind. Short of a class action case, that kind of litigation is a tall and expensive order for any consumer. The way most false advertising appears in court is in civil suits by one company against another. Like Häagen-Dazs vs. Frusen Glädjé, for example. And of course the legal restrictions we're talking about extend only to commercial advertising. Political advertising isn't included.

You don't need to say something false in order to get someone to believe something false. One way of misleading someone is to tell them the truth but to be careful not to tell them the whole truth. Consider just a few of the phrases that tend to show up in advertising: "None proven more effective." That claim is true even if all competing brands have been shown to be equally effective. Indeed "none proven more effective" is true if none were proven effective at all.

"Faster acting (or more effective, or longer lasting) than the leading brand." It's not always clear what the leading brand is. It might be true that my product is faster acting than the leading brand, and still true that it's slower acting than any of the other contenders. "Recommended by doctors." It doesn't say recommended by all doctors, or even recommended by most doctors. It just says recommended by doctors. It would be hard to find a pharmaceutical that wasn't recommended by some doctors. It may be literally true, then, that a product is recommended by doctors. But any implication of general medical consensus is a false implication. You get the idea. If the advertiser can speak a limited truth and get you to make false inferences, he's within the letter of the law and yet still doing a great job of deception. Everything he says may be literally true. What you are led to believe may be false nonetheless. More recently, "recommended by doctors" has given way to a similar implication carried by something that isn't even a statement: "Ask your doctor whether Bonitranq might be right for you"

Here's a final category of advertising tricks, the magic word. There are a whole set of magic words that reappear in advertising. Some may not mean what you think they mean. Some may not mean much of anything at all.

Consider, for example, foods labeled low fat, reduced fat, light, 98 percent fat free or simply fat free, low calorie, reduced calorie, lower sodium, lite. All of that sounds great, but what do those words mean? When you check the Food and Drug Administration guidelines, some of those terms carry absolute standards. A main dish counts as low fat if it contains less than three percent fat and not more than 30 percent of its calories come from fat. Some of those terms are relative terms instead. A potato chip can boast of reduced sodium if it has at least 25 percent less sodium than your standard run-of-the-mill potato chip. The standard is higher for light. You need 50 percent fewer calories from fat than the standard can of beer to qualify as a light beer.

A big proviso for standards like that is that they often have no legal teeth; they are merely

non-binding industry recommendations. There is also another problem here. If a product is labeled low fat, that means it contains a certain percentage of fat. It doesn't mean it won't make you fat. A product can be labeled low fat despite the fact that it is packed with calories, sugar, carbohydrates, and salt. Indeed Twizzlers candy boasts of being a low-fat food. They don't mean it will keep you low-fat. So here again we have selective mention. It may be true that a product has low fat, but that should make you wonder about the other aspects that might be important, calories, carbohydrates, and salt, for example. The label that says low salt may speak the truth, but what it labels may still be unhealthy in other ways. There is no complete list of magic words to watch out for. Advertisers can be counted on to borrow or invent magic words faster than they can be specified in law or in non-binding industry recommendations. It took the USDA 10 years to settle on a definition for organic.

Many of the magic words sound good. But when you stop to ask yourself what they mean, it's not clear that they mean much of anything at all: new, improved, new and improved, original, vitamin enriched, with six anti oxidants, healthy, economy sized, heart healthy, nutritious, better tasting, longer lasting, home style, all natural, as seen on TV.

We certainly haven't exhausted the topic of fallacious advertising. That topic is big enough for a course of its own. What I have tried to do is to give you a feel for two major categories of fallacious advertising. One thing to look out for is everything that isn't information—everything that is just out to push your buttons. The other thing to look out for is the misleading use of any information that does appear, to think about what that information really means, and what information you are not being given.

Now it's your turn. We're going to end by giving you two clips for analysis. These aren't real ads, but they'll work as samples, and you can go on to dissect the real ads on your own. In each case, listen to the ad, pause the tape, and think about it for a minute. I want you to name what is wrong or suspicious about each ad. You could do this in terms of categories—standard categories like appeal to authority, for example. But it will do just as well to ask just three questions. What psychological buttons are being pushed? What is the precise claim that is being made? What relevant information may have been left out? If you can do that in each case, you are well on your way to outwitting the advertiser.

[Video start.]

Male voiceover: You are alone upstairs in the dark. You hear a window break downstairs, and the phone being ripped from the wall. Someone is starting up the stairs. They are coming closer to your bedroom door. Closer … closer. You can see the doorknob turn. Don't you wish you had a gun?

Gun control? Tell them "no" on Proposition 12.

[Video end.]

Professor Grim: What is happening in that ad? If you said they were pushing emotional buttons, you're right. That's a classic appeal to fear. Gun control is a complicated topic with both facts and important principles at issue. Principles about the right to self protection, for example, but also statistics on the likelihood of shooting friends or family rather than an intruder. It's a tough issue worthy of careful and detailed rational discussion, deliberation, and debate. Did the ad address any of that? No. It merely stimulated fear. It dealt with a complex and controversial social issue merely by manipulating your emotions.

Okay. We have time for one more. Let's see if you can name the main fallacy committed in this commercial, and at least two other points. Let's see if you can outwit the advertiser.

[Video start.]

Lab Technician: Hard day at the lab? Do what I do. Energize with Ergo-Sip. Physicians will tell you that fatigue is a feeling of weariness, tiredness, or lack of energy. Physicists will tell you that energy is the ability to do work.

Energy … that's what energy drinks are made for. But did you know that most other energy drinks are mostly water? That's right—water!

Ergo-Sip. Made with the finest ingredients. And the taste? Well, there's nothing quite like it.

Had a hard day at the lab? Do what I do. Reach for a tall cold can of Ergo-Sip.

[Video end.]

Professor Grim: Okay, what about that one? Some of the finer points: "Most other energy drinks are mostly water." That's true enough. The major ingredient in all energy drinks is water. The implication is that Ergo-Sip is somehow different, but of course the ad never said that. "Made of the finest ingredients." What does that mean? "And the taste…well, there's nothing quite like it." That's true for just about everything. No real information there. The major fallacy here, however, is appeal to dubious authority. It actually appears in two different ways. There's an appeal to the authority of both physicists and physicians. "Physicians will tell you that fatigue is a feeling of weariness, tiredness, or lack of energy. Physicists will tell you that energy is the ability to do work." Those are really just definitions in the fields, of course. They aren't talking about Ergo-Sip. Appeal to dubious authority is built into the whole setup, test tubes, lab coat, blackboard full of equations.

Who is this person? What lab? This has appeal to dubious authority written all over it.

In order to outwit the advertiser, you have to see through attractiveness attracts, appeal to prestige, appeal to emotion, appeal to dubious authority, misleading half-truths and magic words. Learn to see through the spin. Next time we'll talk about spin again, not in advertising, but in putting a spin on statistics.

Putting a Spin on Statistics
Lecture 17

People love statistics, and well they should. Statistics give us the straight facts, in numbers and graphs. We want our information manageable and easy to understand. But while statistics can give us the facts we need in ways that are easy to understand, they can also give us the facts in ways that are prone to mislead and can be manipulated. We should all be a bit more critical of statistics on all sides of an issue, our own included.

Statistics versus Facts

- The appeal of statistics is they are hard-edged and factual. Information conveyed numerically is, however, just like any other information: It may or may not be true. The unfortunate psychological fact is that people seem to be more inclined to believe a claim simply because it comes with precise numbers. That's why numerical information spreads fast. But often the numbers get even more impressive—and even farther from the truth—as the information spreads.

- Consider this statistic: Every year since 1950, twice as many American children have been killed with guns. Joel Best is a sociologist at the University of Delaware and the author of a book on misleading statistics. He nominates this as the most inaccurate social statistic ever.

- Suppose it were true that child deaths due to guns had doubled each year. Start with just 1 child killed with a gun in 1950. If the number doubles each year, that means there would have been 2 children killed in 1951, 4 in 1952, 8 in 1953. By 1960, the number would have been 1,024. It would have passed 1 million in the year 1970 alone. By 1980, there would have been 1 billion children killed with guns—more children killed than there were people living in the United States in 1980.

- Best first encountered this claim in a dissertation prospectus. But the dissertation cited a source, and he found the statistic in the source, too. It came from a 1994 report by the Children's Defense Fund. But what the source actually said was: "Since 1950, the number of American children gunned down every year has doubled."

- Consider those two sentences closely. Do you see the difference? What the original report actually said was that the number of gun-related child deaths doubled between 1950 and 1994.

- Between 1950 and 1994, the U.S. population almost doubled, so the change in gun-related child deaths per capita wasn't particularly dramatic. Just by a different placement of the phrase "every year," that statistic mutated into the outrageous claim that the number of deaths doubled every year.

© iStockphoto/Thinkstock.

Evaluating Statistics Critically:
Sampling Size, Survey Bias

- Statistics represent some kind of counting, and counting is about as objective as things get. But in evaluating statistics, you have to pay attention to who did the counting and how. It's also important to pay attention to the sample and where it came from. The perennial problem is that the basic sample may turn out to be unrepresentative. The bigger a percentage in the sample, the more confident you will be that the sampling reflects the true reality.

Because statistics are explicitly factual, they can be checked, but checking may be difficult, and a bogus number may spread and mutate faster than it's possible to check.

- The classic case here was Al Gore versus George W. Bush in the election of 2000. Around 8 pm on election night, most stations had called Florida for Gore. Given that only a small portion of the voting precincts had officially reported by then, how was that early projection made? The answer is exit polling. In exit polling, pollsters pick a certain percentage of precincts to sample. Out of tens of thousands of voting precincts in the United States, at best, a few thousand are sampled, with perhaps 100 voters per precinct.

- We can see how that could go wrong. By 9:30 pm, with more precincts in, using their own exit polling, the Bush team challenged the network results. By 9:55, all the networks had changed their verdict from "Gore wins" to "too close to call."

- A more important problem, however, was that the sampling space for the exit surveys might have been unintentionally biased. Pollsters didn't sample all precincts in Florida, and the choice of which precincts to sample may have slanted the results. Asking for volunteers may itself have slanted the results.

- Questionnaires often come with their own biases. A major issue is the leading question. Consider the way these questions are phrased: "Do you agree with most Americans that it's time we got a handle on decay in our inner cities?" "Do you think it's important to back our troops by giving them the support they need?"

- Think what it would take to answer those questions in the negative. If the statistics we're dealing with are taken from that kind of questionnaire, the data are biased simply by the way the question was asked. And yet, when results are reported, you may have no way of knowing that the original questions were loaded.

Test Your Skills
- In a graph of the number of drivers in fatal crashes by age group for a typical year, we see there were 6,000 fatal crashes for 16- to 19-year-olds. For 20- to 24-year-olds, the number jumps to almost 9,000. From there, it tapers down by five-year intervals. Those over 79 counted for fewer than 1,000 fatal crashes. These data suggest that people between 20 and 24 are a major hazard on the road.

- The data in the chart may be entirely accurate, but the conclusions we're tempted to draw from them are not. There aren't very many people over age 79 who drive, and thus, there are fewer drivers of that age to be involved in fatal crashes. The same thing is true at the bottom of the scale. Because not all 16- to 19-year-olds drive, there are fewer of them on the road. The sampling, then, is uneven.

- A better graph is the number of fatal crashes per miles driven by people in different age categories. Those numbers give us a very different graph. It becomes clear that people on the extremes of the age spectrum aren't the safest.

Mean, Median, and Mode

- In order to condense data into a single number, some information is inevitably left out. But what information to emphasize and what information to leave out is always a choice. The difference between mean and median is a good example.

- The mean of a series of numbers is what most people think of as the average. Statistics are often reported in terms of the mean. But that can be misleading.
  - Consider the salaries of 1985 graduates of the University of North Carolina immediately upon graduation. Which graduates could boast the highest salaries, those in computer science or business?

  - Actually, the graduates with the highest mean salary were those in cultural geography. The fact is that there were very few people who graduated with a degree in cultural geography, but one of them was the basketball star Michael Jordan.

  - A major problem with using average or mean as a measure is that it is vulnerable to the problem of **outliers**. One enormous salary in a group can lift the mean in misleading ways.

  - Here's a simple example: Imagine that there are nine people in a company. Four of them earn $20,000 a year; one makes

$30,000; and three earn $40,000. The boss gives himself a salary of $200,000. In this sample, the **mode** is $20,000, the mean is almost $48,000, and the median is $30,000.

- Means, modes, and medians may give you different numbers, and people on different sides of an issue, pushing a different line, will make different choices about what numbers to use.

Statistical Distribution
- The truth is that anything that compresses a complex social reality into a single number loses information. If you want more information, you want to see how the numbers distribute across the sample. That demands something more like a graph. But here again, there are possibilities for misrepresentation.

- Consider two graphs that claim to show the changes in violent crime between 1993 and 2010. One shows a slight decrease over time, and the other shows a dramatic decline. Which one is right? The answer is that they're both right—they portray precisely the same data—but the difference is in how they portray it. The choice of scale on a graph can exaggerate or minimize an effect.

Correlation
- Sometimes statistics concern not just one thing but a relationship or **correlation** between two. Consider a graph with two lines, one representing the number of homicides per 100,000 people between the years 1960 and 1970 and one representing the number of people executed for murder in that period. The graph shows that as the number of executions between 1960 and 1970 steadily falls, the number of homicides steadily rises.

- The two lines go in opposite directions. It is precisely as the execution rate falls that the homicide rate increases. In one of the earliest sophisticated statistical studies of deterrence, Isaac Ehrlich used data on executions and the homicide rate to conclude "an additional execution per year…may have resulted in seven or eight fewer murders."

- But this is where an important distinction comes in. What the graph shows is correlation. Ehrlich's conclusion, on the other hand, is phrased in terms of **causation**: Murder rates go up because execution rates go down.

- Correlation and causation are not the same. At best, such statistics yield a correlation. That is important in judging causality, but correlation alone never provides causal information.

Terms to Know

causation: The act of producing an effect. In a cause-and-effect relationship, the presence of one variable, the effect, can be established as a direct result of the other variable, the cause.

correlation: A direct or inverse relationship between two variables. In a direct correlation, the strength or frequency of both variables increases proportionately to each other; in an inverse correlation, the strength or frequency of one variable increases as the strength or frequency of the other decreases. However, correlation does not establish that one variable causes the other; one variable is not necessarily the cause and the other the effect.

mode: The most frequently occurring value in a set of numbers.

outlier: An extreme value; something that is out of the ordinary. In statistics, a number that is extremely divergent from all the others in a set of numbers; outliers have misleading impact on the mean, or average, of a set of numbers.

Suggested Reading

Best, *Damned Lies and Statistics*.

———, *More Damned Lies and Statistics*.

Huff, *How to Lie with Statistics*.

1. From your own experience, is there a case where your views were particularly shaped by statistics? What were the crucial statistics, and how did they influence you?

2. What precisely is the difference between these two sentences: (1) Every year since 1950, the number of American children gunned down has doubled. (2) Since 1950, the number of American children gunned down every year has doubled.

3. Why do you think it is that an appeal to statistics seems so persuasive? What aspect of the power of statistics is deserved? What aspect of their power is not?

Eleven children take a quiz, with scores between 1 and 10. Construct a series of scores so that each of these is importantly different:

    (a) The mean score (the average)

    (b) The median score (there are as many scores above as below)

    (c) The mode (the score that most children got)

(See "Answers" section at the end of this guidebook.)

"If you take the drug Doomleva, your chances of catching Doomlitis are cut in half. A very small portion of the population has been shown to carry a genetic allergy to Doomleva, which can cause sudden death." What further statistics do you need in order to know whether you should take Doomleva?

Putting a Spin on Statistics
Lecture 17—Transcript

Professor Grim: Statistics are an important form of information. That's why proponents on any side of an issue will try to use statistics, even abuse statistics, in order to win arguments. But here is an informal statistic on statistics, most people are far more critical of statistics offered for a view they don't support than they are of statistics offered for a view they agree with. I'd like that to change a bit. In this lecture I'll be using lots of examples. They will come from many sides of many issues. What I'd like is for us all to be a bit more critical of statistics on all sides of an issue, our own included.

Let me give you a couple of statistics. I don't want you to think about them in terms of whether they support your view or not. I want you to think about them in more critical terms than that. I want you to think about them critically as statistics. Statistic number one can be easily put in a single sentence: Every year since 1950 twice as many American children have been killed with guns. Statistic number two is trickier. This one is in the form of a graph for the years between 1960 and 1970. One line on the graph represents the murder rate between those years, the number of homicides per 100,000 population. The other line represents the number of people executed for murder each year. What the graph shows is that as the number of executions between 1960 and 1970 steadily falls, the number of homicides steadily rises.

People love statistics, and well they should. Statistics give us the straight facts. They give us those facts in numbers and graphs that are easy to understand. I want to start by making it clear that I think statistics are great. We want the straight facts, inflexible and uncompromising, and statistics can give us those. We want our information manageable and easy to understand. Statistics can give us that too. The problem, of course, is that statistics give us the straight facts, except when they don't. Statistics can give us the facts we need in ways that are easy to understand, but they can also give us the facts in ways that are prone to mislead and be manipulated.

Let's start with statistics and the facts. The appeal of statistics is that they are so hard-edged and factual. It's great to be able to settle an argument with simple numbers. The influence of environment on language and thought? Eskimos have 200 different words for snow. You think handguns aren't a problem? Then how about that statistic from a minute ago? Every year since 1950 twice as many American children have been killed with guns. Unfortunately, those so-called statistics are factoids, rather than facts. They are truthy rather than true. It's not true that the Eskimo have 200 different words for snow, or at least not simply true. The other factoid has taken on a life of its own, but it's not true that every year since 1950 twice as many children have been killed with guns.

Information conveyed numerically is just like any other information. It may or may not be true.

The unfortunate psychological fact is that people seem to be more inclined to believe a claim simply because it comes with precise numbers. Consider Senator Joe McCarthy and the Communist witch hunts of the 1950s. It's because of Senator Joe that we have the term McCarthyism. He loved precise numbers. He loved the hard-edged power they conveyed. In one address he claimed to have in his hand a list of 205 names, names of people working in the State Department known to be members of the Communist party. At another point the list was said to have 207 names; later it was 57 and at one point 81. He gave precise numbers every time, numbers that seemed to convey credibility just because they were numbers. But the numbers on the list kept changing, and, in fact, there were no such numbers, because McCarthy had no such list. The lure of precision, that's why numerical information spreads fast, far, and wide. But the lure of numbers may help it spread whether it is true or not. Often the numbers get even more impressive—and even farther from the truth—as the information spreads.

That is what happened with the Eskimo words for snow. In 1911, the anthropologist Franz Boas reported that the Inuit use four apparently distinct roots for snow on the ground, falling snow, drifting snow, and a snow drift. That comment was picked up and used by Edward Sapir and Benjamin Lee Whorf, who wanted to argue that the character of our language determines the character of our thought. In Whorf's article the number had increased to seven. With repetition, it continued to grow. A 1984 editorial in the New York Times put it at one hundred. I have often heard it said that the

Eskimo have 200 words for snow. When you track it down, the facts are very different. They're actually much more interesting. The Inuit languages are polysynthetic, which means that they form compound nouns in just about any way you want. As the German author Kathrin Passig says in one of her short stories, "that means that even seldom-used expressions like snow that falls on a red T-shirt are combined into one word." Even in English we have lots of different roots related to snow: powder, slush, snowflake, blizzard.. Somehow Franz Boas's comment about four snow-related roots in Inuit languages mutated into that widely-circulated, but misguided, claim about 200 distinct words for snow.

Something similar happened with the claim that twice as many American children have been killed with guns every year since 1950. Joel Best is a sociologist at the University of Delaware and the author of a great book on misleading statistics. He nominates this as the most inaccurate social statistic ever. Suppose it were true that child deaths due to guns had doubled each year. Start with just one child killed with a gun in 1950. If the number doubles each year, that means there would have been two kids killed in 1951, four in 1952, eight in 1953. Doubling each year. By 1960, the number would have been 1,024. It would have passed one million dead kids in the year 1970 alone. By 1980, there would have been a billion kids killed with guns. That would mean that more kids were killed in the United States in 1980 than there were people living in the United States in 1980. Joel Best first encountered the claim in a dissertation prospectus. But the dissertation cited a source, and he found it in the source, too. There it was, in academic black and white: "Every year since 1950, the number of American children gunned down has doubled." And where did that come from? It came from a 1994 report by the Children's Defense Fund, but what they actually said was, "Since 1950, the number of American children gunned down every year has doubled." Consider those two sentences closely. Do you see the difference? What the original report actually said was that the number of gun-related child deaths doubled between 1950 and 1994. "Since 1950, the number of American children gunned down every year has doubled." The U.S. population grew by 72 percent between 1950 and 1994. It almost doubled, so the change in gun-related child deaths per capita wasn't particularly dramatic. What that statistic mutated into—just by a different placement of the phrase "every year"—was the outrageous claim that the number of deaths doubled every year. What a difference the placement of that phrase makes. One of the great things about statistics is that they are explicitly factual. That means they can be checked. But checking may be hard, and a bogus number may spread and mutate faster than it's possible to actually check it. Moreover, a bogus number is often dramatic—front page news. The retraction doesn't make the front page. Mark Twain said that "There are three kinds of lies: lies, damned lies, and statistics." He attributes the phrase to the British Prime Minister Benjamin Disraeli. It's interesting that when you track that down, the phrase doesn't actually appear in Disraeli.

Statistics represent some kind of counting, by someone, somewhere, and counting is about as objective as things get. But in evaluating statistics you have to pay attention to who did the counting and how. The foundation of all statistics is the sample—the group you count from. In evaluating statistics it's always important to pay attention to that sample and where it came from. The perennial problem is that the basic sample may turn out to be unrepresentative. I'm sure you've watched the network projections of the winner in a presidential election, moving east to west and state to state. I'm sure you've also noted that the networks try to scoop each other by getting their results out first. The classic case was Al Gore vs. George W. Bush in 2000. At 7:48 on election night, first the Associated Press, then CNN, and then all the networks called Florida for Gore. Many commentators started to speak of Gore as the next President. How was that early projection made? Only a small portion of the voting precincts had officially reported. The results at 7:48 were based predominantly on exit polling. How is that done? They pick a certain percentage of precincts to sample, and they ask people leaving the polls whether they would fill out a 30- to 40-item questionnaire. You don't do it for all precincts, and you don't do it for all voters. There are tens of thousands of voting precincts in the United States. At best a few thousand are sampled, with perhaps 100 voters per precinct. You can see how that could go wrong, and it did. By 9:30 on election night, with more precincts in, using their own exit polling, the Bush team challenged the network results. By 9:55 all the networks had changed their verdict, from Gore wins to too close to call.

Here one issue is sampling size. You want to know something about all voters. The bigger a percentage you actually get results from, the more confident you will be that the sampling reflects the true reality. Here the sampling space was small. A more important problem, however, was that the sampling space for the exit surveys might have been unintentionally biased. They didn't sample all precincts in Florida, and Florida is known for being a very heterogeneous state. The choice of which precincts to sample may, therefore, have slanted the results. Asking for volunteers may itself have slanted the results. What if the Florida voters who had more time to spend answering 40 questions were also those more likely to have voted for Gore? There may also be a feedback loop. What if people thought one candidate was more likely to win because of what they had heard over the media, perhaps, and were more inclined to fill out a volunteer survey if they had just voted for that candidate?

I don't think it applies in this case, but questionnaires often come with their own biases. A major issue is the leading question. If you're like me, you've gotten unsolicited telephone calls that say you've been chosen to participate in a survey. But the questions in the survey may be things like this: Do you agree with most Americans that it's time we got a handle on decay in our inner cities? Do you think it's important to back our troops by giving them the support they need? Think what it would take to answer those questions in the negative. No, I don't agree with most Americans. I think decay in the inner cities is just fine as it is. No, I don't think we should back our troops, even if they need it. Who would really say that? If the statistics we're dealing with are taken from that kind of questionnaire, the data is biased simply by the way the question was asked, and yet when results are reported, you may have no way of knowing that the original questions were loaded.

Okay. It's your turn for some critical statistical analysis. Here, for a typical year, is a graph of the number of drivers in fatal crashes by age group.

For 16- to 19-year olds, there were 6,000 fatal crashes. For 20- to 24-year-olds, the number jumps to almost 9,000. From there it tapers down by five-year intervals. Those over 79 counted for less than 1,000 fatal crashes. That suggests that people between 20 to 24 are a major hazard on the road. I knew it. But it also suggests that people under 20 to 24 are safer drivers and that people become progressively safer from that age on. A glance at the chart will tell you that people over seventy-nine are the safest drivers of all. Here's the challenge for you: Is there something wrong with the chart? Is there something wrong with drawing that conclusion from it? Press pause if you want, think about it, and then come back.

The data in the chart may be entirely accurate, but the conclusions it's tempting to draw are not. What the chart shows is the absolute number of drivers in certain age categories involved in fatal crashes. But there aren't very many people over 79 who drive and, thus, fewer of those to be involved in fatal crashes. The same thing happens at the bottom end of the scale. Because not all 16- to 19-year-olds drive, there are fewer of them on the road than 20- to 24-year-olds. The sampling, then, is uneven. Because of that, the graph doesn't show you what you might think it does. In order to know what age categories are safest, we need the number of fatal crashes relative to the number of people in that category who are actually driving. Even better, we need a number of fatal crashes per miles driven by people in different age categories. When you take that into account, you get a very different graph.

It becomes clear that people on the extremes of the age spectrum aren't the safest. They're the ones with the most accidents per mile. They're the ones to look out for.

Suppose we have tracked down a statistic, established that it is genuinely factual, and have established that the basic samples are representative. If we have that, we have a pretty solid statistic. But there is still room for misrepresentation. Statistics can package important and complex information in a simple and understandable way. That's one of the great things about statistics. It's also one of the potential pitfalls. In order to condense something to a single number, you inevitably have to leave some information out. That's what makes it tidy. But what information to emphasize and what information to leave out is always a choice. The difference between means and medians is a good example. The mean of a series of numbers is what most people think of as the average. If you want the average SAT score for students from North Dakota, you add up all the scores and divide them by the number of North Dakota students who took the test, presto, the average, or more technically the mean.

Statistics are often reported in terms of the mean, but that can be misleading. Consider the salaries of 1985 graduates of the University of North Carolina immediately upon graduation by department. Which graduates could boast the highest salaries? Those in computer science, maybe? Business? Actually, the graduates with the highest mean salary were those in Cultural Geography. The figure in most departments was below 40,000 or 50,000, but the mean salary for graduates in Cultural Geography was in the hundreds of thousands. Cultural geography? How come those guys are making so much? The fact is that there were very few people who graduated in Cultural Geography from the University of North Carolina in 1985, but one of them was the basketball star Michael Jordan. He was making enough already, but 1985 was also the year that Nike introduced their Air Jordan line. A major problem with using average or mean as a measure is that it is vulnerable to the problem of outliers. One enormous salary in a bunch can lift the mean in extremely misleading ways.

There are really three concepts to consider. If we're talking salaries, the mean is the average salary. The mode is the typical salary—what most people got. The median is something like the middle salary—the salary such that about half got more than that, and half got less. Let's look at all of those in a simple example. There are just nine people in the company. Four of them work on the shop floor, each earning 20,000 a year. There is one shop foreman, who makes 30,000. There are two managers and a salesman, each of whom make 40,000 each. The boss gives himself a salary of 200,000. What is the average salary in that company? Add up four times 20,000, plus 30,000, plus three times 40,000, plus 200,000. Divide by 9. That's the mean salary. It comes out a little shy of 48,000. But that's not the typical salary. In this case the mode—the category with the most people in it—is just 20,000. The mean salary of 48,000 is more than anybody in the company gets except the boss, just like the Michael Jordan case. In this case the median is 30,000, the shop foreman's salary. There are four people getting less than that. There are four people getting more. That's the median. So the mode is 20,000, the mean is almost 48,000, the median is 30,000. Means, modes, and medians all give you simple numbers, but they may give you different numbers. Which you choose is always that, a choice. It's to be expected that people on different sides of an issue, pushing a different line, will make a different choice. In the debate over a federal tax cut, proponents of the bill claimed that the average family tax reduction would be more than $1,000. That makes it sound like a great deal for the taxpayer. Opponents claimed that more than half of all families wouldn't see a tax cut that hit even $100. That doesn't make it sound like such a great deal. In fact, both statistics were solid, factual, and well-sampled. It's just that one result was reported in terms of the mean while the other was reported in the spirit of the median.

The truth is that all of those measures lose information. Anything that compresses a complex social reality in to a single number is going to lose information. If you want more, you want to see how the numbers distribute across your sample. That demands something more like a graph, but here again there are possibilities for misrepresentation. There are two problems that happen with statistical graphs. Consider two graphs that claim to show the changes in violent crime between 1993 and 2010. Look at the difference in these graphs.

One shows a very slight decrease over time. One shows a dramatic decline. Which one is right? They're both right. They portray precisely the same data. The difference is just in how they portray it. The choice of scale on a graph can exaggerate or minimize an effect. If crimes fell from 120,000 to 100,000 and we put that on a graph that shows just the range from 100,000 to 120,000, the decline is going to look huge; it will fill the entire graph. If we put that same information on a graph that ranges from 0 to 1 million, the decline is going to look almost unnoticeably tiny. In trying to represent data with statistics, you have to choose which statistics to use. In trying to represent data with graphics, you have to choose which graphics to use. Today, anyone with a spreadsheet program can experiment with different graphics for their data. That makes it very easy for someone to shop around for that form of graph that seems to make their point most dramatically. That may not be such a great thing for the consumer of statistics, and we're all consumers.

Sometimes statistics concern not just one thing but a relationship or correlation between two. Consider the graph I described at the beginning of the lecture. For the period between 1960 and 1970, one line shows the number of people executed for murder in the United States each year. The other line shows the homicide rate.

The dramatic thing is that those lines go in directly the opposite directions. It is precisely as the execution rate falls that the homicide rate increases.

In one of the earliest sophisticated statistical studies of deterrence, Isaac Ehrlich used data on executions and the homicide rate from 1933 to 1969, including this data: Do homicide rates go up when execution rates go down? Ehrlich's answer was yes. He phrased his conclusion this way: "an additional execution per year...may have resulted in seven or eight fewer murders." That sounds pretty impressive. That statistic was quickly put to work by the Attorney General in a case before the Supreme Court. But this is where an important distinction comes in. What the graph shows is correlation. It's a fact. Between 1960 and 1970, murder rates go up and execution rates go down. Ehrlich's conclusion, on the other hand, is phrased in terms of causes. An additional execution per year may have resulted in seven or eight fewer murders. That claim is not just murder rates go up and execution rates go down. That claim is murder rates go up because execution rates go down. Correlation and cause are not the same. What you get from statistics like that is, at best, correlation. That is important in judging causality, but correlation alone never just gives you causal information. Remember *post hoc ergo propter hoc*? A and B might be correlated because A causes B, or because B causes A, or because some third element C causes both of them. Here's a correlation, for example. Lung cancer rates are correlated with the number of ashtrays in someone's house. But that's not because ashtrays cause cancer. It's because smoking causes both lung cancer and the accumulation of ash trays, and sometimes correlation is just coincidence. So what is it in this case?

Ehrlich's was a sophisticated statistical study and was subjected to sophisticated statistical critique. The central issue, however, has to do with information that doesn't appear on the graph. If you look at all crime between 1960 and 1970, you'll see that all crime rose on the same curve that murders did. Burglaries, robberies, car thefts all went up in the sixties and went up in the same way. But none of those other crimes carried the death penalty. So perhaps murders were on the rise for some other reason, or some host of reasons, the reasons that were driving all crime up. Perhaps other factors were in play. Perhaps the correlation between declining executions and increasing murders isn't enough to show a cause. Ehrlich's study spans the period between 1933 and 1970. It's on that data that he concludes that the death penalty does deter murder. But it's that ten-year period between 1960 and 1970 that turns out to be crucial. The same study done on just the period from 1933 to 1960 showed no deterrent effect at all. Don't get me wrong. I am arguing neither for nor against a deterrent effect. I think the question is complicated. What I am saying is that dramatic graphs that correlate a rise in murders with a decline in executions suggest a short and immediate proof of causality that doesn't really exist. Lots of factors may be at play, including factors that don't show up in the graph. A single correlation like this won't tell us what we really want to know about cause.

Sometimes we have statistical information on correlation but just don't know about cause. Here's a statistic for you. Among college graduates, it turns out that left handers earn more than right handers. Among the general population, there's no difference. That's strange. I wonder what it means. Here are some other intriguing numbers. Of seven recent U.S. presidents, four have been left-handed, while a fifth was said to have been ambidextrous. Gerald Ford was left-handed and Ronald Reagan was left-handed at birth. Reagan's VP, left-handed George H.W. Bush, later become President too. Left-handed Bill Clinton won against him in a three-way race that also included left-hander Ross Perot. More recently, it was the left-handed Barack Obama who defeated the left-handed John McCain. Those statistics offer a surprising correlation between left-handedness and success, or at least ambition. Is there some causal connection? The answer is we simply don't know. You can spin some plausible causal stories, but the correlations—even if real—don't tell you those causal stories are true.

Here, then, is the take-home message on statistics and spin. Statistics are great. They give us factual information in an easily understandable way, except when they don't. Beware of Joe McCarthy and the lure of precise numbers. Statistics are great because they can be checked, and they should be. Who did the counting, and how? The sample is the first place to look. Is that sample representative? If we're dealing with a questionnaire, was there bias built into the questions asked? Statistical representation always represents a choice. Why did they use mean rather than some other statistic? Why did they use that graph, scaled in that way? And finally, remember that correlation isn't cause.

In the next lecture we'll use the kinds of techniques we developed here to look at poker, probability, and everyday life.

Poker, Probability, and Everyday Life
Lecture 18

W e gamble all the time. Although we may not call it gambling—we reserve that term for high-risk games in casinos—we are always playing the odds. For example, we play the odds as individuals when we decide what job to take, or where to live, or whether to marry a particular person. As a culture, we gamble all the time, too. Science stands as one of the greatest intellectual achievements of all time. But outside of mathematics, none of our scientific conclusions is a certainty. Our confidence in those predictions is all based on a batch of evidence—we hope good enough—and a best guess on the basis of that evidence.

Being Rational about Probability

- As long as we live and breathe, we're forced to play the odds, to gamble. Our concern is whether or not we gamble rationally, and we often don't. **Probability** is another name for what we're talking about here. We often don't handle probabilities rationally. For example, which is more dangerous, a handgun or a swimming pool? Say that you have one neighbor with a swimming pool and another neighbor who keeps a gun in the house. Which is the safer place for your child to play?

- If we take the statistics seriously, swimming pools are far more dangerous. About 550 children drown in swimming pools each year— 1 per every 11,000 residential pools. About 175 children are killed by guns each year—1 for every 1 million guns in the United States. The probability of a child drowning in a swimming pool is more than three times greater than the probability of a child being killed with a gun.

- Why do we tend to think otherwise? Imagine kids playing in a swimming pool: That's a wonderfully positive image. The image of kids with guns, on the other hand, is frightening. When it comes to probabilities, we tend to think in terms of images, and it's those images that often lead us astray.

Law of Large Numbers

- Calculating probability works like this: Imagine you have a quarter in your hand. As we know, the chance that it will come up heads when you flip it is 50/50, or a probability of 1/2. The probability is the ratio of the number of possible outcomes in your chosen category (heads) over the total number of possible outcomes.

A grand-prize winner for the lottery might be a news item for 30 seconds; a news show of 1-second pictures of every person who bought a ticket but didn't win the grand prize would take more than six years to watch.

- The basic principle of probability, however, demands **equiprobable** outcomes. That works well for standard games of chance. Cards, dice, and roulette wheels are made for easily calculated equiprobable outcomes. But we also want to use probability in real life, and real life isn't so straightforward.

- Here's an illustration with coin flips: We know that the probability of a quarter coming up heads is 1/2. If you flip a coin 20 times, how many heads will you expect to get? If you said about 10, you're right. You shouldn't expect to get exactly 10 heads. The perfect division of half and half is unlikely with small numbers. But as the number of flips gets larger, the probability increases that the result will be close to a 50/50 split.

- This is known as the **law of large numbers**. In the long run, frequencies of heads and tails will tend to even out.

Linking Probability and Frequency

- The law of large numbers matters, especially in real life, because it links probability with frequency. If the probability of heads is 1/2, then in the long run, we can expect about half the flips to come up heads. In other words: If the probability is 1/2, the long-range frequency will approach 1/2.

- In real life, we estimate probabilities by reasoning in the other direction—from frequencies to estimated probabilities. For example, what's the probability that a hurricane will hit New Orleans this year? By one reckoning, New Orleans has been hit by hurricanes 61 times in the last 139 years. That's a frequency of 61/139, or about 43/100. Reading our probability off that frequency, we'd say the probability of a hurricane hitting New Orleans this year is about 40 percent.

Combined Probabilities

- After an airplane crash, the FAA wants to know not just the probability of hydraulic system failure in a plane, not just the probability of landing gear problems, not just the probability of pilot error, but the probability of any of those things. For that kind of case, we need to be able to calculate **combined probabilities**.

- Suppose you're playing stud poker. You have to stick with the five cards you're dealt. If we know the probability that you'll be dealt a straight of some kind, and we know the probability that you'll be dealt two pairs, we calculate the probability that you'll be dealt one or the other by simply adding those probabilities together. The probability of a straight is a little more than 0.39 percent. The probability of being dealt two pairs is significantly better—just shy of 5 percent. Thus, the probability of getting one or the other is just shy of 5.39 percent.

- There is an important proviso to the simple addition rule, however. Adding probabilities this way works only if the two events are **mutually exclusive**—only if we know they can't both happen.

- That's the simple rule for "or." There's an equally simple rule for "and." If we know the probability of drawing a red card from deck 1, and we know the probability of drawing a black card from deck 2, can we calculate the probability of doing both? In that case, we multiply. The probability of drawing a red card from deck 1 is 1/2; the same is true for drawing a black card from deck 2; thus, the probability of doing both is 1/2 x 1/2 = 1/4.

- The simple rule for "and" comes with a proviso, just like the simple rule for "or." You can figure the probabilities for two things happening by multiplying their individual probabilities as long as they are **independent events**.
  - What does it mean to say that two events are independent? They are probabilistically isolated. There are no strings of influence between them. The fact that one happens doesn't affect whether the other will.

  - For example, successive flips of a coin are independent. If you flip a coin once, the chance of heads is 1/2. If you flip it a second time, the chance of heads is again 1/2. What comes out on the second flip doesn't depend on what came out on the first flip. The two flips are independent events. The quarter doesn't remember what it did last time. As far as it knows, each flip is a new day—a clean slate.

Gambler's Fallacy

- To gamble rationally, you must know whether or not you're betting on independent events. But we know that people often don't bet rationally. Treating events that are independent as if they aren't is called the **gambler's fallacy**.

- For example, Mr. and Mrs. Smith have five children, all boys. They would like to have a girl, and they believe that with five boys in a row, their next child is bound to be a girl. Of course, that's not true. The reproductive system doesn't have a memory any more than the coin does. The gender of a couple's next child is an independent event.

- Sometimes, people offer a little more reasoning: With five boys in a row, the next one's bound to be a girl. What are the chances the Smiths would get six boys in a row?

- It's true that the chances of having six boys in a row are small. After all, if the odds are 50/50 between having a boy and having a girl, the chance of having six boys is $1/2 \times 1/2 \times 1/2 \times 1/2 \times 1/2 \times 1/2 = 1/64$.

- But that's not the question. The question is what are the chances that the next child will be a boy? And the chances of that are precisely the same as the chance that the first one was going to be a boy. Each birth is an independent event. Mr. and Mrs. Smith have committed the gambler's fallacy, treating independent events as if they aren't independent.

A Classic Example

- If you use the full canon of the more complicated laws of probability, it is possible to calculate combined probabilities even when events are not mutually exclusive, or not independent. The following is a classic example.

- You find yourself with 30 or 40 other people in a restaurant, office, or gym. What are the chances that two of those people were born on the same day of the year? In fact, it's very probable you would have two people born on the same day of the year.

- The calculation might go like this: We pick person 1, who was born on October 15. We then pick person 2. The probability that person 2 was born on a different day is fairly high: 364/365. With person 3, the odds are down to 363/365. As we add people, we keep multiplying: $364/365 \times 363/365 \times 362/365 \times 361/365\ldots$ and so on.

- With just 23 people, the multiplication boils down to 50/50 odds that those people will all have different birthdays. In a room with only 23 people—the size of a small classroom—the odds are 50/50 that two people share the same birthday. At 30 people, the probability reaches 70 percent that two birthdays match. At 35, it becomes an 80 percent probability; with 50 people, the probability of a match hits 97 percent.

- Life is full of decisions, and rational decision making demands knowing the odds. But good decisions also demand a lot more than that. In the next lecture, we'll explore some of the lessons and some of the limits of decision theory.

Terms to Know

combined probability: The rules for calculating the odds that two or more events will happen. The probability of either one or another of two mutually exclusive events happening can be calculated by adding their individual probabilities. The probability of two independent events both happening can be calculated by multiplying their individual probabilities. Other rules apply for dependent events.

equiprobable: Having an equal mathematical or logical probability of occurrence.

gambler's fallacy: Treating independent events as if they were dependent events. Someone who thinks black is "bound to come up" in roulette because of the appearance of a string of blacks (or reds) has committed the gambler's fallacy.

independent events: Events that are isolated probabilistically; the fact that one event happens is not linked to and will not affect the fact that the other event happens.

law of large numbers: In probability theory, the fact that as the number of trials increases, the outcome will approach the mathematically expected value. For example, in a situation where there are two equiprobable outcomes, such as heads or tails when flipping a coin, the longer the run, the more likely the outcome will be a 50/50 split.

mutually exclusive events: Events are mutually exclusive if the presence of one categorically excludes the presence of the other; both cannot occur.

probability: The ratio calculated by dividing the number of possible outcomes in a particular category by the total number of possible outcomes.

Aczel, *Chance.*

David, *Games, Gods and Gambling.*

Questions to Consider

1. You draw a card from a standard deck. What is the probability that it is both red and has an even number on it? Convince yourself that these two approaches give you the same answer:

   (a) Count the cards that fit the bill: 2 of hearts, 4 of hearts, and so on. Put that total over 52.

   (b) The probability of drawing a red card is 1/2. The probability of drawing a card with an even number is 5/13. Thus, the probability of drawing a card that is both red and has an even number is 1/2 x 5/13 = 5/26.

2. Do you know anyone who has committed the gambler's fallacy? Have you?

3. What do you think that John Maynard Keynes meant by "In the long run, we're all dead"?

Exercises

Your cat just had 4 kittens. Which of the following do you think is more likely?

   (a) All are of the same sex.

   (b) Half are of each sex.

   (c) Three are of one sex, and one is of the other.

First, give your intuitive answer. Then, calculate the answer by laying out all 16 possibilities. We can lay them out like this, left to right in terms of the first kittens to appear:

MMMM	FMMM
MMMF	FMMF
MMFM	FMFM
MMFF	FMFF
MFMM	FFMM
MFMF	FFMF
MFFM	FFFM
MFFF	FFFF

In how many of those cases are the kittens all of the same sex? In how many cases are there half of each sex? In how many cases are there three of one sex and one of the other?

Poker, Probability, and Everyday Life
Lecture 18—Transcript

Professor Grim: We gamble all the time. It may not be called gambling, we reserve that term for high-risk games in casinos, but we are always playing the odds. We play the odds as individuals when we decide to drive a few hundred miles to have Thanksgiving with family or friends despite the fact that we might have an accident on the highway. We play the odds as individuals when we decide what job to take, or where to live, or whether to marry a particular person. In none of these cases is the outcome a matter of certainty. We know lots of things can happen. We try to figure out what is most likely to happen and make our choices accordingly. Life is chancy, but we try to deal with chance rationally. As a culture we gamble all the time too. Science stands as one of the greatest intellectual achievements of all time. But outside of mathematics, none of our scientific conclusions is a certainty. We use what we call scientific laws to calculate trajectories to the moon, safety margins in factories, procedures for public vaccination against disease. But our confidence in those predictions is all based on a batch of evidence—we hope good enough—and a best guess on the basis of that evidence. Our culture, and perhaps any culture, is based on that kind of gambling. So the question isn't whether to gamble or not. As long as we live and breathe, we're forced to play the odds. The question is whether we'll gamble rationally or not. That's what this lecture is all about.

The problem is that we often don't gamble rationally. Probability is another name for what we're talking about. The problem is that we often don't handle probabilities rationally. Which is more dangerous, a handgun or a swimming pool? Your eight-year-old has two friends, living two houses away on each side of you. You know that one family has a swimming pool. You know that the other family keeps a gun in the house. Which is the safer place for her to play? If we take the statistics seriously, swimming pools are far more dangerous. About 550 children drown in swimming pools each year; that's about one per every 11,000 residential swimming pools. About 175 children are killed by guns each year. That's about one for every 1 million guns in the United States. Both statistics are horrible, but the probability of a child drowning in a swimming pool is over three times greater than the probability of a child being killed with a gun.

Why do we tend to think otherwise? Imagine kids playing in a swimming pool. That's a wonderfully positive image. The image of kids with guns, on the other hand, is frightening. When it comes to probabilities, we tend to think in terms of images instead. It's those images that often lead us astray. Here's another question for you. In terms of annual deaths, which of these is the most dangerous? Asthma? Drowning? Tornados? Fireworks? What do you think? Asthma is the big killer, accounting for about 4,000 deaths a year. Drowning is up there too, claiming about 3,500 lives. Tornados? We had a particularly bad year recently. In a typical year the rate is less than 100 people. Fireworks? Three or four deaths each year. Death by tornado is a vivid image, as is death from fireworks. Both are the kind of thing that shows up on the news. Asthma isn't. Yet people are about 1,000 times more likely to die from asthma than from fireworks. The relative vividness of the images makes us miscalculate probabilities. Vividness of images and what shows up on the news may also be what reinforces buying lottery tickets. The happy grand prize winner for the multi-state lottery might be a news item for about thirty seconds. If we compiled a news show of one-second pictures of every person who bought a ticket but didn't win the grand prize, we'd be watching that show for over six years.

Let's talk about how to be more rational about probabilities. The probabilities we're really concerned with are the important probabilities we have to calculate as individuals and as a society: probabilities of economic downturns, of natural disasters, of the impact of airbags and seat belts on child fatalities, for example. But here, as elsewhere, it helps to start with the simple cases. Here's the simplest case of all. I've got a quarter in my hand.. What is the chance that it will come up heads? That's right: 50/50. A 50 percent. chance of heads. Probability, one over two. That is the simplest case. It's a perfect instance of the general rule. In all calculations like this, we want to know how many ways can it come up, two, leaving out edges. How many of those ways count as heads? The probability of heads is the ratio of the number of possible outcomes in your chosen category, heads, over the total number of possible outcomes.

That is the central concept of probability. All probability is a calculation of that ratio. Everything else we'll talk about is merely a shortcut for calculating that fundamental concept. Remember that you can always come back to that. When we're talking about probability, we're talking about the number of possible outcomes in a chosen category over the total number of possible outcomes. So take another simple case. You have a complete, shuffled deck of 52 cards. Let me give you some questions. Listen to them, press pause, and then figure out the answers. You draw a card at random. What is the chance that it's the ace of spades? What's the chance that it's a red card? What's the chance that it's a face card—a jack, queen, or king? What's the chance that it's a red face card? In each case we're applying the basic concept of probability. How many different ways can a single draw come out? As many ways as there are cards in the deck. If we don't have any jokers, that's 52. In calculating each of these probabilities, we're just counting how many of those 52 possible draws count as in our category. How many of the 52 possible draws are in the category ace of spades, just one. The odds are 1/52. How many of the 52 are the category a red card? Half the cards are red. 26/52. How many count as face cards? There are three face cards in each suit: three clubs, three hearts, three diamonds, and three spades are face cards. That makes 12. The odds? 12 in 52. And what's the probability of a red face card? Half of that, six in 52. Yes, I could have reduced the fractions. But there's no need to. What we're interested in is the ratio, and that remains the same. The number of possible outcomes in a chosen category over the total number of possible outcomes. That's the core concept of probability. But there's an important proviso. In order for that to work, we have to be talking about equiprobable outcomes.

In the questions I just asked you, I specified a well-shuffled deck. Suppose I had done this instead. Suppose I had taken a brand new deck, opened it up, and shuffled it. Suppose I had then asked you something like "I draw a card. What are the chances that it is the ace of spades?" If I pull the card out from the upper half of a face-down deck, like I just did, I can tell you exactly what the chances are of drawing the Ace of Spades, zero. Why? Every new deck has the ace of spades as the first card you see, the bottom card of a face-down deck. If you cut and do a riffle shuffle, alternating cards like I just did, all of the spades are still going to be on the bottom half of the deck. The ace is going to be on the bottom of the spades. It turns out that six or seven riffle shuffles can be a pretty good randomizer. It turns out that less than five shuffles aren't very good. True randomization demands that all possibilities be equiprobable. It has been shown that with just five shuffles it's absolutely impossible to get all the cards in the reverse order. With just five shuffles, then, all the orders aren't equiprobable. The basic principle of probability demands equiprobable outcomes. Given that, we can always calculate probability as the number of outcomes in our chosen category over the total number of outcomes.

That works great for standard games of chance. Cards, and dice, and roulette wheels are made for easily calculated, equiprobable outcomes. But we also want to use probability in real life, and real life isn't made that way. What's the probability that a hurricane will hit New York next year? Or New Orleans? What's the probability of a tornado taking out Omaha? Or Sacramento? In real life cases like that we don't count up possible outcomes. How many equiprobable outcomes are there, anyway? You can't count possible hurricane tracks the way you can count face cards. Nonetheless we do often calculate probabilities in something like the same way. Let me illustrate with coin flips. We know that the probability of a quarter coming up heads is one half. If you flip a coin 20 times, how many heads will you expect to get? If you said about 10, well you're about right. You shouldn't expect to get exactly 10 heads. The perfect cut of half and half is pretty unlikely with small numbers. But here's the fundamental fact: As the number of flips gets larger, the probabilities rise that the result will be increasingly close to a 50/50 split. This is known as the Law of Large Numbers. In the long run, frequencies of heads and tails will tend to even out. And how long is the long run? It doesn't have an exact number, 100,000, or a million, or 100 million flips. What it really means is, the longer the better. The longer the run, the better the bet that frequencies will be close to our calculated probabilities.

John Maynard Keynes was one of the great economists of the early twentieth century. As Keynes noted, that long run can be a misleading guide to current affairs. In the long run, he said, we are all dead. Here's why the law of large numbers does matter, even in current affairs and in real life. It links probability with frequencies. If the probability of heads is one half, then in the long run we can expect about half the flips to come up heads. In other words, if the probability is half, long-range frequency will approach half. In real life, we estimate probabilities by reasoning in the other direction— from frequencies to estimated probabilities. If we see a die come up sixes with a frequency of 90 times in 100 throws, we figure it's a loaded die. Reasoning from the observed frequencies to an estimated probability, we figure that the probability of a six on the next throw will be about 90 percent . We do the same thing with natural events. What's the probability of a hurricane hitting New Orleans this year? There are several ways of doing that, and of course it depends on what you count as a hit. But here's a simple way. By one reckoning New Orleans has been hit by hurricanes 61 times in the last 139 years. That's a frequency of 61/139, or about 43/100. Reading our probability off that frequency, we'd say the probability of a hurricane hitting New Orleans this year is about 40 percent. Probability is the insurance companies' business. They use all the sophisticated statistics they can find. At base, however, their calculations are grounded in precisely this kind of calculation, an attempt to estimate probabilities on the basis of frequency data.

If we want to deal with the real probabilities, we have to pay attention to real frequencies. Images aren't enough. A shark attack is a vivid image—that's what Jaws is all about. You can practically hear the music, can't you? But in order to live rationally in a chancy world we have to do more than think in images. Otherwise we'll be scared of the wrong things and scared when we don't need to be. We have to look at the real frequencies—at the real statistics. Those tell us that we should worry a lot less about having sharks around and a lot more about having lifeguards.

Cards and dice are neat and tidy. We project our probability calculations to estimate frequencies in the long run. Tornadoes and hurricanes aren't nearly so tidy. There we think backwards from observed frequencies to estimates of background probabilities. Once we have probability estimates for basic events of either type, however, the rules for calculating combined probabilities are the same. We often want to know not just the probability of hydraulic system failure in a plane, not just the probability of landing gear problems, not just the probability of pilot error, but the probability of any of those things going wrong. For that kind of case we need to be able to calculate combined probabilities.

We'll start simple. Suppose we're playing stud poker. I have to stick with the five cards I'm dealt. If we know the probability that I'll be dealt a straight of some kind, and we know the probability that I'll be dealt two pairs, can we calculate the probability that I'll be dealt one or the other? Sure, it's easy. We just add those probabilities together. A straight? The probability is less than one percent. In fact it's a little over .39 percent . The probability of being dealt two pairs is significantly better, just shy of 5 percent. So the probability of getting one or the other? Add those two probabilities. A straight or two pairs? The probability is just shy of 5.39 percent. By the way, the probability of getting nothing in stud poker, just the value of your highest card? About 50 percent. Half the hands dealt will be nothing more.

That same simple rule works for some of the probabilities we estimate empirically. If we know the probability of a tornado hitting Omaha and not Sioux Falls, and if we know the probability of a tornado hitting Sioux Falls and not Omaha, can we calculate the probability of a tornado hitting one or the other but not both? Yes, we just add them. If the probability of a tornado hitting Omaha and not Sioux Falls is one percent, and the probability of a tornado hitting Sioux Falls and not Omaha is 1 percent, the probability of a tornado hitting Omaha or Sioux Falls, but not both, is two percent .

You may be able to tell from my careful phrasing that there is an important proviso on the simple addition rule. Adding probabilities this way works only if the two events are mutually exclusive—only if we know they can't both happen. If you deal me just five cards, those cards can't be both two pairs and a straight. If it's a straight, no two cards are alike. If I have two pairs, that can't be a straight. Those outcomes are mutually exclusive. It's when the two outcomes are mutually exclusive that we can find the probability of one or the other just by adding the probabilities. If a tornado hits Omaha and not Sioux Falls, that tornado can't also hit Sioux Falls and not Omaha. Those are mutually exclusive. In that case we can just add the probabilities.

That's the simple rule for or. There's an equally simple rule for and. If we know the probability of drawing a red card from deck number one and we know the probability of drawing a black card from deck number two, can we calculate the probability of doing both? What's the probability of drawing red from deck one and black from deck two? In that case we just multiply. The probability of drawing a red care from deck one? One half.

The probability of drawing a black from deck two? One half. The probability of doing both? One half times one half, one fourth.

We can do it for real probabilities too. If we know the probability of a tornado hitting Sacramento and a hurricane hitting Atlanta in a given year, can we calculate the probability of both happening—a tornado in Sacramento and a hurricane in Atlanta? Sure, we just multiply them. A tornado in Sacramento? Let's say one percent. A hurricane in Atlanta? Let's say two percent. What's the probability of both happening? It's .01 times .02, one percent of two percent, a mere two hundredths of a percent. The simple rule for and comes with a proviso just like the simple rule for or did. You can figure the probabilities for two things happening by multiplying their individual probabilities as long as they are independent events.

Whoa, let's take a break. Where did all this probability theory come from, anyway? It came from gambling. You probably know that's where the sandwich came from, from John Montagu, the Fourth Earl of Sandwich, known as an avid gambler. Late one night in 1762, the legend goes, Montagu was too busy gambling even to stop to eat. "Just bring me my beef between two pieces of bread," he shouted, and thus the sandwich was born. Probability theory came from the ponderings of a French nobleman a century earlier, the avid gambler Antoine Gombaud, Chevalier de Méré. The Chevalier was trying to figure out what he should bet on a double six coming up in 24 throws of the dice. So he did what any well-heeled aristocrat might do. He wrote to two of the best mathematicians in the world. He wrote to Blaise Pascal, who we'll talk more about next lecture, and Pierre de Fermat. That got the mathematicians talking about rational betting odds. As they say, the rest is history.

Okay, back to work. We just introduced the concept of independent events. You can figure the probabilities for two things happening by multiplying their individual probabilities as long as they are independent events. What does it mean to say that two events are independent? They are probabilistically isolated—no strings of influence between them. The fact that one happens doesn't affect whether the other will. Successive flips of a coin are independent. If I flip a coin once, the chance of heads is one half. If I flip it a second time, the chance of heads is, again, one half. What comes out on the second flip doesn't depend on what came out on the first flip. They're independent events. The quarter doesn't remember what it did last time. As far as it knows, every flip is a new day, a clean slate. That's what it means to say that coin flips are independent events. So are rolls of the dice. So are spins of a roulette wheel.

Card games are often a mixture. Take traditional Blackjack, in which the dealer plays several rounds before shuffling the deck. Because the hands played last round don't go back into the deck, you know that those cards won't appear on this round. The probabilities this time around are not entirely independent of the last round. That observation is at the core of card-counting schemes, in which players keep count of high cards on the one hand—aces and tens—and low cards on the other, adjusting their play accordingly.

If someone is going to gamble rationally, it's important that they know whether what they're betting on are independent events or not, but we know that people often don't bet rationally. One way they don't is called the gambler's fallacy. You commit the gambler's fallacy if you treat events that are independent as if they aren't. Mr. and Mrs. Smith have five children, all boys. They sure would like a girl. And they say, "Oh, don't worry, with five boys in a row the next one's bound to be a girl." Well, of course it's not bound to be a girl. Your reproductive system doesn't have a memory any more than the coin does. The sex of your next child is an independent event. Sometimes people offer a little more reasoning. "With five boys in a row, the next one's bound to be a girl. What are the chances you'd get six boys in a row, tiny, right?" Well, the chances of six boys in a row are tiny. After all, if odds are 50/50, the chance of six boys is one half, times one half, times one half, times one half, times one half. The chance of having six boys is one in 64. But that's not the question. The question is, given that you already have five boys, what are the chances that the next child will be a boy? And the chances of that are precisely the same as the chance that the first one was going to be a boy. Each birth is an independent event. Mr. and Mrs. Smith have committed the gambler's fallacy, treating independent events as if they aren't independent. Some people have even worked the fallacy into what they think is a system. If the roulette wheel comes up with a string of reds, they'll bet black because it's got to come up black now. There are, of course, other ways that irrationality creeps into gambling. It has been shown that sleep deprivation increases the chance of making a risky gamble. Perhaps that's why many casinos are open twenty-four hours a day.

In this lecture we have looked at a few simple rules that are special cases of the full laws of probability. Using those more complicated laws, it is possible to calculate combined probabilities even when events are not mutually exclusive or not independent. I'm afraid we don't have time to go into that here. Here's an example, however. You may already know this one. You find yourself with 30 or 40 other people in a restaurant, office, or gym. What are the chances that two of those people were born on the same day of the year? In fact, it's very probable you would have two people born on the same day of the year. The calculation might go like this. We pick a person number one to start with. Of course there was a particular day they were born on, October 15th, maybe. We then pick a person number 2. The probability that they were born on a different day? Pretty likely. There are 364 days that aren't October 15th, so the odds that person two had a different birthday from person one are 364/365. Take person number three. What are the odds that they have a different birthday from the first two? We've eliminated two days—person number one's birthday and person number two's birthday. The chances of number three having a different birthday are now down to 363/365. As we add people we keep multiplying by those probabilities of having a different birthday than the previous ones, so it's. 364 out of 365, times 363 out 365, times 362 over 365, times 361 over 365 times. You can prove it for yourself, but at just 23 people that multiplication boils down to just 50/50 odds that those people all have different birthdays. That means 50/50 odds that two of them do have the same birthday. In a room with only 23 people—a small classroom—the odds are 50/50 that two people share the same birthday. At 30 people the probability reaches 70 percent that two birthdays match. At 35 it becomes an 80 percent. probability that two people are alike. With 50 people the probability of a match hits 97 percent. Next time you're at a party with 35 people you should feel free to bet that two have the same birthday. Have them announce birthdays in turn, and have anyone else shout out if they have the same birthday. Life is chancy. With any fewer than 365 people it's not absolutely guaranteed, but 80 percent is a pretty safe bet. That's rational gambling.

I wish I had room for another lecture on probability. What I've given you are just a few simple principles, but they will have to do for now. I hope you'll remember the simple principles of probability. There are really only three of them. The basic concept, in probability we're always talking about the number of possible outcomes in a chosen category over the total number of possible outcomes. You can always come back to that. The simple rule for the probability of X or Y happening if they're mutually exclusive—if they can't both happen—the probability of one or the other happening is just the probability of one plus the probability of the other, addition. The simple rule for the probability of both X and Y happening, if they're independent, the probability of them both happening is just the probability of one times the probability of the other, multiplication. Finally, and above all, beware the gambler's fallacy.

Life is full of decisions, and rational decision-making demands knowing the odds. But good decisions also demand a lot more than that. Next time we'll explore some of the lessons and some of the limits of decision theory.

Decisions, Decisions
Lecture 19

Sometimes there are just too many options, as in the cereal aisle of the supermarket. Choice is a good thing, but more options may not necessarily make us happier. We obsess over a perfect or optimal solution and get analysis paralysis. When more important decisions are at stake, ignorance is a major problem. What will happen if we take this or that course of action? Are we making things better or worse? If the context didn't start out emotional, it may end up that way. This lecture is about how we can make decisions more rationally.

Decision Theory
- **Decision theory** builds on the probabilities we discussed in the last lecture. Here's an example: As a gift, someone has given you your choice of tickets in two local lotteries. Lottery A offers a 2 percent chance of winning $100. Lottery B offers a 1 percent chance of winning $150. Which one should you take?

- Decision theory tells you to take lottery A. In lottery A, you multiply the 2 percent probability of winning by the $100 you might win: $0.02 \times \$100 = \$2$. That $2 is called the **expected utility** of picking lottery A. In lottery B, you multiply your 1 percent chance of winning by the $150 you might win: $0.01 \times \$150$ gives you an expected utility of $1.50.

- Decision theory comes with a slogan: Maximize expected utility. An expected utility of $2 beats an expected utility of $1.50; thus, you should choose lottery A.

Desirability
- According to decision theory, the rational way to make decisions is on the basis of expected utility, calculated by multiplying probability by desirability. However, in making decisions, what we

consider desirable is also influenced by tags or labels that make us think we should consider certain things more desirable.

- In a study by researchers at Stanford and the California Institute of Technology, volunteers were asked to taste a series of wines. The volunteers weren't wine experts but moderate wine drinkers. The wines were labeled by price: $5, $10, $35, $45, and $90 per bottle. But because this was a psychological experiment, the wine in the bottle labeled $10 and in the bottle labeled $90 was exactly the same.

- Desirability followed the labels rather than the wine itself: Subjects preferred a wine when it came with a more expensive label. In fact, in brain scans of the participants, pleasure centers in the brain showed a stronger response to the wine labeled $90 than to the wine labeled $10, even though they were exactly the same wine.

Decision Theory by the Numbers

- Let's look at a more complicated example, in which the values aren't purely monetary, and our options have downsides, as well as potential winnings. Imagine that you're invited to someone's house for dinner. You volunteer to bring the wine, but should you bring white or red?

- According to decision theory, you need to multiply probability by desirability. We'll do that with a matrix, working with desirability first.

Type of Wine	Fish or Chicken	Beef or Pork
White	Right wine	Wrong wine
Red	Weird choice	Right wine

- In order to apply decision theory, we need to translate those values into numbers. The right wine will be worth +10 points; the wrong wine will be −10. We'll score "weird choice" in the middle, at 0. If we plug in those numbers, we have a desirability matrix:

Type of Wine	Fish or Chicken	Beef or Pork
White	+10	−10
Red	0	+10

- Decision theory demands that we now add probabilities. What is the probability that your hosts will serve fish or chicken versus beef or pork? Perhaps you know enough about your friends to know that they're health-conscious and might favor fish or chicken for that reason. But then again, it's a special occasion; they might serve beef. You end up guessing that the probability is 70 percent that your hosts will serve fish or chicken.

- The probability of fish or chicken is the same, no matter what kind of wine you buy. Thus, we multiply both entries in that column by 0.70. You estimated only a 30 percent probability of beef or pork; thus we multiply both entries in that column by 0.30. Remember, decision theory demands that we multiply probability by desirability. The results give us a matrix of values from +7 to −3.

Type of Wine	Fish or Chicken	Beef or Pork
White	0.70 x +10 = +7	0.30 x −10 = −3
Red	0.70 x 0 = 0	0.30 x +10 = +3

- For the final stage, we look at the problem from your perspective. You don't have any control over what's for dinner. You have control only over the choice of the wine. What we want to do is combine your potential gains and losses for each of your options.

- If you chose white wine, you add together the +7 points from the case of fish or chicken with the −3 points for the case of beef or pork. Your final expected utility for the white wine option is 4 points.

- If you choose red wine, you add the 0 for fish or chicken with +3 for beef or pork, giving you a final expected utility of 3 points.

- Again, the goal is to maximize expected utility, and 4 beats 3. Decision theory says to go with the white wine.

Tversky and Kahneman
- Some of the case studies in this lecture come from the work of Daniel Kahneman and Amos Tversky, for which Kahneman received the Nobel Memorial Prize in Economic Sciences for his work on decision theory (Tversky had died by the time of the award). Interestingly, both Kahneman and Tversky were psychologists, not economists.

- Why do people make the decisions they do? Tversky and Kahneman put it this way: Choices involving gains are often **risk averse**. Choices involving losses tend to be **risk taking**. That holds true even though the only difference is in how a question is phrased. Phrase the question in terms of gains, and we decide one way. Phrase it in terms of losses, and we decide a different way.

- Kahneman received the Nobel Prize for **prospect theory**. Decision theory is all about how people should make decisions. It's a normative theory: It tells you what you rationally should do. Prospect theory is a descriptive theory: It's about how people actually make decisions.
  - Decision theory dictates a straight line on a graph, going through the 0 point at 45 degrees. Losses and gains on this graph are perfectly symmetrical; a gain of 2 perfectly counterbalances a loss of 2, for example.

  - With prospect theory, the same graph has a curvy line, intended to track how people actually view gains and losses. Losses count negatively more than correlate gains count positively. The graph indicates that this is particularly true when smaller rather than larger gains are at stake.

- Sometimes the decisions we make are irrational, for precisely the reasons that prospect theory points up. Sometimes we miscalculate probabilities. Sometimes we think in images instead of taking statistics seriously. Sometimes context leads us astray about what we really value.

- But decision theory, at least as we know it, may also have important limits. Tversky and Kahneman are very careful not to say that the gain-loss differences in valuation are necessarily irrational.

Diminishing Marginal Utility
- Economists have long recognized the concept of **diminishing marginal utility**. Additions of one unit are not always additions of one unit of value. As the units mount up, their individual value may diminish. One dollar is probably not worth as much to a billionaire as to a homeless person.

- Diminishing marginal utility doesn't threaten the foundations of decision theory. All the calculations we've given are written in terms of utilities: probabilities multiplied by what is really valued, to the extent it is valued. If dollars have diminishing marginal utility, we won't always be able to use a straight monetary scale as if it were a straight value scale.

- The lesson of Tversky and Kahneman may be that we have to do something similar for gains and losses. In at least some contexts, it may be that losses are rationally feared more than gains measured in the same units.

Pascal's Wager
- Decision theory goes in a surprising direction envisaged by one of its founders, Blaise Pascal, who provided an argument for the existence of God. Pascal's argument is interesting because it isn't really an argument that God exists; rather, it's a decision-theoretic argument. Whether God does or does not exist, Pascal argues, it is rational to believe in him. The argument is known as **Pascal's wager**.

- Pascal's argument can be framed as a matrix, exactly like the ones we used before. Across the top, we have two possibilities—that God exists and that he doesn't. Down the side, we have two other possibilities—that you believe or that you don't. That gives us four spaces to fill in. What will happen in each of those combinations?

- Suppose you don't believe that God exists, and you're right. We put a value of 100 in the lower right box. Suppose that you believe that God exists, and you're wrong. We enter a value of –10 in the upper right box. Now suppose that God exists, and you believe in him. You stand to gain infinite bliss, so we put an infinity sign in the upper left box. Suppose that God exists, and you don't believe in him. We enter a negative infinity sign in the lower left box.

Personal Belief	God Exists	God Doesn't Exist
I believe	$+\infty$	–10
I don't believe	$-\infty$	+100

- The expected utility of the "I don't believe" option is $-\infty + 100$, which still equals $-\infty$. That's the expected utility of disbelief. The expected value of the "I believe" option is $\infty - 10$, which still equals ∞. That's the expected utility of belief.

- If you are rational, Pascal surmises, you will maximize expected utility. In this case, you have an infinite everything to gain in the case of belief and an infinite everything to lose in the case of disbelief. The rational option is belief.

Terms to Know

decision theory: A theory of how to make rational decisions by maximizing expected utility.

diminishing marginal utility: The concept that units of value may not amass in equal increments; additions of one unit may not always be the same as additions of one unit of value. For example, one dollar may not be worth as much to a billionaire as to an individual who has no savings at all.

expected utility: In economics, calculated by multiplying the potential benefit of an outcome by its probability.

Pascal's wager: A decision-theoretic argument offered by Blaise Pascal in the 17th century with the conclusion that it is rational to believe in God. If you believe in God and he exists, your payoff is eternal bliss; if you believe in God and he does not exist, your payoff is being wrong, a small loss; if you do not believe in God and he does not exist, you have the pleasure of being right, a small gain; if you do not believe in God and he does exist, your payoff is eternal damnation. Given the potential gains and losses, expected utility dictates that the most rational thing to do is believe in God.

prospect theory: In the work of Daniel Kahneman and Amos Tversky, a descriptive theory concerning how people make decisions in terms of prospective loss and gain rather than final outcome alone.

risk aversion: In economics, an investor is considered risk averse if he or she avoids risk as opposed to being risk taking. People's choices involving possible losses are often more risk averse than similar choices regarding possible gains.

risk taking: In economics, an investor is considered risk taking if he or she is not opposed to taking risks. Choices involving possible gains are often more risk taking than similar choices regarding possible losses.

Suggested Reading

Kahneman and Tversky, "Prospect Theory."

Priest, *Logic: A Very Short Introduction*, chap. 13.

Resnik, *Choices: An Introduction to Decision Theory*.

1. Suppose you really were invited to someone's house for dinner and asked to bring the wine. But you don't know what will be served, can't contact your host by phone, and don't have enough money for two bottles. Should you bring red or white? Use your own estimate of probabilities and desirabilities in order to explain your answer.

2. How good would you feel if you found a $100 bill on the street? How bad would you feel if you found that a $100 bill had accidently slipped out of your wallet? It's $100 either way, but people tend to rate potential gains differently from potential losses. Do you think it's rational or not to have different reactions in these two cases?

3. Here is the matrix for Pascal's wager. Do you think that it should convince someone to believe in God?

Personal Belief	God Exists	God Doesn't Exist	Expected Utility
I believe	$+\infty$	-10	$+\infty$
I don't believe	$-\infty$	$+100$	$-\infty$

Your company has the option of acquiring two new lines out of three options. Each will involve a $100,000 investment, and the company can afford only two of the three. You have to choose which two to take:

Option 1: You estimate that this option has a 20 percent chance of success and an 80 percent chance of failure but a $1 million payoff over your investment if it succeeds.

Option 2: You estimate that this option has a 40 percent chance of success, a 60 percent chance of failure, and a $500,000 payoff over your investment if it succeeds.

Option 3: You estimate that this option has a 90 percent chance of success and only a 10 percent chance of failure but will return only $150,000 over your investment if it succeeds.

Which two options would you take? Explain your reasoning in terms of probabilities multiplied by desirabilities, plus any other considerations you think are relevant. (See "Answers" section at the end of this guidebook.)

Decisions, Decisions
Lecture 19—Transcript

Professor Grim: This lecture focuses on rational decision making. In the second lecture of the course we talked about decision making in emotional contexts, but decision making is hard enough even without that. It's easy to hit analysis paralysis in which we defer, or even abandon, making a decision because it's just too difficult. Sometimes there are just too many options, like in the cereal aisle of the supermarket. Choice is a good thing, but more options may not necessarily make us happier. We hit analysis paralysis. We obsess over a perfect or optimal solution when we should satisfice instead. Sometimes good enough is good enough. You can probably guess how I feel about breakfast cereal. When more important decisions are at stake, ignorance is a major problem. We just don't know what's going to happen if we take this course of action. We don't know what's going to happen if we take that one. Are we making things better or just making them worse? If the context didn't start out emotional, it may end up that way. Decision making under uncertainty is inherently stressful. How should we make decisions like that rationally? That is what this lecture is about.

Decision theory is devoted to the problem and builds on the probabilities we discussed last time. The core concept is simple. I'll start with an example. As a gift, someone has given you your choice of two tickets in small local lotteries. It would be impolite to back out. You have to choose either lottery one or lottery two. Lottery one offers a two percent chance of winning $100. Lottery two offers a one percent chance of winning $150. Which one should you take? Decision theory tells you to take Lottery number one. Here's why. Consider lottery number one. Take the two percent probability of winning and multiply it times the $100 you might win. Multiply .02 times $100 and you get $2.00. That $2.00 is called the expected utility of picking Lottery number one. Consider lottery number two. Take your one percent chance of winning and multiply it by the $150 you might win. That's .01 times $150, giving you an expected utility of $1.50. So the expected utility of choosing lottery number one is $2.00. The expected utility of choosing lottery number two is $1.50. Decision theory comes with a slogan: Maximize expected utility. An expected utility of $2.00 beats an expected utility of $1.50, so you should choose lottery number one. All the concepts of decision theory are built into that example. Expected utility always involves multiplying the probability that something will happen times the value or desirability of that thing. Money calculations are easy, but the same principles apply when the values aren't monetary. You still need to calculate the desirability of all the things at stake and the probabilities that various options will get you those things. You pick the option that maximizes probability times value. You maximize expected utility.

There are two parts to the basic formula: probability and value. Unfortunately, people are often pretty bad at calculating both of them. Which is greater, the probability of getting mugged in Detroit or of getting mugged in Michigan? When asked that question, a surprising number of people immediately say Detroit. But, of course, Detroit is in Michigan. Every Detroit mugging is a Michigan mugging. You can't have a higher probability of being mugged in Detroit than being mugged in Michigan. Why do people say Detroit? It's the image problem again. Despite the fact that Detroiters have made some very positive changes, we have this image of Detroit as a dark and dangerous city, the kind of place where muggers lurk. Our image of Michigan? Farmland and lakes, no muggers in sight. The problem is that people often estimate probabilities based on vivid images. But images are often a bad guide to real probabilities. People miscalculate. They ignore the fact that Detroit is in Michigan.

Here's another one. I'll describe someone and then ask you to rank probabilities. Linda is 31 years old, single, outspoken, and very bright. She majored in philosophy. As a student, she was deeply concerned with issues of discrimination and social justice. Where is Linda today? Here are three possibilities. I want you to rank these one, two and three in terms of most probable, second most probable, third most probable. Linda works in a bookstore. Linda is an insurance salesperson. Linda works in a bookstore and takes yoga lessons. Which is most probable? The surprising fact is that many people say the third is, that Linda works in a bookstore and takes yoga lessons. Of course there's no way that can be more probable than she works in a bookstore. Everyone who works in a bookstore and takes yoga is already guaranteed to work in a bookstore. The probability of both is bound to be less than the probability of one or the other. Why do people get the probabilities wrong? They're thinking in images, and they're trying to fill in gaps. Works in a bookstore, takes yoga, that's consistent with what we've been told and is more vivid because more gaps are filled in. Images are a bad guide to real probabilities, but at least we'll get our calculations of value and desirability right, won't we? It turns out we are vulnerable on that side too.

To begin with, our relative valuations tend to be inconsistent in terms of time. When people have to make a decision now regarding things that will be used over time, their decisions favor variety. Go to the supermarket and pick what snacks you want in your lunches next week. You'll probably pick a variety, potato chips, and barbecue potato chips, Fritos, Doritos. But if you buy a snack from the machine each day next week, the odds are you won't choose such a variety. Day after day you push the buttons and out come Fritos. Desirability judgments are also alarmingly sensitive to context. We tend to value things in terms of comparisons, and those comparisons are made in context. This isn't news to people in sales and marketing. Real estate agents may show you a very bad house first. It makes the pretty good house they show you next look a lot better.

What we consider desirable is also influenced by tags—labels that make us think we should consider certain things more desirable. In a study by researchers at Stanford and Cal Tech, volunteers were asked to taste a series of wines. These weren't wine experts, just moderate wine drinkers. The wines were labeled by price: $5.00, $10.00, $35.00, $45.00, and $90.00 a bottle, but of course, this was a psychological experiment. The wine in the bottle labeled $10.00 and the bottle labeled $90.00 was exactly the same wine. The wine in the $5.00 bottle was the same as the wine in the $45.00 bottle. But desirability followed the labels rather than the wine itself. Subjects preferred a wine when it came with a more expensive label. That wasn't just putting on airs. In brain scans of the participants, pleasure centers in the brain showed a stronger response to the wine labeled $90.00 than to the wine labeled $10.00, even though those were exactly the same wine. Decision theory says the rational way to make decisions is on the basis of expected utility, calculated by multiplying probabilities by desirabilities. That doesn't guarantee that we're going to be very good at estimating either one.

The lottery example was a particularly simple case, just two options, known probabilities, and an easy monetary measure of value. Let me give you a slightly more complicated example. Here the values aren't purely monetary, and our options have downsides as well as potential winnings. Here's the story. You're invited to someone's house for dinner. "Can I bring something?" you say. "Sure. You bring the wine." "Great." You're on your way to their house for dinner, just in time. You've forgotten their phone number, but luckily you know where they live, and you're stopping at the liquor store to buy the wine. The guy behind the counter says "Would you like white, or red?" Uh-oh, you have no idea what's for dinner. You know that white goes best with fish and chicken, and red goes best with beef and pork. You know that red wine with fish and chicken is pretty weird, but white with beef is absolutely awful. You only have enough money for one bottle, unless you get the really cheap stuff, but you're out to impress these people, so that won't work. So which is it to be? Would you like white, or red?

Decision theory says we have to calculate probabilities times desirabilities. We'll do that with a matrix, working with desirabilities first. It will start something like this. Across the top we have dinner possibilities. We'll limit them to fish or chicken and beef or pork. Down the left side we have the two wine options, white or red. We'll fill in the matrix with our desirabilities. The upper-left corner is white with fish or chicken. That's the right wine. The lower right is red with beef or pork. That's another case where we have the right wine. The upper-right corner is white with beef or pork. That's the wrong wine. The lower left is red wine with fish or chicken. We can just write in pretty weird. In order to apply decision theory, we have to translate those values into numbers. So, here's a proposal. Let's just score things negative ten to positive ten. The right wine will be worth positive ten points. The wrong wine will be negative ten. We'll score pretty weird in the middle at zero. That zero is admittedly a little arbitrary, but luckily, this is just an example. If we plug in those numbers, we have a desirability matrix.

Decision theory demands that we now add probabilities. So what is the probability that they're going to serve fish or chicken? What are the odds that they're going to serve beef or pork? Perhaps you know enough about your friends to know that they're health conscious and might favor fish or chicken for that reason. But then again, it's a special occasion. You end up guessing that the probability is 70 percent that they'll be serving fish or chicken. The probability of fish or chicken is the same, there in the left column, no matter what kind of wine you buy. So we put a times .70 for our probability in both entries on the left. You've estimated only a 30 percent probability of beef or pork, so we put .30 for our probability in both entries on the right. Decision theory demands that we multiply probabilities times desirabilities. When we multiply those out, we get a matrix with values from +7 to -3.

Now for the final stage. We have to look at it from your perspective. You don't have any control over what's for dinner. You only have control over the choice of the wine. What we want to do is combine your potential gains and losses on each of your options. If you choose white wine, you have to add together the +7 points from the lucky case of fish or chicken with the -3 points for the unlucky case of beef or pork. Your final expected utility on the white wine option, 4 points. If you choose red wine, we add the zero for fish or chicken with 3 for beef or pork. Maximize expected utility. Four beats three. Decision theory says go with the white wine.

We've said that we are vulnerable in both the probability and the desirability estimates that decision theory demands. We're also vulnerable when it comes to putting those together. Let me give you a problem setup and then ask a couple of questions. We find out that the United States is vulnerable to an outbreak of a particularly deadly type of influenza. If left unchecked, 600 people will die. There are two alternative proposals for combating the disease. Tell me which one we should go for. All scientific indicators tell us that If program A is adopted, 200 people will be saved. If program B is adopted, there is a one-third probability that 600 people will be saved and a two-thirds probability that no-one will be saved. Which would you go for? Program A or program B? Seventy-two percent of those asked went for option A. The first interesting thing to note is that in terms of decision theory those options come out exactly the same. In program A we have 200 people saved, an expected utility of 200 lives saved. In program B we calculate expected utility by multiplying probability times desirability. A one third probability of 600 saved with two thirds probability of no one saved gives an expected utility of one third times 600. There, again, the expected utility is 200 lives saved. Hmm, the same expected utility for both programs, 200 lives in each case. So why do 72 percent of those people who are asked go for option A?

Consider a case like that in terms of two different programs. As before, we anticipate a particularly deadly influenza. Left unchecked, 600 people will die. We have two programs, program C and program D. If Program C is adopted, 400 people will die. If Program D is adopted, there is a one-third probability that no one will die, and a two-thirds probability that 600 people will die. When people were asked to choose between Programs C and D, 78 percent chose D. That makes things even more interesting. C and D are, again, identical in terms of expected utility. With program C, we have 400 people dead—expected utility, 400 dead. With program D, we have a one-third probability of no deaths, and a two-thirds probability that 600 will die—expected utility, two thirds times 600, or 400 deaths. Hmm, the expected utility is 400 dead in each case. So why did people go so strongly for option D?

The final thing to note is that A, B, C, and D are all identical in terms of decision theory. With 600 people at risk, the A option of 200 alive is the same as the C option of 400 dead, which in terms of expected utility are the same as options B and D. This example and the case of Linda in the bookstore both come from the work of Daniel Kahneman and Amos Tversky. Tversky had died by that time, but Kahneman received the Nobel Prize in economics for that work. The interesting fact about that is that Kahneman isn't an economist. He's a psychologist, so was Tversky.

Why do people make the decisions they do in these cases? Programs A and B are phrased in terms of lives saved. Given a choice between lives saved for sure and just a chance of saving lives, we go for the sure thing. Options C and D, in contrast, are phrased in terms of lives lost. If one option involves a certainty of lives lost, we'll take our chances instead. Tversky and Kahneman put it this way: Choices involving gains are often risk averse. Choices involving losses tend to be risk taking. That holds in this case even though the only difference is how the question is phrased. Certain gains and losses are the same in two cases. Risks of gains and losses are the same, but phrase the question in terms of gains and we decide one way. Phrase it in terms of losses and we decide a different way.

What Kahneman received the Nobel Prize for was prospect theory. Decision theory is all about how people should make decisions. It's a normative theory; it tells you what you rationally should do. Prospect theory is a descriptive theory. It's about how people actually do make decisions. The difference between the two can be put in terms of two graphs. One graph shows the straight line dictated by decision theory. It goes through the zero point at 45 degrees. Losses and gains are perfectly symmetrical. A gain of two perfectly counterbalances a loss of two, for example. A gain of four perfectly counterbalances a loss of four. The other graph shows Tversky and Kahneman's curvy line, intended to track how people actually do view gains and losses. Losses count negatively more than correlate gains count positively. Their graph indicates that's particularly true when it's smaller rather than larger gains that are at stake.

Sometimes the decisions we make are irrational for precisely the reasons that decision theory points up. Sometimes we miscalculate probabilities. Sometimes we think in images instead of taking statistics seriously. Sometimes context will lead us astray as to what we really value, but decision theory, at least as we know it, may also have important limits. Tversky and Kahneman are very careful not to say that the gain-loss differences in valuation is necessarily irrational. They may be right. Economists have long recognized the concept of diminishing marginal utility. Additions of one unit are not always additions of one unit of value. As the units mount up, their individual value may diminish. One dollar is probably not worth as much to a billionaire as it is to a homeless person. That is known as diminishing marginal utility. Diminishing marginal utility doesn't threaten the foundations of decision theory. All the calculations we've given are written in terms of utilities—probabilities times what is really valued, to the extent it is valued. If dollars have diminishing marginal utility, we won't always be able to use a straight monetary scale as if it were a straight value scale.

The lesson of Tversky and Kahneman may be that we have to do something similar for gains and losses. In at least some contexts, it may be that losses are rationally feared more than gains measured in the same units. Suppose you're living comfortably on an income of $50,000 per year. The prospect of losing that $50,000 is awful. What about the prospect of gaining an extra 50,000 dollars? Is that to be desired as much as losing 50,000 is to be feared? My guess is no. There are cases in which potential losses are things we are rational to treat differently than the same potential gains. If decision theory is interpreted in such a way that that's always irrational, we have hit an important limitation. Either the theory or its interpretation needs to be changed. Just as dollars have diminished marginal utility in terms of a comparison base—that of the billionaire or the pauper—dollars may have a variable utility in terms of gain and loss.

What I hope you'll take home from this lecture are the basics of decision theory. Expected utility is calculated by multiplying the probability of an outcome by its value. How should you make decisions rationally? Calculate probabilities of various outcomes on various options. Choose that option with the highest expected utility. I hope you'll also remember some of the ways we go wrong in applying such a principle. We mistake vivid images for real probabilities. Even our desirabilities can be irrationally influenced by context.

I want to close by taking decision theory into a surprising direction envisaged by one of its founders. Blaise Pascal was born in 1623, one of those wonderful combinations of mathematician and philosopher. He was one of the founders of probability theory and one of the first to outline the fundamentals of decision theory. He did amazing mathematical work while still in his twenties. On the 23rd of November, 1654, between 10:30 and 12:30 at night, Pascal had an intense religious experience. The rest of his life was devoted to theological reflection. Left incomplete at his death was the philosophical work for which he is most remembered, Pascal's *Pensees*— the "Thoughts."

Let me set the stage for what I want to show you by saying that infinity is a major theme in Pascal's thought. He says "Nature is an infinite sphere of which the center is everywhere and the circumference nowhere." Perhaps most memorably, "The eternal silence of these infinite spaces terrifies me." I've given you that background on Pascal—decision theory, religion, and infinity—in order to introduce his argument regarding rationality and belief in God. The attempt to offer logical proofs for the existence of God has a long and interesting philosophical history. Pascal's argument is interesting because it isn't really an argument that God exists. It's a decision-theoretic argument. Whether God does or does not exist, Pascal argues, it's rational to believe in Him. The argument is known as Pascal's Wager. Pascal's argument can be framed as a matrix, exactly like the ones we used before. Across the top we have two possibilities, that God exists, and that he doesn't. Down the side we have two other possibilities, that you believe, or that you don't. That gives us four spaces to fill in. What's going to happen in each of those combinations? Suppose that you don't believe that God exists, and you're right. He doesn't. That's the lower-right box. Aren't you smart? I guess we'll have to put you down for lots of smart points. We'll put a value of 100 in that box. Suppose that you believe that God exists, and you're wrong. That's the upper-right box, all that work for nothing. So we should probably put a zero in that box. Maybe we'll even mark it down a little for disappointment. We'll put in a value of -10. Now suppose that God exists, and you believe in him. That's the upper left. What do you stand to gain? Infinite bliss. We'll put down the lazy eight for an infinite gain. Suppose that God exists and you don't believe in him, that's the lower left. If what they say about Hell is true, the payoff there is eternal torment—eternal torment. We'll put down a negative lazy eight. You remember how we calculated the wine case. We took the options under your control, choosing red or white, and added up the values across the rows in order to calculate expected utility. Pascal has us do the same thing here for the options of belief or disbelief.

Take the don't-believe row on the bottom. What do you stand to gain and lose? You might gain 100 smart points, but you might lose an infinite amount. What's the expected utility of that option? A negative infinity plus 100 smart points is still a negative infinity. That's the expected value of disbelief. What's the expected value across the top row, the "belief" option? You've got those negative 10 points, but you add those to a positive infinity. Infinity minus any finite amount is still infinity, positive infinity. That's the expected utility of belief. What happened to the probability part? Weren't we supposed to multiply the expected value by the probability of the event? Interestingly, because we've got infinity in the matrix, it doesn't matter what the probabilities are that God exists or that he doesn't. However small a probability you assign to His existence, that probability times infinity will still give you an expected utility of infinity. If you are rational, Pascal thinks, you will maximize expected utility. In this case you have an infinite everything to gain in the case of belief and an infinite everything to lose in the case of disbelief. The rational option is belief.

It's an interesting argument, both because of the use of infinity and the application of decision theory where you might least expect it. Pascal's argument is something like gambling on God, But that would be a pretty cheap ticket into heaven. I expect the cost of admission is a little higher than that. It also looks like a similar argument might justify belief in Santa Claus. He brings toys only to good little boys and girls who believe in him. An argument that justifies belief in Santa Claus is an argument to watch out for.

Most philosophers don't take Pascal's Wager seriously as a vindication of religious belief. They think it offers some warnings regarding the limits of decision theory. The first warning is that infinity is a difficult concept to deal with in many contexts. We should probably not be surprised if it's difficult to deal with in decision theory too. The second warning is this: Be careful how you set up a decision problem. As Pascal sets up the wager, he has just two options across the top: God exists or he doesn't. Given his historical context, that may not be too far fetched. But many critics since then have said, what about the other possibilities? What about all those other possible gods of ancient Greece and Egypt, some of whom might jealously penalize us for belief in this one? If you add the right infinite counter values in those additional cases, the infinities cancel out.

That warning has a practical application, too. Decision theory is made to calculate the expected utility of your various options, probability times desirability, but in order to make any decision between options, it's important, first of all, to have a good grasp of what all your options really are. In the next lecture we'll take on another aspect of rationality, thinking scientifically.

Thinking Scientifically
Lecture 20

W hy is it important to think scientifically? Thinking scientifically means putting claims rigorously and systematically to the test. It means holding beliefs about the world up to the world in order to see whether those beliefs are true or not. Just because someone calls a discipline a science doesn't make it so. If it's real science, it's science because of the procedures of rational test.

Science versus Pseudoscience

- We can easily differentiate between science—physics, chemistry, microbiology, neurophysiology—and **pseudoscience**—astrology, UFOlogy, phrenology. In certain cases, however, people disagree about these categories; an example is parapsychology. What precisely is the difference between science and pseudoscience?

- The 20th-century philosopher most responsible for concentrating on this problem—and for presenting one of the most durable and well-known answers to it—was Karl Popper. Popper wrote, "I wished to distinguish between science and pseudoscience; knowing very well that science often errs, and that pseudoscience may happen to stumble on the truth."

- How do we distinguish between science and pseudoscience? There were two answers to Popper's question that were in circulation at the time: an old answer and a new one. The old answer was that what distinguishes science is that it is inductive; it proceeds from experience to theoretical generalizations. That does seem to be true of science: It builds from individual experiences to general theories. But it's not clear that that is what distinguishes science from pseudoscience.

- Popper was dissatisfied with the old answer. But there was also a new answer current in Vienna during his time, an answer associated with a group of scientists and philosophers called the Vienna Circle. The young mathematician Kurt Gödel was a member, and Wittgenstein's *Tractatus Logico-Philosophicus* was a highly respected source. Their answer to Popper's question was that the mark of science was **verifiability**.

Falsifiability Is the Key

- Popper didn't think that that answer was any better. His objection was that it seemed all too easy to verify a theory. Popper's conclusion was that this new answer wasn't adequate to distinguish real science from its pretenders either. Popper proposed

The history of the pseudoscience astrology demonstrates that the issue is not always falsification of theories but how people use theories—in this case, to make money.

an alternative: What a theory needed in order to be scientific wasn't verifiability but **falsifiability**.

- A genuinely scientific theory, according to Popper, is one that takes risks. A theory that nothing would disconfirm is pseudoscientific. A genuinely scientific theory isn't like that. It sticks its neck out. Here, Popper's favorite candidate is Einstein's theory of relativity.

  o Popper was working at just about the time of the Eddington expedition, intended to test Einstein's theory. Einstein predicted that light would be bent by gravity. If the theory was true, the position of stars photographed during an eclipse should differ from their positions when photographed in the night sky, without the gravitational effect of the sun.

○ If the Eddington expedition came back with photographs that didn't show the effect, Einstein's theory was dead in the water: falsified. It was precisely because Einstein's theory was so clearly falsifiable that Popper took it as the paradigm of genuine science.

Unfalsifiable: The Opposite of Science

- Suppose you want to create a pseudoscience, one that is protected against falsifications precisely because it is unfalsifiable. Perhaps you claim that there is a very strange thing hovering in the distance: a "cereotopic IV (invisible) sphere." There is nothing anyone can say or do to refute the theory. Unfalsifiability is built right in.

- Another way to create a pseudoscience is to make your theory a moving target. Every time someone tries to refute it, you say, "Oh, that wasn't exactly my theory. More precisely, it is…," and you change the theory a bit.
  ○ That was the history of **Ptolemaic astronomy**. In simple form, the Ptolemaic theory held that the planets and the sun circle the earth. If that were true, all the planets would move uniformly across the night sky. But, in fact, they don't. If you plot the positions of Mars against the constellations night by night, you'll find that Mars moves in the same direction for a while but then starts moving backwards and then forward again. This movement is called **retrograde motion**.

  ○ Thus, the theory was modified a bit; what is circling the earth aren't planets but invisible pivots around which the planets turn. What you see in retrograde motion is the planet going backwards around that invisible point, called an "epicycle."

  ○ Such additions to a theory as epicycles are called *post hoc* modifications—modifications after the fact. If you add those often enough, you make your theory unfalsifiable by making it a constantly revised moving target.

- Yet another way to make a theory unfalsifiable is to build on ambiguous phenomena. Use something that is hard to see or that can be interpreted in any way you want. In this category fall Percival Lowell's reports of canals on Mars.

- Another tip for building a pseudoscience: Build in a stipulation that only the properly trained or the truly initiated can see at work.

- The point of these examples is to alert you to what pseudoscience is so that you will recognize it. Pseudoscience isn't a problem only outside the scientific establishment; it is something to guard against even within the hallowed halls of established science—perhaps especially within the hallowed halls. Even the best of scientists can fool themselves by falling for the traps of pseudoscience.

Putting Theories to the Test

- When it comes to influential bits of 20th-century philosophy of science, Popper's falsifiability criterion may top the list. Scientists themselves use it as a mark for good science. No theory of any breadth, Popper emphasizes, can ever be conclusively confirmed. There is no point at which we say "confirmed," close the book, and abandon all future testing. We should continue to look for falsifications. We should continue to put our theories to the test. That's the only way we'll find better ones.

- The core demand of scientific rationality is that we test theories. How precisely do we do that? Let's look at a simple case in what's known as the **Wason selection task**: Imagine that you have four cards in front of you, labeled E, K, 4, and 7. Here's the theory you want to test: If a card has a vowel on one side, it has an even number on the other. What is the minimum number of cards you have to turn over to test this theory?

- The most popular answer is to start with the E card. And, of course, that's right. If we turn over the E and see an even number there, that's in accord with the theory. If there isn't an even number on the E card, the theory is refuted: falsification. That's a good test. If it comes out one way, we get a confirmation. If it comes out another, we get a falsification.

- Are there any other cards that would test the theory? The second most popular answer is the 4 card. But that is definitely not right. Suppose we turn over the 4 card and find a vowel. Then we have a card that fits the theory. But what if we don't find a vowel? If we turn over the 4 and don't find a vowel, we still haven't falsified the theory. All the 4 card can give us is confirmation. Turning over the 4 card can't give us falsification; thus, it can't really test the theory.

- The second card we should turn over is the 7 card. Suppose we find a vowel on the other side. Then we have a vowel on one side, an odd number on the other, and we've falsified the theory. That gives us a real test.

The Limits of Experimentation
- In this lecture, we've emphasized "putting it to the test," or experimentation, as the core concept of scientific rationality, with an emphasis on the importance of seeking falsifications of a theory. There are a few warnings and provisos regarding the role of experimentation in science, however. There are many different sciences, and the role of experiment differs in each one.

- If an organic chemist forms a hypothesis regarding the composition of a specific protein, he or she can design an experiment and test it. But if a paleoanthropologist forms a hypothesis regarding the origins of *Australopithecus afarensis*, there is no clear and obvious experiment that would confirm or refute it. There are rival scientific theories about the formation of the moon, but there is no clear moon-forming experiment we can perform to decide between them.

- The role of experiment in science is crucial, but it is often unavailable. Sometimes, it takes years to invent an experiment that would confirm or disconfirm aspects of a theory. This was true both for parts of Einstein's relativity and for quantum mechanics.

- When we do have experimental evidence for a theory, it is often indirect. We cannot afford, either economically or ethically, to perform wide-scale social experiments that would answer some of our questions. The experimental evidence we have regarding such questions must be extrapolated from social psychological experiments involving small groups of people in relatively artificial circumstances.

- Scientific rationality demands that we put hypotheses to the test—any theory that waives away empirical test as irrelevant would clearly be an unscientific theory. But the role of experiment is often more elusive and indirect than it might seem.

Terms to Know

falsifiability: Karl Popper's criterion for distinguishing real science from pseudoscience was that real scientific theories can be shown to be false. Real science takes risks in that theories are formulated so that they are open to disconfirmation by empirical testing or empirical data.

pseudoscience: A set of practices or tenets that is made to resemble science and claims to be scientific but violates scientific standards of evidence, rigor, openness, refutation, or the like.

Ptolemaic astronomy: The view of astronomy put forward by Ptolemy (A.D. 90–168); based on a geocentric (earth-centered) model of the universe.

retrograde motion: In astronomy, the apparent movement of a body opposite to that of other comparison bodies. Plotted against constellations night by night, Mars moves in one direction for a while but then appears to move backwards before it again seems to reverse directions. Retrograde motion does not fit with the predictions of a simple geocentric theory, which holds that the sun and planets circle around the earth.

verifiability: The criterion for what distinguishes science from non-science, proposed by a group of 20th-century scientists and philosophers called the Vienna Circle. Science, they said, is that which can be confirmed by empirical or experiential data.

Wason selection task: A psychological task used to examine the psychology of reasoning. If shown four cards, one displaying the letter E, one the letter K, one the number 4, and one the number 7, using a minimum number of cards, which cards do you turn over to test the theory "If a card has a vowel on one side, it has an even number on the other"? It has been demonstrated that people tend to seek confirmation when working with unfamiliar or abstract concepts, such as "vowels" and "even numbers," but will seek falsifiability in similar tasks using familiar situations, such as underage drinking.

Suggested Reading

Gardner, *Fads and Fallacies in the Name of Science.*

Grim, ed., *Philosophy of Science and the Occult.*

Shermer, *Why People Believe Weird Things.*

Questions to Consider

1. You certainly don't want your theory to be false. Why then should you want it to be falsifiable?

2. One of the points made in the lecture is that people make different decisions in different forms of the Wason selection task.
   (a) Which cards should you turn over to put this theory to the test: If a card has a vowel on one side, it has an even number on the other?

   | E | K | 4 | 7 |

   (b) Which cards should you turn over to put this theory to the test: If you are drinking alcohol, you must be 18?

   In the first case, people tend to get the answer wrong; they say. "Turn over the E and the 4." In the second case, they tend to get the answer right: "Turn over 'Drinking beer' and 16."

The cases are logically identical. Why do you think people get the answer right in one case and wrong in the other?

3. Karl Popper's criterion of falsifiability states that a theory is scientific only if it is falsifiable—only if there is some experience that would prove it wrong. Is Popper's theory falsifiable? Is the criterion of falsifiability itself falsifiable? What conclusion should we draw if the answer is no?

Exercise

Below are some examples of theories. From your own familiarity, or with a bit of research, decide which of these you think belongs in which category:

    (a) A theory that is falsifiable but hasn't yet been falsified.

    (b) A theory that is falsifiable and has, in fact, been falsified.

    (c) A theory that is unfalsifiable.

- Erich Von Daniken's theory that a small group of aliens was stranded on Easter Island in ancient times and taught the natives to make robot-like statues.

- Thor Heyerdahl's theory that the Polynesian islands were populated not from Asia to the west but from ancient Peru to the east.

- The theory in Aristotle and earlier Greeks that everything is made of four essential elements: earth, air, fire, and water.

- The Ptolemaic theory that the sun orbits around the earth.

- The Lamarckian theory of evolution that an animal can pass on to its offspring characteristics that it has acquired during its lifetime.

- Erich Von Daniken's theory that the stone from which the Maoi statues on Easter Island are made is not found on the island.

Thinking Scientifically
Lecture 20—Transcript

Professor Grim: In this lecture I want to talk about thinking scientifically. Why is that important? Thinking scientifically means putting claims rigorously and systematically to the test. It means holding beliefs about the world up to the world in order to see whether those beliefs are true or not. It's not rational because someone calls it science. If it's real science, it's science because of procedures of rational test.

Let me start here. If I asked you to give me a list of scientific disciplines, you'd have no trouble, physics. Chemistry. Microbiology. Neurophysiology. If I asked you to give me some scientific theories, you'd have no trouble. The theory of relativity in physics, evolutionary theory in biology, maybe string theory. But we are all familiar with things that look like science, that present themselves as science, but that have a whiff of the fake and phony, astrology. UFOlogy, phrenology. ancient astronauts. Those are classified as pseudo sciences. When it comes to what side of the fence something is on—science or pseudo-science—people are going to disagree in particular cases. The hardest case I know of is parapsychology. What I want to press here, however, isn't what side of the fence something like parapsychology should fall on. The question I want to press is, what, precisely, is this fence? What precisely is the difference between science and pseudo-science? That's called the problem of demarcation. How do you draw the line? How do you demarcate science from pseudoscience?

The twentieth century philosopher most responsible for concentrating on this problem and for presenting one of the most durable and well-known answers to it was Karl Popper. Popper died in Britain, as a British citizen, in fact as Sir Karl Popper. He was 92. But Popper was born in Vienna, Austria, and grew up in the ferment that was Vienna between the late 1800s and the 1920s. Sigmund Freud was working in Vienna developing the science of psychoanalysis. Popper himself was among a number of people attracted to Marxism, explicitly put forward by Karl Marx and Friedrich Engels as a scientific theory of historical forces. Vienna was home to the young philosopher Ludwig Wittgenstein and the young mathematician Kurt Gödel. And there was this new theory of space and time, Einstein's theory of relativity. That was certainly an intellectually exciting time but also a confusing time. Here, side by side, were Freudian psychoanalysis, Marxist theory of history, Einsteinian physics, and some pretty strange philosophy and mathematics. Was it all genuinely scientific, or just some of it? How do you tell the real science from the pseudo-science?

That was Popper's problem. Thinking back on that period, Popper wrote later that what troubled him was neither when is a theory true nor when is a theory acceptable. "My problem was different. I wished to distinguish between science and pseudo science knowing very well that science often errs, and that pseudo science may happen to stumble on the truth." How do you distinguish science from pseudo science? There were two answers to Popper's question that were in circulation at the time, an old answer and a new one. Neither of those was good enough for him. So let me give you all three, the old answer, the new answer, and Popper's answer.

The old answer was this. The thing that distinguishes science is that it is inductive; it proceeds from experience to theoretical generalizations. That does seem to be true of science. It builds from individual experiences to general theories. But it's not clear that that's what distinguishes science from pseudo science. One problem is that generalization from experiences seems to characterize many of the things dismissed as pseudo sciences. Von Däniken's books on ancient astronauts are filled with specific experiences, visits to Stonehenge, to the pyramids of Egypt, to the plains of Nazca. He proceeds from these specific examples to a general theory, that we were visited by ancient astronauts. But is that enough to make it science? Perhaps generalization from experience is a necessary condition for something to be scientific but not a sufficient condition.

Popper was dissatisfied with the old answer. But there was also a new answer current in Vienna at the time, an answer associated with a group of scientists and philosophers called the Vienna Circle. The young mathematician Kurt Gödel was a member, and Wittgenstein's *Tractatus Logico-Philosophicus* was a highly respected source. Theirs was the new answer to Popper's question. The mark of science, they said, was verifiability. What makes something scientific is that it's verifiable; there is some piece of empirical or experiential data that could confirm or verify the theory. Popper didn't think that answer was any better. His objection? That verifications seemed to come too cheap. It seemed all too easy to verify a theory.

The study of Marx or Freud seemed to have the effect of an intellectual conversion or revelation. "Once your eyes were thus opened, you saw confirming instances everywhere. The world was full of verifications of the theory." A Marxist couldn't open a newspaper without finding confirming evidence for a Marxist interpretation of history, particularly in the things that weren't said. A Freudian would think his theory confirmed if you said you hated your father and also confirmed if you said you didn't. That was precisely the problem, said Popper. These theories were all verifiable, all too verifiable. Anything that happened, no matter what, could be taken as verification. Popper's conclusion, this new answer wasn't adequate to distinguish real science from its pretenders either. Popper proposed an alternative. What a theory needed in order to be scientific wasn't verifiability, but falsifiability. A genuinely scientific theory, Popper says, is one that takes risks. A theory that nothing would disconfirm is pseudo scientific, and Freud and Marx are Popper's prime candidates. A genuinely scientific theory isn't like that. It sticks its neck out. Here, Popper's favorite candidate is Einstein's theory of relativity.

The time we're talking about was just about the time of the Eddington expedition intended to test Einstein's theory. Einstein predicted that light would be bent by gravity. If the theory was true, the position of stars photographed during an eclipse should differ from their positions when photographed in the night sky without the gravitational effect of the sun. The Eddington expedition set out to make precisely that comparison. If they came back with photographs that didn't show the effect, Einstein's theory was dead in the water, busted, falsified. It was precisely because Einstein's theory was so clearly falsifiable that Popper took it as the paradigm of genuine science.

What about evolution? Is that falsifiable, or should we put Darwin next to Freud and Marx on the shelf of pseudo science? Popper initially thought Darwinism was unfalsifiable too: "not a testable scientific theory but a metaphysical research programme." But Darwin himself lays down a potential falsification in the Origin of Species. He says, "If it could be demonstrated that any complex organ existed which could not possibly have been formed by numerous, successive, slight modifications, my theory would absolutely break down." J.B.S. Haldane, when asked what would disprove evolution, said "fossil rabbits in the Precambrian." Popper's later view was much more nuanced. There are ways to falsify major aspects of evolutionary theory, though, "really severe tests of natural selection are harder to come by than comparable theories in physics in chemistry," he said. When it comes to influential bits of twentieth-century philosophy of science, Popper's falsifiability criterion may top the list. Scientists themselves use it as a mark for good science. And yet it still causes confusion.

Everyone knows that false is bad. You certainly don't want to brag that your theory of the universe is false. But if false is bad, how can falsifiable be good? What Popper ends up with is an evaluative hierarchy of theories. At the bottom are unfalsifiable theories. Those are theories which no empirical or data or experience would refute. Their defenders will defend them no matter how things come out. Those are theories that have been insulated against the world that they're supposed to be about. They're not even good enough to be bad scientific theories, says Popper. Those are mere pseudo science.

Go up one step you have falsifiable theories that have, in fact, been falsified. Phlogiston theory postulated a special substance in things that explained why they burned. Lamarckian evolutionary theory postulated the inheritance of acquired traits. Newtonian physics postulated absolute space and time. Scientific theories? Absolutely. They took a genuine risk. They could be disproven by experience, and in fact they were. We now know that all those theories are false, but better that than unfalsifiable. On the highest rung we have falsifiable theories that haven't been falsified, well, at least not yet. Those are the best of the lot. No theory of any breadth, Popper emphasizes, can ever be conclusively confirmed. There is no point at which we say confirmed, close the book, and abandon all future testing. We should continue to look for falsifications. We should continue to put our theories to the test. That's the only way we'll find better ones.

There are, however, two ways to read the falsifiability criterion. One of them has to do with theories. One of them has to do with people. Popper writes as if he's talking about the theories, as if the theories themselves came intrinsically falsifiable or unfalsifiable, scientific or unscientific. I think it's more important to emphasize how people use theories—whether the people involved are treating the theory as falsifiable or not, whether the people involved are treating it scientifically or not. Take astrology, the theory that our destinies are controlled by the stars under which we're born. Could such a theory be put forward in a falsifiable form? My guess is that it could be. My guess is also that it would be falsified pretty quickly. But at least it would be on the second rung next to Phlogiston theory and Lamarck—scientific but unfortunately falsified. But that's not what the history of astrology looks like. The history of astrology is not a history of people looking for falsifications in order to formulate better theories. It's a history of people making money from fortune telling. The problem then may not be in the theories themselves. It may be in how people handle those theories.

Suppose you want to build a pseudo science. You want a genuinely bogus claim at the core, but one that sounds great, gets a lot of attention, but is protected against falsifications precisely because it is unfalsifiable. Suppose you want to build a pseudo science. How would you do that?

You've probably got some good ideas. Lay them out, and you'll know what to watch out for when you're on the receiving end of pseudo-science. Some claims are going to be unfalsifiable just because they come that way. What if I claimed that there is a very strange thing hovering just by your left shoulder. It's a cereotopic IV sphere, roughly 4.5 inches in diameter. It's there, alright. But don't expect to see it. The IV of IV sphere stands for invisible. There's an invisible cereotopic IV sphere hovering over your left shoulder. Cereotopic? I just made that up, but it means undetectable by any means that can be manipulated by mammals. There is nothing you can do or say to refute my theory. I can deflect your every objection. You don't see it? Of course not, it's an IV sphere. You can't detect it by touch, or with x-rays, or a CT scan? Of course not, it's a cereotopic IV sphere, and you're a mammal. That's an air-tight, pseudo-scientific theory. Unfalsifiability is built right in.

There are other ways to make something pseudo scientific. One is to make your theory a moving target. Every time someone tries to refute it, you say,

"Oh, that wasn't exactly my theory. More precisely..." and you change the theory a bit. That was the history of Ptolemaic astronomy. In simple form, the Ptolemaic theory held that the planets and the sun circle the earth. If that were true, all the planets would move uniformly across the night sky. But in fact they don't. If you plot the positions of Mars against the constellations night by night, you'll find that Mars moves in the same direction for a while but then starts moving backwards, then moves forward again. It's called retrograde motion. The astrologers, in fact, know all about it and figure it into their predictions for what will happen in your life.

Retrograde motion is an observable fact that doesn't fit the simple Ptolemaic theory. Busted. Falsified. But we could modify the theory a bit, and that's what happened. Perhaps what circles the earth aren't planets but invisible pivots around which the planets turn. What you see in retrograde motion is the planet going backward around that invisible point. It's called an epicycle. Well, it turns out that theory doesn't fit observations of Mars very well either. Oops. You get a better match if you have the planet circling a point which, itself, circles a point which goes around the earth. Well, it turns out that doesn't quite work either, but if you add a further epicycle...Epicycles upon epicycles.

Things like epicycles are called post-hoc modifications, modifications after the fact. Do that often enough and you make your theory unfalsifiable by making it a constantly-revised moving target. Ptolemaic astronomers were never conclusively refuted because they just kept adding epicycles. But it turns out there is a much simpler explanation, a sun-centered universe with elliptical orbits. With that simpler explanation, planets are like runners on an elliptical track, and retrograde motion happens when one planet laps another.

Built-in undetectability, post-hoc modifications, how else could you make your theory unfalsifiable? A good way is to build on ambiguous phenomena. Use something that is hard to see or that can be interpreted any way you want. In this category falls Percival Lowell's reports of canals on mars. Lowell was a primary astronomer of the early 1900s. It was he who founded the famous Lowell observatory in Arizona. Night after night Lowell peered at the tiny and wavering image of Mars through his optical telescope. He convinced himself that he could see a pattern of canals and that those canals must have been constructed and navigated by intelligent beings on the red planet. He published Mars and its Canals in 1906, followed two years later with Mars as the Abode of Life.

Another tip for building a pseudo science, build in a stipulation that only the properly trained or the truly initiated can see it work. There is, for example, the famous case of N-rays at the University of Nancy, in France, in 1903. Shortly after the discovery of X-rays, the discovery of N-rays was announced by René Blondlot, highly respected member of the French Academy of Sciences. N-rays were emitted by certain metals and blocked by lead, as X-rays were. They increased the luminosity of a spark and made luminescent paint glow slightly and could be bent by metal prisms. No one outside of France seemed able to duplicate Blondlot's experiments. The French claimed they just weren't properly trained. After all, it did take some training to be able to see the increase in spark intensity or paint luminosity. The American physicist Robert Wood volunteered to participate in Blondlot's experiments, but when it was his job to position the metal prism that would bend the N-rays, he palmed it instead. Amazingly, the reports given by Blondlot and his students were the same as if the prism had been in position. The same thing happened when Wood lied about whether a lead screen was in place. Blondlot's students were trained, alright. They were trained to see what they were supposed to see.

If you set out to build a pseudo science, those would all be good ways to do it. Be prepared to shift the theory by adding post-hoc modifications. Build on ambiguous phenomena, and insist that only the properly trained or the true believers can see the results. All of those will help to make your theory unfalsifiable. Of course you aren't setting out to build a pseudo-science—at least I hope you're not. But by thinking how you'd do it, you can see the tricks from the inside. That will help you recognize them from the outside. That will help vaccinate you against pseudo science.

But I want you to note that something very strange has happened in the course of this lecture. We started out talking about the demarcation between science on one side—physics, chemistry, microbiology—and so-called pseudo sciences on the other—astrology, phrenology, ancient astronauts. The strange thing is that many of the examples I just gave came from the science side. Ptolemaic astronomy was genuine astronomy. Percival Lowell was a well-respected scientist. René Blondlot's scientific credentials couldn't have been higher. That means that pseudo-science isn't just a problem outside the scientific establishment. Pseudo science is something to guard against even within the hallowed halls of established science—perhaps especially within the hallowed halls. Let me also point out that neither Lowell nor Blondlot were con men. Lowell really did believe he was seeing canals on Mars. Blondlot really did believe he could see the increase in spark luminosity. Even the best of scientists can fool themselves by falling for the traps of pseudo science.

The core demand of scientific rationality is that we put things to the test. How precisely do you do that? It sounds easy. You have a theory that says Y happens when X happens. So you make X happen and you see if Y happens. What's so tough about that? Let me give you a simple case. You have four cards in front of you. One has an E on it. Another has a K. One has the number four on it. Another has the number seven. E, K, four, and seven. Here is the theory you want to test: If a card has a vowel on one side, it has an even number on the other. It's your turn to put it to the test. Which cards must be turned over in order to test the theory? We can imagine that each experiment costs a lot of money, but we really want to know whether the theory is true. What is the minimum number of cards you have to turn over to test it? Press pause, take a minute, and think about that.

Which cards do you have to turn over to test the theory? The most popular answer is, turn over the E card. And, of course, that's right. The theory says, if you have a vowel on one side you have an even number on the other. So we turn over the E and see if there is an even number there. If there is, that's in accord with the theory. Great. If there isn't, the theory is refuted. Falsification. That's a good test. If it comes out one way, we get a confirmation. If it comes out another, we get a falsification. Are there any other cards that would test the theory? The second most popular answer is, turn over the four card. But that is definitely not right. If our experiments cost money, turning over the four card is a waste of research dollars. Suppose we turn over the four card and find a vowel, then we have a card that fits if there's a vowel on one side, there's an even number on the other. But what if we don't find a vowel? The theory didn't say if there's an even number there's a vowel. It didn't say only if there is a vowel is there an even number. So if we turn over the four and don't find a vowel, we haven't falsified the theory. All the four card can give us is confirmation. Turning over the four card can't give us falsification. So it can't really test the theory. The second card we should turn over is the seven card. Suppose we find a vowel on the other side. Then we have a vowel on one side, an odd number on the other, and we've falsified the theory. That gives us a real test.

Even in a case this simple, with only two variables, experimental design can be trickier than you might think. What I just showed you is the Wason selection task, one of the most studied examples in the psychology of reasoning. It's often taken as showing that people have a confirmation bias, that they look for confirmation of a theory (just like Popper said that the Marxists and Freudians did), but don't seek out falsifications, the real test of a theory. Exactly why people do that remains a matter of dispute. One of the strange things about the Wason selection task is that people don't make the mistake all the time. It seems to depend on context, and even on how everyday the problem is. In the way I just presented it, the problem is outlined in abstract terms, vowels and numbers. People rightly see that they should turn over the E but wrongly say they should turn over the four when it's really the seven they need to turn over in order to test the theory.

But people don't tend to make that same error when we change the problem just a little. We'll have four cards representing four people that you see on the beach. Their ages are on one side. What they are drinking is on the other. We want to turn over cards in order to test this theory: If you are drinking alcohol, you must be over 18. The cards say drinking beer, drinking Coke, 18, 16. Which cards you do you have to turn over to test if you are drinking alcohol, you must be 18?

People say, "Turn over drinking beer." Right. That's like the E card before. In this context, people see that the other right answer is urn over the 16 card. Then you'll be able to see if there's underage drinking going on. That's a real test of the theory. The interesting thing is that people don't say, "Turn over the 18 card." They know that it doesn't matter what's on the other side. If it's alcohol, fine, but if it's not, no falsification. They   don't go for that card, although, it's doing just what the four card did in the previous test, and they did go for the four card in the abstract case. So do people have a confirmation bias? Or is it just in unfamiliar or abstract contexts that such a bias appears? The psychologists haven't yet reached consensus on that one.

In this lecture we've emphasized putting it to the test as the core concept of scientific rationality with an emphasis on the importance of seeking falsifications of a theory. I hope you'll not only remember Popper's falsifiability criterion but put it to work. But let me close with a few warnings and provisos regarding the role of experiment in science. There are many different sciences, and the role of experiment is going to be different in different sciences. If an organic chemist forms a hypothesis regarding the composition of a specific protein, he can design an experiment and test it. But if a paleoanthropologist forms a hypothesis regarding the origins of Australopithecus Afarensis, there is no clear and obvious experiment that would confirm or refute it.

Here's another example. There are rival scientific theories as to the formation of the moon. But there is no clear moon-forming experiment we can perform to decide between them. The role of experiment in science is crucial, but it's often unavailable. Sometimes it takes years to invent an experiment that would confirm or disconfirm aspects of a theory. That was true for both parts of Einsteinian relativity and for quantum mechanics. Sometimes the results of an experiment are clear, but it may take decades to produce a theory for which those results will serve as confirmation. In *On the Nature of Things*, written about 60 B.C.E., the philosopher Lucretius notes the random motion of motes in sunlight. In 1828 a botanist named Robert Brown noted the same phenomenon in grains of pollen suspended in still water. Those random little jerks became known as Brownian motion. The phenomenon was widely known for decades, easily repeatable and a complete mystery, an experiment in search of a theory. Einstein's 1905 doctoral dissertation developed a statistical molecular theory of liquids. Among other things, it finally offered a theory for Brownian motion.

When we do have experimental evidence for a theory, it is often indirect. We can't afford, either economically or ethically, to perform wide-scale social experiments that would answer some of the questions we'd like to know the answer to: Whether the death penalty deters murder. Whether pornography leads to rape. What experimental evidence we do have regarding those questions has to be extrapolated from social psychological experiments involving small groups of people in relatively artificial circumstances. Scientific rationality does demand that we put hypotheses to the test. Any theory that waived away empirical test as irrelevant would clearly be an unscientific theory, but the role of experiment is often more elusive, harder, and more indirect than it might seem. Putting it to the test is often harder than it might seem. In the next lecture I want to take that further by examining some of the really beautiful experiments in the history of science.

Put It to the Test—Beautiful Experiments
Lecture 21

An understanding of scientific experimentation, and of its limits, is an important part of the contemporary philosopher's toolkit. This understanding is crucial both in backing your own claims and in evaluating the claims of others. In this lecture, we'll talk about the structure of experiments—putting questions to the test.

A Beautiful Experiment

- On January 28, 1986, the Space Shuttle *Challenger* lifted off at the Kennedy Space Center, but 73 seconds into the flight, the *Challenger* disintegrated in the skies over the Atlantic, instantly killing all aboard. What caused the crash?

- To answer that question, President Ronald Reagan set up the Rogers Commission, which included Charles Yeager, Neil Armstrong, Sally Ride, and the physicist Richard Feynman. The most famous moment in the commission's work occurred during a televised hearing, when Feynman performed a simple experiment to show that the O-rings sealing a joint on the right solid rocket booster could have failed under conditions of extreme cold.

The Well-Designed Experiment

- In order to establish an empirical claim, the gold standard is a well-designed experiment. The key, however, is the qualifier "well-designed."
  - Consider again the structure of Feynman's experiment. The commission was looking for a chain of causal events: A causes B, B causes C, and C caused the disaster.

  - It became clear that a blowby of pressurized hot gas and flame could have hit an adjacent external tank, leading to the structural failure. The manufacturer insisted that the O-rings couldn't have been what allowed the blowby. But Feynman's experiment with ice water showed that indeed they could have.

- Note three things about Feynman's experiment: the risk of the experiment, its power, and its limitations. Karl Popper insisted that every good scientific hypothesis takes a risk—a risk of being proven wrong. Feynman took that risk. His hypothesis about the O-rings might have been disproven in the public eye.

- But note also the limitations of the experiment. All that was shown is that the O-rings could have failed. The experiment didn't show that something else might not also have been a factor—even a major factor. The experiment addressed only one link in the causal story that the Rogers Commission finally settled on: the link from O-rings to blowby. Feynman's experiment, like all experiments, functioned only against a broader theoretical background—within a wider theoretical context.

The Controlled Experiment
- All experimentation is about an "if…then" statement. What any experiment really shows is "If this, then that." Feynman's experiment shows that if you put a C-clamped O-ring of this material in ice water, then its resilience fails.

- We need to know "If X, then Y." But we also need to know "If not X, then not Y." We need to know that X is what makes the difference as to whether or not Y happens. The core of any test of causality is, therefore, a double test. We need to have two cases side by side, one with X and one without X. That's the whole idea of a control and a **controlled experiment**.

Randomized Controlled Trial
- Another beautiful experiment was conducted in 1747 by James Lind, surgeon on H.M.S. *Salisbury*, in investigating the problem of scurvy. Lind tested the numerous proposed cures for the disease by selecting 12 men from his ship, all suffering from scurvy and as

The "scientific method" many of us were taught involves defining a question, forming a hypothesis, designing and performing a test with a measurable outcome, and analyzing the results.

similar as he could choose them. He then divided the men randomly into six pairs, giving each pair one of the cures.

- The important points here are that Lind chose men as similar as possible and that he randomized them into pairs. As it turned out, all the men showed some improvement. But the two who were given fruit recovered dramatically.

- Lind's is an early example of what we now call a randomized controlled trial. In constructing a solid experiment with complex subjects, such as people, you want to make sure you have a random enough selection procedure within a large enough group so that your results are unlikely to have happened by chance alone.

- Complex experimental subjects—again, such as people—make randomized testing a necessity. There are also other ways in which people make things complicated. These call for other constraints in experimental design.

Philosophical Understanding of Scientific Procedure
- The philosophical understanding of scientific procedure developed hand in hand with its application. We'll look at three high points of this development: the work of Francis Bacon, John Stuart Mill, and R. A. Fisher.

- In 1620, Francis Bacon published his *Novum Organum*, intending his work to be a new organon, a new philosopher's toolkit. The core of that new organon was an emphasis on "if…then" experimentation.
  - Bacon contrasts experimentation with other ways—what he takes to be erroneous or fallacious ways—of forming beliefs.

  - He pits empirical experiment against appeal to authority, against vague theorizing, and against appeals to experience without experimental manipulation.

- In 1843, the philosopher John Stuart Mill published his *System of Logic*, intended to offer a logic appropriate to scientific investigation. Mill gave a series of "methods," all of which focus on the kind of double "if…then" test we have emphasized for deciding whether X causes Y.
  - **Mill's method of difference** is precisely that double test. If X is the only difference between two cases, with Y a result in one and not in the other, we have an indication that X causes Y.

  - In **Mills method of agreement**, if two cases are alike only in X, and Y occurs in both, we again have evidence that X and Y are causally linked.

  - In **Mill's joint method of agreement and difference**, Y appears whenever we have X and disappears when we don't.

  - We have talked as if X and Y were simply on or off, present or not. Mill also generalizes that to a case in which things might be more or less X or Y. This is **Mill's method of concomitant variation**: If Y varies as X varies, we again have indication of a causal link.

- Beyond Mill, the most important improvements in our understanding of scientific procedure have been in the development of statistical measures for randomized testing. Here, much of current practice traces back to the English biologist R. A. Fisher.
  - Fisher's work was motivated by a beautiful experiment of his own, known as the Lady Tasting Tea example. The topic was trivial, but Fisher's experimental design and analysis are not.

  - Fisher's experiment is essentially what all controlled randomized experimentation looks like today, from political polling to medical experimentation.

The Double-Blind Experiment
- Suppose you want to know whether Coke or Pepsi tastes better. You do a straightforward test: You buy a can of Coke and a can of Pepsi and offer people a sample from each. In order to improve your experiment better, you blind the taste testers so that they are unable to see the cola cans.

- There are cases, however, that call for an even more scrupulous experimental design. When a researcher strongly believes in a hypothesis, there is always the chance that he or she will influence the results by giving unconscious clues to the experimental subjects. In order to guard against that, **double-blind experiments** are used. In a double-blind experiment, neither the investigator nor the subject knows until later what test condition the subject is in.

- In evaluating experimental evidence, you want to know the following: What was the control? If complex subjects are involved, how was **randomization** assured? What is the probability of getting the result by chance? For some questions, you may want to know whether the experiment was blinded or double-blinded.

Scientific Method and Scientific Discovery

- The principles of experimental design we have been discussing are all parts of the toolkit of empirical investigation. The scientific method is as follows: Define a question. Formulate an explicit hypothesis. Design a test with measurable outcome. Perform the test and analyze the results.

- Scientific research, however, also often requires a little bit of luck. It has been estimated that half of all scientific discoveries have been stumbled on. This was the case, for example, with the discovery of penicillin by Alexander Fleming in 1928.

- Scientific discovery can be chancy. Philosophers of science have long distinguished between a **context of discovery** and a **context of justification**. There's no recipe, no algorithm, no scientific method of discovery. Once a discovery is made, however, the process of demonstrating the fact to others can follow a very specific method— the method of repeatable experiment.

Limitations to the Scientific Method

- Thus far, we've seen aspects of experimental design that are important both in presenting and in evaluating experimental evidence, but there are limitations in some aspects of experimental design.

- For example, experimental results can be faked and sometimes have been, both in the fringes of pseudoscience and in core areas of testing.

- Experimental results can be selectively chosen to support the favored case. This is called cherry-picking. There are perennial worries that companies choose to report those experiments that make their products look good and choose not to report those that don't. What we really need to know are the results of all the relevant experiments.

- Not every experimental result is submitted for publication, and not every result submitted for publication is accepted. This is called the file-drawer problem. When you survey the literature to find out whether most experiments support or fail to support a particular effect, you may be surveying a misrepresentative sampling of all the tests that have actually been run. The tests that show a less dramatic or less popular result may be buried, unseen, in someone's file drawer.

Consider the Context

- Experimentation is never pure; it always takes place in a context. All effective experimentation tests one belief against a background of other beliefs. For the moment, at least, those other beliefs are simply assumed.

- One of the earliest cases cited in the history of experimentation is Eratosthenes's calculation of the circumference of the earth.

    ○ The basic observation was this: On June 21, the summer solstice, in the Egyptian town of Syene, the sun appeared directly overhead. Looking down a well, a man's head would block the sun's reflection.

    ○ Back in Alexandria, Eratosthenes knew that on that day, the elevation of the sun was 1/50 of a circle off from being directly overhead—a little over 7 degrees, and he knew that Syene was directly south of Alexandria. Thanks to the royal surveyors, he also knew the distance between the two towns.

    ○ Assuming the earth to be a sphere and the sun to be so far away that its rays could be thought of as essentially parallel, Eratosthenes was able to use those two facts and a little mathematics to conclude that the circumference of the earth was a little more than 25,000 miles—very close to our contemporary estimate.

    ○ But note the assumptions here. If the earth were not a sphere, the calculation would not hold, and the experiment would not show what it was taken to show. If the sun were not so far away that its rays could be treated as parallel, the calculation would not hold.

- o Because no experimental result is established independently of a set of background assumptions, what every experiment really establishes is an alternative: either the claimed result holds or one or more of our background assumptions is wrong.

Terms to Know

context of discovery: Used by philosophers of science to label the process by which a hypothesis first occurs to scientists, as opposed to the context of justification, the process by which a hypothesis is tested. The consensus is that context of justification follows clear, rational rules, but context of discovery need not do so.

context of justification: Used by philosophers of science to label the process by which a hypothesis is tested or established, as opposed to how it is first invented or imagined. The consensus is that context of justification follows clear, rational rules, but context of discovery need not do so.

controlled experiment: A test that demonstrates a causal connection between a variable (X) and an outcome (Y) by establishing both "if X, then Y" and "if not X, then not Y." In essence, a controlled experiment is a double test of two cases side by side, one with and one without X.

double-blind experiment: In experimentation, a way of controlling the possibility that unconscious bias of subjects and investigators may influence the results by assigning subjects so that the experimental condition is hidden from them. In a double-blind experiment, neither the subjects nor the investigators who interact with them know which experimental condition is being tested until after the experiment is complete.

Mill's joint method of agreement and difference: In John Stuart Mill's *System of Logic*, evidence for causal linkage between X and Y is stronger when conditions for both Mill's method of agreement and Mill's method of difference are present; that is, Y appears whenever X is present and disappears when it isn't.

Mill's method of agreement: In John Stuart Mill's *System of Logic*, if two cases are alike only in X, and Y occurs in both, we have evidence that X and Y are causally linked.

Mill's method of concomitant variation: In John Stuart Mill's *System of Logic*, applies to cases in which there is a range of values for X and Y. If the amount or strength of Y varies as the amount or strength of X varies, there is evidence for a causal link between X and Y.

Mill's method of difference: In John Stuart Mill's *System of Logic*, if the presence or absence of X is the only difference between two cases, with Y a result in the first case and not in the second, we have an indication that X causes Y.

randomization: In experimentation, a procedure for assigning groups so that the differences between individuals within groups distribute with equiprobability across experimental conditions.

Suggested Reading

Crease, *The Prism and the Pendulum*.

Peirce, "The Fixation of Belief."

Salsburg, *The Lady Drinking Tea*.

Questions to Consider

1. Do you think that dogs and cats make experiments? Draw on your own experience to explain why or why not.

2. This lecture outlines several ways that an experiment can fail to be well designed. Here are examples of two ways an experiment can fail:
   (a) For a taste test, people are asked to choose whether Post Toasties or Kellogg's Corn Flakes taste the best, with a full view of the product boxes.

   The experiment is not blinded.

(b) You put a deck of cards between you and a friend. You perform an experiment in clairvoyance by picking cards one by one and looking at them with their backs to your friend. Your friend's job is to use extrasensory perception to say whether the card is red or black.

The experiment is not double-blinded.

Now supply your own examples for an experiment that would fail in each of these ways:

(a) The experiment shows that when factor X is present, factor Y is present, too. But it doesn't show that when factor X is not present, factor Y is not. It is not a controlled experiment.

(b) The experiment exposes one group of people to factor X. They develop factor Y. The experiment includes another group of people who are not exposed to factor X. They do not develop Y. But people were not assigned randomly to the two groups. It is not a randomized controlled experiment.

3. The Greeks, for all their brilliance, never developed a full science involving experimentation. Why do you think that was? How might history have been different if they had?

Exercises

Can your friends tell expensive bottled water from inexpensive bottled water and each of those from tap water? Expand R. A. Fisher's Lady Drinking Tea experiment by preparing four samples of each and giving them to a friend "blind," in random order.

How well did your friend do in categorizing each sample, compared with what one might expect by chance alone?

Put It to the Test—Beautiful Experiments
Lecture 21—Transcript

Professor Grim: On January 28, 1986, the Space Shuttle *Challenger* lifted off at the Kennedy Space Center. The launch was broadcast live, with many classrooms watching, because Christa McAuliffe was among the crew as the first member of the Teacher in Space Project. Seventy-three seconds into the flight, the *Challenger* disintegrated in the cold skies over the Atlantic, instantly killing all aboard. What caused the crash? To answer that question President Ronald Reagan set up the Rogers Commission, named after its chair, former Secretary of State William P. Rogers. Also on the commission were Charles Yeager, the first person to break the sound barrier, Neil Armstrong, the first person to walk on the moon, Sally Ride, the first American woman in space, and, with reluctance, the physicist Richard Feynman. The most famous moment in the Commission's work occurred during a televised hearing. The subject was the temperature at which the flight had been launched and the O-rings that sealed a joint on the right solid rocket booster. A gas leak at that point had been seen in footage just before the disaster. But Morton-Thiokol, the company that manufactured the O-rings, insisted that their product could not have been the problem.

In the course of the televised hearing, Feynman offered an experiment. He used a flexible O-ring of the same material, clamped in a C-clamp, and submerged in the ice water on the table in front of him. When he took the O-ring out and removed the C-clamp, it did not resume its previous shape. It was not resilient. For at least a few seconds, long enough to cause a major failure, the material cramped in the cold. In an appendix to the final Rogers Commission report, Feynman wrote "For a successful technology, reality must take precedence over public relations, for nature cannot be fooled."

Feynman's was a beautiful experiment, beautiful first and foremost for its simplicity. Media was at a fever pitch. Within an hour 85 percent of the American public had seen rebroadcasts or heard of the disaster. The reputation, not only of major companies, but of NASA and the United States space program were at stake. What caused the crash? Look at what happened to that O-ring in a glass of ice water.

In this lecture I want to talk about the structure of experiments—putting questions to the test. Feynman's is just one of the beautiful experiments I want to explore. An understanding of scientific experimentation, and of its limits, is an important part of the contemporary philosopher's toolkit. An understanding of experimentation is important both in backing your own claims and in evaluating claims of others. In order to establish an empirical claim, the gold standard is a well-designed experiment. The key, however, is that qualifier: well designed. Many of the claims that come our way—claims in the media regarding medical safety or efficacy, for example—are presented as having scientific experiment behind them. In order to really evaluate these claims, we need to be able to evaluate the experimental work behind them. Are those experiments well designed?

Look again at the structure of Feynman's experiment. The question is what caused the disaster. The commission had studied footage of the event repeatedly, and that footage seemed to show a gas blowby just before the spacecraft broke apart. Could that have been the problem? And what caused that? They were looking for a chain of causal events: A causes B, B causes C, and C causes the disaster. It became clear that a blowby of pressurized hot gas and flame could have hit an adjacent external tank, leading to the structural failure. But Morton Thiokol insisted that the O-rings couldn't have been what allowed the blowby. What Feynman's ice water showed was that indeed they could have, that Morton Thiokol was wrong.

I want you to note three things about that experiment: the risk of the experiment, its power, and its limitations. First of all, the risk. Karl Popper insisted that every good scientific hypothesis takes a risk, a risk of being proven wrong. Feynman took that risk. If the ring had come out of the water and sprung instantly back into shape, his O-ring hypothesis would have been discredited right there on television. But of course it didn't come out that way. Hence, the power of the experiment. It no longer mattered who said that the O-rings couldn't have been the problem. "Nature cannot be fooled," as Feynman said. And nature said, they could have.

But note also the limitations of the experiment. All that was shown is that the O-rings could have failed. Perhaps it shows that they could be expected to fail under the temperatures of the flight—15 degrees colder than any previous launch, incidentally. But the experiment didn't show that something else might not also have been a factor, even a major factor. The experiment only addressed one link in the causal story that the Rogers Commission finally settled on: the link from O-rings to blowby. The rest of the story about released gasses and flame hot enough to impact the exterior tank, about that leading to the break-up of the spacecraft, all that part of the story, had to come from somewhere else. Feynman's experiment, like all experiments, functioned only against a broader theoretical background within a wider theoretical context.

All experimentation is about an if, then. What any experiment really shows is if this, then that. Feynman's experiment shows that if you put a C-clamped O-ring of this material in ice water, then its resilience fails. If X, then Y. It seems to me that humans must have always tested things by asking what happens if you do this, if you hit a tasty bone with a large rock, or extend a spear with a throwing extension. In that sense it seems that humans must have always performed experiments. Animals seem to test things in much the same way. They feign an attack to see if the enemy retreats. What will he do if I do this? They nose prey to see if it will move again. Scientific experimentation is just a more formal extension of that kind of if-then testing applied to general claims at the level of scientific theories. The question, what was the first scientific experiment, is, thus, bound to invite controversy.

Around 200 B.C. Archimedes submerged himself in his bath and saw the displacement of water, and then applied that principle to other cases. Does that count as an experiment? Around 50 A.D. Hero of Alexandria developed some amazing hydraulic devices, which must have required extensive if-then testing. Does that count as experiment? Credit for the beginnings of scientific experimentation is often given to the Islamic School, the school of Jābir ibn Hayyān, around 800 A.D., known for elaborate alchemical equipment and an insistence on practical experience, or Alhazen, who around 1000 A.D. published a work on optics with a number of experimental demonstrations. I think if-then testing is quite natural to both people and animals. Scientific experimentation is just a formalization in a context of large-scale theory. If so, and as long as formalization and large-scale theory remain vague, the answer to what was the first scientific experiment will be vague as well.

The concept of controlled experiment leaves a clearer historical record. We envisage the functioning of the universe as a complicated pattern of causal chains. We want to know what that causal pattern is, what causes what. We want to know whether Xs cause Ys. In order to establish that X causes Y, we have to establish more than just if X, then Y. In order to establish that X causes Y we really have to establish two if, thens. We need to know that if X, then Y. But we also need to know that if not X, then not Y. We need to know that X is what makes the difference as to whether or not Y happens. The core of any test of causality is, therefore, a double test. We need to have two cases side by side, one with and one without X. That's the whole idea of a control, and of controlled experiment.

Here's another beautiful experiment as an example. The theory of spontaneous generation had the authority of Aristotle, who claimed that some animals developed, not from kindred stock, but "from putrefying ... matter." It was the decaying meat itself that was thought to generate maggots, for example, which then develop into flies. Imagine yourself in an Italian city in the 1600s. You can imagine the crowded streets, the tiny butcher shops, and the clouds of flies around those butcher shops. Put a piece of raw meat out, and it will soon have maggots. Aha, spontaneous generation. There's an if, then for you. Francesco Redi had a suspicion that Aristotle was wrong; that all animals developed "from kindred stock," and that it was flies, rather than meat, that generated maggots. In 1687 he published the results of a controlled experiment as a demonstration. He put fresh meat in four jars, exposed to flies and the open air. That's the if-X-then-Y part: if flies, then maggots. But he also put fresh meat in four other jars covered with cloth so as to keep the flies away. That's the if-not-X-then-not-Y part. That's the control. Maggots developed in the uncovered jars, but not in the covered ones. It wasn't the meat but the flies that bred maggots.

The double if, thens of controlled experiments form the basis of all our causal knowledge. In constructing an experiment, you have to make sure that you have a clear conception of both the experimental condition and the control. In evaluating an experiment you need to ask, what's the control? But take a minute to think about what that demands. What we want is two cases that are exactly alike except for the presence of X. If in one of those cases we observe Y, and in the other we don't, and the two cases are otherwise exactly alike, in that setup it must be X that made the difference.

The double if, thens form the basis of all our causal knowledge. The demand that our two conditions be identical except for X is the major source of all experimental difficulty. The problem is that there is always the possibility of hidden differences between our two conditions. Redi's covered and uncovered jars were supposed to differ only in the presence and absence of flies. But of course they didn't. Air circulation would be different with and without a cloth over the top. Temperature was probably affected. Maybe the type of cloth he used suppressed spontaneous generation. Asking for just one difference between two cases may be asking for the impossible. Any two cases will differ in hundreds of ways. One jar will be north of another, or east, or closer to the window, or the one I put the meat in first. We think we know that those are differences that don't make a difference. But that's just the point. All experiment tests things we don't know against a background of other things we think we do know. The conditions of controlled experiment are no different.

The difficulties of getting two cases otherwise identical, except for some factor X, become even more pressing when we are dealing with complex experimental subjects—people, for example. We want to know whether a given treatment prevents scurvy, or cures cancer. We need two groups of people, those who are given the treatment and those who are not. But we want the treatment or the lack of it to be the only difference. We want the groups to be identical in all other respects. Good luck. People are complicated. Groups of people are still more complicated. You will never get two groups of people identical except for factor X. So what do we do?

Here's another beautiful experiment. Scurvy had long been a problem for men at sea. It was a problem in 1747 when James Lind served as a surgeon on Her Majesty's Ship *Salisbury*. There were a number of proposed cures for scurvy, and he tested them with a controlled experiment. The cures, a quart of cider every day; twenty-five drops of sulphuric acid in water three times a day; one half pint of seawater every day; a lump of garlic, mustard, and horseradish; two spoonfuls of vinegar three times a day; two oranges and one lemon every day. Lind selected 12 men from the ship, all suffering from scurvy, and as similar as he could choose them, he said. He then divided the men randomly into six pairs, giving each pair one of the cures, and he watched what happened. The important point is that Lind chose men as similar as he could, and that he randomized them into pairs. As it turned out, all the men showed some improvement, but the two given oranges and lemons recovered dramatically. One was back to duty after six days. The other helped care for the rest.

Lind's is an early case of what we now call a randomized, controlled trial. When it comes to experimental subjects as complicated and as various as people, we can never be confident that the factor X we are testing for is the only difference at issue. Indeed, we know that it won't be; there will always be other differences between people and between groups of people. What we do instead is to assign our groups randomly. With large enough groups, with tests repeated often enough, with strong enough results on one side and not the other, it becomes increasingly implausible that there is some hidden factor other than X that is producing the result. In constructing a solid experiment with complex subjects like people, you want to make sure you have a random enough selection procedure within a large enough group that your results are unlikely to have happened just by chance. You want to know precisely those same things in evaluating experimental claims from other sources—claims from the media, for example. Unfortunately, those crucial details often don't fit within a sound bite. In order to evaluate those claims you often have to do a little more research as to how the research was conducted.

The philosophical understanding of scientific procedure developed hand in hand with its application. I'll give you three high points: the work of Francis Bacon, John Stuart Mill, and R. A. Fisher. In 1620, about 60 years before Redi's experiments refuting Aristotle, Francis Bacon published his *Novum Organum*. That means the "new organon". The title is a direct reference to Aristotle's logic in *The Organon*. Francis Bacon intended his work to be a new organon—a new philosopher's toolkit. The core of that new organon is an emphasis on if-then experimentation. Bacon contrasts experimentation with other ways—what he takes to be erroneous or fallacious ways—of forming beliefs. He pits empirical experiment against appeal to authority, against vague theorizing, against appeals to experience without experimental manipulation.

In 1843, the philosopher John Stuart Mill published his *System of Logic*, intended to offer a logic appropriate to scientific investigation. Mill gave a series of methods, all of which focus on the kind of double if-then test we have emphasized for deciding whether X causes Y. Mill's method of difference is precisely that double test. If X is the only difference between two cases, with Y a result in one and not in the other, we have an indication that X causes Y. Mill's other methods complement that structure. Here is his method of agreement. If two cases are alike only in X, and Y occurs in both, we, again, have evidence that X and Y are causally linked. Our evidence is even stronger, of course, if we have both of these. That's Mill's joint method of agreement and difference. Y appears whenever you have X, and disappears when you don't. We have talked throughout as if X and Y were just on or off, there or not. Mill also generalizes that to a case in which things might be more or less X or Y. That's the method of concomitant variation. If Y varies as X varies, we, again, have indication of a causal link.

Beyond Mill, the most important improvements in our understanding of scientific procedure have been in the development of statistical measures for randomized testing. The philosopher Charles Sanders Peirce was important in early development, but much of current practice traces back to the English biologist R. A. Fisher. It has been said that Fisher almost single-handedly created the foundations for modern statistical science. Fisher's work was motivated by a beautiful experiment of his own, known as the Lady-tasting-tea example. The topic is trivial. Fisher's experimental design and analysis are not. The British take their tea with milk. At a summer tea party, a Lady claimed that tea tasted differently if one first put in the milk and then added the tea, or first put in the tea and then added the milk. Preposterous, said the scientifically minded men around her. Fisher put the question to the test with a randomized, controlled experiment. Four cups were prepared with the milk first. Four cups were prepared with the milk second. The Lady was fully informed of the procedure and was then given the cups in a random order. Her task, to pick the four that were prepared with the milk first. Fisher formalized the case in terms of two hypotheses. He actually thought the Lady wouldn't be able to detect any difference. He called that the null hypothesis. The alternative hypothesis was that she could tell the difference. If she picked the four cups correctly, what confidence should we put in that alternative hypothesis? Fisher's calculation of what we now call statistical significance was based on the probability that she would have made the right choice by chance alone. That's essentially what all controlled, randomized experimentation looks like today, from political polling to medical experimentation. Oh, and the result? She picked the four cups correctly. We can say with a high degree of confidence that there really is a difference in how tea tastes depending on whether you put in the milk first or second, a difference, at least, in how it tastes to a lady.

Complex experimental subjects, like people, make randomized testing a necessity. There are also other ways in which people make things complicated. These call for other constraints in experimental design. Suppose you want to know whether Coke or Pepsi tastes better. You do a straightforward test. You buy a can of Coke and a can of Pepsi and offer people a sample from each. Which tastes better? You can see the problem with that experiment. Coke drinkers may be loyal to their Coke, the same for Pepsi people. Because they can see where each sample is coming from, you may not be getting a pure taste test. In order to make that experiment better, you'd like your taste-tasters to be unable to see the cola cans. You want them to be blind. Now we're talking about a blinded experiment. There are cases, however, that call for an even more scrupulous experimental design. We've been talking about experiments that involve human subjects. All of those experiments also involve human experimenters. When a researcher strongly believes in a hypothesis, there is always the chance that he or she will influence the results by giving unconscious clues to the experimental subjects. In order to guard against that, double-blind experiments are used. In a double-blind experiment, neither the investigator nor the subject knows until later whether the randomized case they are dealing with is from the if-X group or the if-not-X group.

In evaluating experimental evidence, then, you want to know what was the control? If complex subjects like people are involved, you want to know how was randomness assured? What is the probability of that result just by chance? For some questions, you may want to know was this experiment blinded or was it double-blinded?

Is there a scientific method? In a sense, there surely is. The principles of experimental design we have been talking about are all parts of the toolkit of empirical investigation. The scientific method taught to many a junior high school student includes aspects of investigation emphasized since Francis Bacon: define a question, formulate an explicit hypothesis, design a test with measurable outcome, perform the test and analyze the results. But the notion of a scientific method that can be followed robotically, like a recipe for truth, is a myth. Alhazen said it in his experimental work on optics: "Finding the truth is difficult, and the road to it is rough." William Whewell's *Philosophy of the Inductive Sciences*, published just before John Stuart Mill's *System of Logic*, emphasized that science requires not mere method but "invention, sagacity, genius."

Scientific research also often requires luck. It has been estimated that half of all scientific discoveries have been stumbled on. Often the process starts with what appears to be a problem with the scientific apparatus, a bug in the program. When that bug doesn't go away, it starts to dawn on the investigator that the unexpected result may not be an error but precisely that, an unexpected result. Aha! Here's a beautiful example. On the morning of September 28th, 1928, Alexander Fleming came back from vacation to his lab in the basement of St. Mary's Hospital in London. As he tells the story, he noticed a Petri dish containing the bacterium staphylococcus. He had left it open by mistake, and it was now contaminated by a blue-green mold. Around the mold was a halo in the bacteria. Fleming concluded that the mold was releasing a substance that inhibited bacterial growth. He had discovered antibiotics. After a few months of calling the active substance mold juice, he decided to name it after the blue-green Penicillium mold. He called it penicillin, and as he often had to remind people, "I did not invent penicillin. Nature did that. I only discovered it by accident."

Much of scientific discovery is a matter, not of method, but of invention, sagacity, genius, and luck. All of those can go together, Louis Pasteur is credited with the saying that "Chance favors the prepared mind." Philosophers of science have long distinguished between a context of discovery and a context of justification. Scientific discovery can be chancy. There's no recipe, no algorithm, no scientific method of discovery. Once a discovery is made, however, the process of demonstrating the fact to others can follow a very specific method, the method of repeatable experiment.

Once Fleming had his initial hunch, however fortuitous, it was by controlled experiment that he demonstrated its truth.

I've tried to emphasize aspects of experimental design that are important both in presenting and in evaluating experimental evidence. But there are limitations. Unfortunately, there are cases in which the information we need in order to evaluate experimental evidence is somehow hidden or unavailable. Experimental results can be faked, and sometimes have been, both in the fringes of pseudo science and in core areas of medical testing. Experimental results can be selectively chosen to support the favored case. It's called cherry picking. There are perennial worries that drug companies choose to report those experiments that make their products look good and choose not to report those that don't. What you really need to know is the results of all the relevant experiments. The file-drawer problem. Not every experimental result is submitted for publication, and not every result submitted for publication is accepted. When one surveys the literature searching for whether most experiments support or fail to support a particular effect, one may be surveying a misrepresentative sampling of the tests that have actually been run. The tests that show a less dramatic or less popular result may be buried, unseen, in someone's file drawer.

There is another limitation to experimentation that I have tried to emphasize throughout. Experiment is never pure. It's always in context. All effective experimentation tests one belief against a background of other beliefs. For the moment, at least, those other beliefs are simply assumed. One of the earliest cases cited in the history of experimentation is Eratosthenes's calculation of the circumference of the earth, sometime around 225 B.C. The basic observation was this: On June 21, the summer solstice in the Egyptian town of Syene, the sun appeared directly overhead. Looking down a well, a man's head would block the sun's reflection. Back in Alexandria, Eratosthenes knew that on that day the elevation of the sun was 1/50th of a circle off from being directly overhead, a little over seven degrees. He knew that Syene was south of Alexandria, and thanks to the royal surveyors, he knew the distance between the two towns. Assuming the earth to be a sphere and the sun to be so far away that its rays could be thought of as essentially parallel, Eratosthenes was able to use those two facts with a little mathematics to conclude that the circumference of the earth was a little over 25,000 miles, very close to our contemporary estimate. But note the assumptions at issue. If the earth were not a sphere, the calculation would not hold, the experiment would not show what it was taken to show. If the sun were not so far away that its rays could be treated as parallel, the calculation would not hold.

The *Huainanzi* is a Chinese text from approximately the same date. It also notes that vertical sticks at different points north and south cast shadows of different length on the same day. From that fact, and the assumption that the earth is flat, the text concludes that the vertical sticks with shorter shadows are directly under the sun. Those with longer shadows are farther away—the sun's rays hit those at an angle. From the length of those shadows, the *Huainanzi* concludes we should be able to calculate the height of the sun. Because no experimental result is established independently of a set of background assumptions, what every experiment really establishes is an alternative. Either the claimed result holds, or one or more of our background assumptions is wrong.

In the next lecture I want to take return to a theme that we've touched on at various points: thinking rationally in a social context.

Game Theory and Beyond
Lecture 22

In a previous lecture, we talked about decision theory. In a nutshell, we said, when choosing between two options, calculate the probability and value of possible outcomes. Then pick the option that has the highest probability multiplied by value. Maximize expected utility. Decision theory, however, is made for contexts in which there is just one decision maker: you. Where decision theory leaves off, game theory begins. Game theory is designed to address, in an abstract way, rational decision making in the social context—the context in which multiple players are making interlocking decisions. An understanding of game theory and an appreciation of its limitations should both be part of the philosopher's toolkit.

Game Theory

- **Game theory** originated with John von Neumann, who was also instrumental in the development of computers and the atomic and hydrogen bombs.

- What does von Neumann's involvement with nuclear weapons have to do with game theory? Not long after World War II, the Soviet Union acquired nuclear capability. The United States could not navigate in that world as if it was the only decision maker that mattered. It was necessary to think through chains of action and response, and multiple decisions on both sides mattered significantly. With the possibility of mutual annihilation looming, the stakes could hardly be higher.

- It was in anticipating a world with two nuclear superpowers that von Neumann developed game theory. In 1944, von Neumann and Oskar Morgenstern published *Theory of Games and Economic Behavior.* The subject sounds innocent enough: games. But the real target was rational strategies for conducting a cold war, rather than a hot one.

Game theory is designed to address, in an abstract way, rational decision in the social context—the context in which multiple players are making interlocking decisions.

Classic Game Theory: Prisoner's Dilemma

- Games in von Neumann's sense aren't just checkers and chess. They include every situation involving cooperation and competition in which the outcome depends on the multiple actions of multiple players. The game that has become the central exhibit of game theory is called the **Prisoner's Dilemma**. That particular game—that particular tradeoff between potential competitors or cooperators—has been the most studied game in game theory.

- The story behind the game is one of two bank robbers, Al and Bruiser, caught and held in separate cells. The district attorney admits that the evidence in the case is pretty thin. If both bank robbers refuse to cooperate, they will both get off with light sentences. But if Bruiser confesses and Al doesn't, Bruiser's evidence will be used against Al; Bruiser will go free; and Al will be put away for years. The same is true if Al confesses and Bruiser doesn't. If both confess, they will both go to jail, but they may receive some time off because they pled guilty.

- The question is: Should one accomplice give evidence against the other? The important point is that the outcome doesn't all depend on one prisoner or the other. It depends on how the decisions by both prisoners play out. That's game theory.

- As in previous lectures, we can use a matrix to represent possible outcomes, and we see that there are four ways this game could play out, reflected in the four quadrants of the matrix: Both players cooperate with each other, both defect against each other, A cooperates and B defects, or A defects and B cooperates. (Note that "cooperate" here means to cooperate with the other prisoner, not the police.) In the matrix, payoffs for A are listed in bold and italics in each box, and those for B are listed in standard type.

		B	
		Cooperate	**Defect**
A	**Cooperate**	*3*, 3	*0*, 5
	Defect	*5*, 0	*1*, 1

- Game theory is about how decisions interact. The matrix shows potential gains and losses for each side given different combined decisions. If one prisoner cooperates with the other, while the second player defects against the first, the defector walks away with five points—a full five years of freedom. The cooperator gets no years of freedom. If both players choose to cooperate with each other, both end up with three points—three years of freedom. If both defect, they both end up with only one year of freedom.

- The rational thing to do is to defect. This is a first major result in game theory. In a one-shot Prisoner's Dilemma, defection is strongly dominant.

Tit for Tat
- In real life, however, interactions are almost never a one-shot deal. Social life involves repeated interactions with the same people and the same organizations. When we step up to this level, we're dealing

with what is called an iterated game. In an iterated Prisoner's Dilemma, the payoffs on each round stay the same, but there is more than one round and the players remember what happened in previous rounds.

- In the iterated Prisoner's Dilemma, there's one strategy that's of particular interest: **tit for tat**. With this strategy, one player mirrors the play made by the opponent in the previous round; the player cooperates if the opponent cooperated in the previous round but defects if the opponent defected.

- Robert Axelrod, a political scientist at the University of Michigan, tried to understand the iterated game by running a computer-based Prisoner's Dilemma tournament. He set it up so that every submitted strategy played 200 rounds against every other strategy. He also made sure that every strategy played against itself and against an opponent that cooperated and defected at random. Axelrod added up the total points that each strategy gained in all those competitions. The winner was the strategy that got the most points over all.

- The strategy that came out with the most points—the most rational strategy to play against that whole field of competitors—was the simple strategy tit for tat.

- In the one-shot Prisoner's Dilemma, defection is the rational choice. But in repeated games played with many different players—as in Axelrod's tournament and more like real life—the strategy that does best is one that starts by cooperating and returns like for like.

- There's an important lesson to be learned from the tit for tat result: Cooperation may triumph even in situations where we might think it impossible.

The Limitations of Game Theory
- The limitations of game theory have lessons to teach us about social rationality.

- The Prisoner's Dilemma is clearly an artificial setup—artificial enough to be captured in a matrix. That in itself isn't bad—**simplification** is one of our most important conceptual tools—but there are several artificialities in that setup of which we should be particularly wary.
  - The first is that everybody knows the rules, everybody knows the payoffs, and it's always clear to both sides whether an action was a cooperation or a defection. Such a situation is called a game of perfect information. But real life is often not like that.

  - Second, note that all moves are either full cooperation or full defection. There's no halfway here.

  - Third, also note that players have to play on each round. They can't just opt out.

- For real social rationality, then, it's not enough to act as one would in game theory. For real social rationality, one also has to keep in mind that actions of cooperation and competition may not always be clear and don't have to be all or nothing; further, opting out may be an important possibility.

Behavioral Economics

- Decision theory and game theory have one core characteristic in common. Both gauge rationality in terms of self-interest. In game theory, each player is calculating how to maximize his or her own gain on the assumption that other players are doing the same. But do people actually make decisions in the way that game theorists—and economists in general—regard as rational?

- The answer is that they don't. That's a major finding of the developing field of behavioral economics, which puts the standard assumptions to the test. The truth is that people systematically violate the economist's canonical assumptions of rationality.

- Consider the **Ultimatum Game**. Two people, the proposer and the responder, meeting for the first time and have to split an amount of money. The proposer gets to propose how the money is to be split—50/50, perhaps, or 90 percent to the proposer and 10 percent to the responder. The responder then gets to choose whether to accept that division or not. If the responder accepts, the money gets divided as proposed. But if the responder declines the division, neither side gets any money at all.

- Here is what economic rationality says that you should do: If you're the responder, you should take any division that gives you any money at all. Why? Because something is better than nothing.

- What is the rational thing to do as a proposer? If the responder is rational in the way just outlined and you're just out for the money, you should propose as much as possible for you and as little as possible for the responder. Because the responder is rational, he or she will take the money offered. Because you're rational, you'll walk away with a bundle.

- But in the real world (at least in industrialized societies), proposers tend to offer about 44 percent to the other side. In sampling across a range of societies, across all continents, virtually no proposer offers less than 25 percent to the other side. Interestingly, the more that a culture incorporates a market economy, the higher the offer is to the other side.

Social Rationality

- As of now, the best formal tool we have for studying social rationality is game theory. It at least recognizes the issue that decisions made jointly may have a different rationality than those made individually. Game theory supplies some simple models. It even offers some hope for emergence of cooperation in egoistic situations, such as the Prisoner's Dilemma.

- Its limitations when applied to real life carry lessons as well. It may not always be clear whether our real-life "moves" are cooperations, defections, or something in between. The "game" we're playing in real life—the payoffs to each side—may be far from obvious. Indeed, the game itself may be open to negotiation. The motivations of the multiple players won't be as predictable as purely rational models might lead us to believe and might well depend on differences in cultural background.

Terms to Know

game theory: The mathematically idealized study of rational interaction in competitive and cooperative situations.

Prisoner's Dilemma: In game theory, a two-person game in which the value of mutual cooperation is greater than the value of joint defection, the value of defecting against a cooperator is greater than the value of mutual cooperation, but the value of cooperating against a defector is lower than the value of mutual defection, as illustrated in the following matrix:

B

		Cooperate	Defect
A	Cooperate	*3*, 3	*0*, 5
	Defect	*5*, 0	*1*, 1

Prisoner's Dilemma Matrix

In the Prisoner's Dilemma, the real loser is cooperation in the face of defection.

simplification: Conceptual replacement of a complex concept, event, or phenomenon with a stripped-down version that is easier to understand or to manipulate mentally.

tit for tat: In game theory, a strategy for playing an iterated game in which a player mirrors the play made by the opponent in the previous round; the player cooperates if the opponent cooperated in the previous round but defects if the opponent defected in the previous round.

Ultimatum Game: In game theory, a game in which two participants are given the task of splitting a payoff, for example, a certain amount of money. One participant is randomly selected to be the Proposer of how to make the split, and the other, the Responder, must decide whether to accept the offer. If the Responder accepts, the payoff is split between them as proposed; if the Responder rejects the offer, no one gets a payoff.

Suggested Reading

Axelrod, *The Evolution of Cooperation.*

Henrich, Boyd, Bowles, et al., "In Search of Homo Economicus."

Poundstone, *Prisoner's Dilemma.*

Questions to Consider

1. From your own experience, can you think of a situation in which you tried to produce one result but something very different happened, not because of what you did but because of what someone else did independently?

2. Suppose you really were involved in the Prisoner's Dilemma. You and your partner in crime are in separate cells and cannot communicate. If you both stonewall, you'll both get off with just one year in prison. If you both confess, you'll both get two years in prison. If you confess and your partner doesn't, you'll go free and he'll do five years in prison. If he confesses and you don't, he'll go free and you'll do the five years. If this were a real case, which option would you take? Would you confess or stonewall?

3. The empirical evidence shows that people from different cultures play even artificial games, such as the Prisoner's Dilemma, differently. Does that show that cultural influence can override rationality or that rationality is somehow culture-relative?

Spend a little Monopoly money to find out what people actually do in the Prisoner's Dilemma.

(a) For the one-shot game, explain the payoffs to two friends, emphasizing that you want them to think rationally and that this game will be a one-shot deal: Each is to choose independently whether to cooperate or defect. If they both cooperate, they get $3 each. If they both defect, they get $1 each. If one defects and the other cooperates, the defector gets $5 while the cooperator gets nothing.

(b) For the iterated game, play 10 rounds with the same payoffs. Who got the most money over all? A question for further thought: Was the play different on the last round?

Try the same experiment with payoffs from one of the other games, Chicken or Stag Hunt. Do you get different results in those cases?

		B	
		Cooperate	Defect
A	Cooperate	3, 3	1, 5
	Defect	5, 1	0, 0

Chicken Matrix

		B	
		Cooperate	Defect
A	Cooperate	5, 5	0, 1
	Defect	1, 0	1, 1

Stag Hunt Matrix

Game Theory and Beyond
Lecture 22—Transcript

Professor Grim: In a previous lecture we talked about decision theory. In a nutshell, when choosing between two options, calculate the probability and value of possible outcomes; then pick that option that has the highest probability times value—maximize expected utility. Decision theory, however, is really made for contexts in which there is just one decision maker, you. There you are, standing at the fork in the road. In that picture, the road taken and the outcome which eventuates are treated as it if was all about you. Most real-life decisions are not all about you. What happens may depend, not only on what you decide to do, but on what you, Bill, Mike, the guy next door, Apple, IBM, and Microsoft all decide to do. Where decision theory leaves off, game theory begins. It's designed to address, in an abstract way, rational decision in the social context, the context in which multiple players are making interlocking decisions. An understanding of game theory and an appreciation of its limitations should both be part of the philosopher's toolkit.

Let me start with some history. Game theory goes back to John von Neumann, my candidate for the smartest man of the 20th century. Von Neumann was instrumental in the development of computers, the atomic and hydrogen bombs, and game theory. Along with Einstein, he was one of the first professors appointed to the Institute for Advanced Study in Princeton. The standard configuration used in today's computers is still called the von Neumann architecture. He played a major role in the Manhattan project and was responsible for the sphere-implosion design used in the first atomic explosion at Trinity, New Mexico, and in Fat Man, the bomb dropped at Nagasaki. After the war, many of those involved wanted nothing to do with that kind of work again. Not von Neumann. He was even more instrumental in the development and testing of the Hydrogen Bomb. There are photographs of von Neumann at ground zero at the Bikini Atoll within a few months of the blast. It's still debated whether radiation exposure contributed to the cancer that killed him at the age of 54. What does his involvement with nuclear weapons have to do with game theory? A lot, It wasn't long after World War II that the Soviet Union had nuclear capabilities too. How do you conduct war, how do you negotiate your way out of it, in a world in which both sides have that kind of killing capacity? You can't navigate that world as if you are the only decision-maker who matters. You have to think through chains of action and response: If we do this, and they do this, and we respond like this. Multiple decisions on both sides are going to matter, and are going to matter a lot. With the possibility of mutual annihilation looming, the stakes could hardly be higher. It was in anticipating a world with two nuclear superpowers that von Neumann developed game theory. The year was 1944. The book was *von Neumann and Morgenstern, Theory of Games and Economic Behavior*. The title sounds innocent enough, how amusing, games. But the real target was rational strategies for conducting a cold war rather than a hot one.

Games in von Neumann's sense aren't just Checkers and Chess. They include every situation involving cooperation and competition in which the outcome depends on the multiple actions of multiple players. There is one game, one situation regarding cooperation and competition, that has become the central exhibit of game theory. You may know it. It's called the Prisoner's Dilemma. The story behind the game is the story of two bank robbers caught and held in separate cells. Call them Al and Bruiser. You're Al. The DA comes in and says, "Look, I've got to admit our evidence is pretty thin. If you guys both stonewall, you'll both get off pretty light, but if Bruiser confesses, Al, and you don't, we'll use his evidence against you. We'll let him walk and put you away for years. The same thing goes the other way around: "If you confess and Bruiser doesn't, we'll let you walk. We'll nail Bruiser instead." And if you both confess? "I have to admit we'd love that. You'd both do some time, alright, but with a little discount for the guilty plea." The question is, do you rat on your accomplice or not? The important point is that the outcome doesn't all depend on you. It depends on how the decisions by both prisoners play out. That's game theory. As in previous lectures, we can use a matrix to represent possible outcomes. I'll build it using positive values rather than negative. We'll talk about a maximum of five positive points. Those would be the number of years of freedom out of the next five.

Al is player A. Bruiser is player B. If you're player A, you have two options. You can defect against Bruiser by turning state's evidence, or you can cooperate with Bruiser by stonewalling. We'll call your two options defect and cooperate. Note that cooperate means to cooperate with your partner in crime, not with the police. B has the same two options, just like in the story. There are four ways this could come out, reflected in the four quadrants of our matrix. Both players cooperate with each other, both defect against each other, A cooperates and B defects, or A defects and B cooperates. We'll lay those out in a matrix with a box for each possibility. Payoffs for A are listed in red in each box, those for B listed in blue.

Game theory is about how decisions interact. What the matrix shows is potential gains and losses for each side given different combined decisions. If one player plays nice by cooperating with the other player and the other stiffs him by defecting against him, the defector walks away with five points, a full five years of freedom, say. The cooperator gets no years of freedom. You can see that in the upper right and lower left corners of the matrix. If both players choose to cooperate with each other, both end up with three points, three years of freedom. Pretty good. If both defect, they both end up with only one year of freedom, not so good. So I put it to you, player A. Should you cooperate with B or defect against him? What's the rational thing to do in the Prisoner's Dilemma? Press pause, think about it a minute, and then come back.

So what are you going to do? Stay true and cooperate with your pal or defect against him? If you said cooperate, you're an incredibly nice person. But the rational thing to do is to defect. This is a first major result in game theory .Look at it this way. You have no control over what the other player does, but no matter what the other player does, you're better off defecting. Suppose the other player cooperates. If you cooperate, you get three points. If you defect, you get five. So if he cooperates, you're better off defecting. Suppose the other player defects. If you cooperate, you get zero points. If you defect, you at least get one. So you're better off defecting in that case too. So no matter what the other player does, you're better off defecting. That's what it means for a strategy to be strongly dominant. In a one-shot Prisoner's Dilemma, defection is strongly dominant.

In real life, however, interactions are almost never a one-shot deal. Social life involves repeated interactions with the same people and the same organizations. Your relationship with your baker, your banker, or your partner in crime is almost never a one-shot deal. The parties to the transaction expect to interact again. That's going to make a difference in what behavior is rational. You and I will interact again, and how I treat you next time may depend on how you treated me this time. When we step up to this level we're dealing with what is called an iterated game. In an iterated Prisoner's Dilemma, the payoffs on each round stay the same, but there isn't just one round. A and B play once, get their payoffs, and then play again. But, of course, they remember what happened last time.

Suppose you know that you and Bruiser will keep playing an iterated game. Suppose he burns you this time. You cooperated, but he defected against you. He walked away with five points. You ended up with zero. How is that going to affect your future play? In an iterated game we're talking about the rationality, not just of a single cooperate or defect, but of strategies of cooperation and defection over time. A really simple strategy might be always cooperate. Simple, but probably not very rational. On the opposite end is always defect. But there are lots of more complex possibilities. There's one strategy that's of particular interest. It's called Tit for Tat. I'll start off cooperating, but whatever you do to me on a given round, I'll give back to you the next time. If you cooperate, I'll cooperate on the next round. If you defect on me, I'll defect on you next time.

We know what it's rational to do in a one-shot Prisoner's Dilemma. Defect. But in an iterated game we're dealing with patterns of play on both sides. You might be playing against people with very different strategies on the other side. You might not know what their strategy is. What is the most rational strategy? Robert Axelrod is a political scientist at the University of Michigan. In 1980, he tried to get a handle on that question by running a computer-based Prisoner's Dilemma tournament. Anyone who wanted to was invited to submit a strategy. Submissions came in from game theorists in economics, psychology, sociology, political science, and mathematics. No philosophers, unfortunately. He set it up so that every submitted strategy played 200 rounds against every other strategy. Strategy one played two hundred iterated rounds against strategy two, then 200 iterated rounds against strategy three, and so forth. He also made sure that every strategy played against itself and also against an opponent that cooperated and defected at random. Axelrod added up the total points that each strategy gained in all those competitions. The winner was the strategy that got the most points over all. Some of the strategies submitted were very complicated: strategies that kept track of the opponent's play, looking for weaknesses. Strategies that tested and probed the opponent's play, forming hypotheses regarding future action. What strategy would you suggest for Axelrod's tournament? When the dust cleared, there was a clear winner. The strategy that came out with the most points, the most rational strategy to play against that whole field of competitors, was Tit for Tat. Axelrod was flabbergasted. All those complicated submissions, and the one that did the best was one of the very simplest, Tit for Tat. Start off cooperating. If the other guy cooperates, do the same for him next time. If the other guy defects, do the same to him next time.

What was even more surprising was the result of a second tournament. The winner of the first competition was announced. People were invited to submit strategies again. Clearly, Tit for Tat was the strategy to gun for, so, lots more complicated strategies, designed this time to exploit weaknesses and to stump Tit for Tat. All strategies are pitted against all others in 200 iterated rounds. The winner is that strategy with the most points over all. The winner that second time? Tit for Tat. Here's what's particularly amazing about all this. Tit for Tat doesn't win because it smashes so many opponents. In fact, Tit for Tat never smashes an opponent. It never gets more points than the opposition. Tit for Tat just mirrors the other guy's play, and the best it can do is as good as the other guy's play. So why did Tit for Tat win? Because it did better against the whole field of opponents than any other strategy did. It was particularly important that Tit for Tat did well against other cooperative strategies. In that case both sides ended up cooperating, racking up three points every time. All those competitive strategies designed to exploit an opponent's weakness, and the one that does the best overall is a strategy of reciprocal cooperation.

In the one-shot Prisoner's Dilemma, defect is the rational choice. But in repeated games played with lots of different players, like in Axelrod's tournament, and more like real life, the strategy that does best is one that starts by cooperating and returns like for like. That lesson from game theory sounds a little like a relative of the golden rule, doesn't it? It says that the rational thing is to do unto your neighbor as he does unto you. The golden says that the ethical thing is to do unto your neighbor as you would have him do unto you.

There's a lesson from the Tit-for-Tat result. Cooperation may triumph even where one might have thought it wasn't possible. That's important. The limitations of game theory also have lessons to teach. Let me turn to those limitations and what they have to teach us about social rationality. I want to give game theory its due, but also look beyond it. The Prisoner's Dilemma is clearly an artificial setup, artificial enough to be captured in a matrix. That in itself isn't bad. Simplification is one of our most important conceptual tools, but there are several artificialities in that setup of which we should be particularly wary. First, note that everybody knows the rules, everybody knows the payoffs, and it's always clear to both sides whether an action was a cooperation or a defection. That's what is called a game of perfect information. Real life is often not like that. You don't always know the payoffs. In real life, your attempts at cooperation might be read as defections instead. Real life is certainly not a game of perfect information. Second, note that all moves are either full cooperation or full defection. There's no half way here. What happens if an action can be somewhat but not fully cooperative? I've actually done some research along those lines, and the simple answer is things get messier fast, more like real life. Finally, note that players have to play on each round. They can't just opt out, I'm just not going to deal with you any more. People have also investigated formal games in which opting out is a possibility. The results, you guessed it, lots more complex, just like real life. For real social rationality, then, it's not enough to act as one would in game theory. For real social rationality, one also has to keep in mind that actions of cooperation and competition don't have to be all or nothing, may not always be clear, and that opting out of certain relationships may be an important possibility.

Here's another limitation of the work I've outlined. All of the results we have been talking about are results in a particular game, the Prisoner's Dilemma. That particular game—that particular trade-off between potential competitors or cooperators—has been the most studied game in game theory. Why? When you look through the literature, it turns out that no one has a very good answer. The game is fascinating. It does seem to capture something about individual gains versus social benefits, but there is no reason to believe that the Prisoner's Dilemma is the only game we play. The matrix values define the game. It's a Prisoner's Dilemma if the value of defecting against a cooperator is greater than the value of mutual cooperation, if that is greater than the value of joint defection, and if the real loser is cooperation in the face of defection. But nothing says that payoffs in all our everyday interactions line up that way. There are other named games with other relative values. Chicken is a game that functions like two teenage boys driving hot rods toward each other. The guy who pulls out first loses, he's the chicken, and everyone makes fun of him. But he doesn't lose as badly as both boys do if neither pulls off. The matrix values for Chicken look different from those for the Prisoner's Dilemma.

Stag Hunt is named after a game that appears in the work of the philosopher Jean-Jacques Rousseau, writing in the 1700s. Rousseau's tells the story of two men hunting, each of whom can choose to cooperatively hunt stag or to go for an easier rabbit when one runs by. The big payoff here is for cooperative stag hunting. Anyone who goes for the rabbit can count on a much smaller payoff. The big loser is the man who is still hunting stag while his companion has gone for a rabbit. The matrix for Stag Hunt is different from both Chicken and the Prisoner's Dilemma. The truth is that we have no reason to believe that any single game captures the balance of payoffs in all our different kinds of interactions. In different contexts, with different people, with different aims in view, we engage in very different games. Payoffs, relative gains, and the possibility of opting out differ across those different contexts. And of course, there are interactions where it's not just the decisions of two players that are important, but three players, or four, or more. The wonderful thing about game theory is that it deals with a very real issue of rationality: how to deal with contexts in which the actions of not one but multiple players determine the result. The not-so-wonderful aspect is that it doesn't deal with that very real issue in a very real way. At least, not yet. The research continues.

Decision theory and game theory have one core characteristic in common. Both gauge rationality in terms of self-interest. In game theory one is calculating how to maximize one's own gain on the assumption that the other players are doing the same. That's going to be applicable to real-life situations, however, only to the extent that people really act that way. Do they? Do people actually make decisions in the way that game theorists and economists in general regard as rational? The answer is that they don't. That's a major finding of the developing field of behavioral economics, which puts the standard assumptions to the test. The truth is that people systematically violate the economist's canonical assumptions of rationality. They violate them in a variety of different ways.

There is a simpler game that is used in many of these studies. It's called the Ultimatum Game. You and another person, meeting for just one time, have to split an amount of money equivalent to a couple days' wages, perhaps. One person is randomly chosen as the proposer. The other person is the responder. Here's the interaction. The Proposer gets to propose how the money is to be split, 50-50, perhaps, or 90 percent to him and 10 percent to the responder. The responder then gets to choose whether to accept that division or not. If he accepts, the money gets divided as the proposer proposed, 50-50, or 90-10, or whatever. But if the responder declines the division, neither side gets any money at all. So think about that for a minute. Put yourself in the place of the proposer. You've got $300 to divide. What would you propose? Now put yourself in the place of the responder. Suppose the proposer suggests that he gets $270 and you get $30. Would you accept that $30, or would you tell him to shove it, with the result that you both walk away with nothing?

Okay. We know what you would do. Here is what economic rationality says that you should do. If you're the responder, you should take any division that gives you any money at all. Why? Because something is better than nothing. If he proposes $30 for you and $270 for him, you should take it. Better something than nothing. Indeed, if he proposes 30 cents for you and $299.70 for him, you should take it. Better something than nothing. What is the rational thing to do as a proposer? If the other person is rational in the way just outlined and you're just out for the money, you should propose as much as possible for you and as little as possible for him. Because he's rational, he'll take it. Because you're rational, you'll walk away with a bundle.

Homo Economicus is the name that has been given to agents who act as the economic game theorist says they should. One of the primary articles on the issue is titled "In Search of Homo Economicus." Is there such a species? Evidently not. In industrialized societies, proposers tend to offer something like a 44 percent cut to the other side. Not close to zero, as Homo Economicus would. Sampling across a whole range of societies, across all continents, virtually no proposer offers less than 25 percent to the other side. But the variability by culture is immense. The Torguud people of Mongolia and the Mapuche of Chile offer about 35 percent to the other side. The Achuar of Ecuador and the Sangu of Tanzania offer 40 percent to the other side. The Aché of Paraguay offer an average of 51 percent. My favorite are the Lamelara of Indonesia. Their Proposers offer a full 58 percent to the other side. We have to split the money? Okay, you take more.

Anthropologists and economists have tried to figure out what it is that explains those variations. The answer seems to be the different cultural patterns of everyday life. The more cooperative work is involved in a culture, like the whale hunting Lamelara of Indonesia, the higher the offer to the other side in this artificial game. It's also true that the more a culture incorporates a market economy, the higher the offer to the other side. What of rejections? In industrialized societies, offers of less than 20 percent tend to be rejected. In other cultures, almost no offer is rejected. But the Au and Gnau of Papua, New Guinea reject both high and low offers. They will reject offers they think are unfair even if it's unfair to their benefit. Offer an Au 58 percent and he'd rather you both walked away with nothing.

What I have tried to emphasize in this lecture is social rationality. As of now, the best formal tool we have for that is game theory. It at least recognizes the issue: that decisions made jointly may have a different rationality than those made individually. Game theory supplies some simple models. It even offers some hope for emergence of cooperation in egoistic situations like the Prisoner's Dilemma. Its limitations, when applied to real life, carry lessons as well. It may not always be clear whether our real-life moves are cooperations, defections, or something in between. The game we're playing in real life—the payoffs to each side—may be far from obvious. Indeed, the game itself may be open to negotiation. The motivations of the multiple players won't be as predictable as purely rational models might lead us to believe and might well depend on differences in cultural background.

I want to close with another aspect of social rationality that goes significantly beyond game theory. In order to make decisions rationally, even individual decisions, the first requirement is information. Often it's social means that make that information available. Wikipedia is a testament to social interaction as a source of information. It's also a testament to the unselfish motivations of thousands of anonymous volunteers, unselfish in a way that seems to defy the assumptions of game theory. It's still true, however, that you should check and double check information that comes from that kind of source. We talked about the wisdom of crowds in an earlier lecture, which is also a way of reading information off social sources.

Here's another. The previous lecture began with Richard Feynman and the *Challenger* disaster. Let me end here with that same case but with a different message. There it was Feynman's experiment that was the focus. Here I want to put the emphasis on information that can be read off social action. In the *Challenger* case, there were four major contractors who manufactured components that might have been at fault: Rockwell International, Boeing, Martin Marietta, and Morton Thiokol. When the spacecraft disintegrated, it might have been any one of those that was to blame. The Rogers Commission made their final report six months later. Feynman was ahead of the curve; he did his experiment in front of television cameras just two weeks after the incident. But there was information of a social sort that beat even Feynman. When the incident happened, stock in all of those manufacturers took an immediate hit on Wall Street. Stock prices for Rockwell International, Boeing, Martin Marietta, and Morton Thiokol all fell immediately. But by the end of that day's trading, three were starting to recover. One of the four just kept falling. It was Morton Thiokol, the company that made the O-rings. My guess is that there was information out there that indicated Morton Thiokol as the probable source of the problem. That information became apparent in the actions of people who knew. Even if you were entirely ignorant regarding the technology involved, then, you might have been able to read off the relevant information from social behavior. You could have beaten even Feynman to the punch by weeks just by watching stock market activity on the day of the disaster.

In the next lecture I want to talk about another tool in the philosopher's toolkit—another type of thinking, of increasing importance, thinking with models.

Thinking with Models
Lecture 23

This lecture covers an important conceptual tool—thinking with models—that incorporates the problem-solving strategies we have been discussing, such as visualization, simplification, and thought experiments. It has a long and distinguished history, but with the advent of computers, it will have an even more significant future.

Schelling's Model: Visualize, Simplify
- Let's start with an example of thinking with models, drawn from the work of the economist and political scientist Thomas C. Schelling.
  - To examine residential segregation, Schelling constructed a model with a checkerboard grid and used dimes and pennies to represent two ethnic groups. He distributed these randomly across the checkerboard, making sure to leave some empty spaces.

  - What Schelling thought of as the "neighbors" of a penny or dime were those in the immediate squares around it. The neighborhood for any square included those eight squares touching it on each side and on the diagonals.

  - Schelling then gave his pennies and dimes some low-level preferences. The dimes wanted to live in a neighborhood where one-third of the people around them were of the same ethnicity, but they were perfectly happy to have two-thirds of their neighbors be different. The same was true of the pennies.

  - What happens as we play those preferences out? Even with those low preference levels, distributions on the board start to form neighborhoods. There are small patches that are almost all pennies and small patches that are almost all dimes.

- One important point to draw from Schelling's model is that thinking is something you can do with your eyes. That is the strategy of visualization.

- A second and obvious characteristic of Schelling's model is that it uses the strategy of simplification.
  - Schelling was out to understand something very complicated—social patterns of housing by ethnicity. Those patterns are the result of innumerable complex decisions by thousands of interacting individuals, influenced by multiple factors.

  - But note that Schelling's model includes none of that detail. It doesn't even really have individuals who belong to ethnic groups—just pennies and dimes. His strategy here is obvious: to try to understand something complicated by understanding something related but much simpler.

Three-Part Model Structure: Input, Mechanism, Output
- In examining the structure of Schelling's thinking, we see that his model works with an **input** and generates an **output** by way of some **mechanism**. The input is the original random distribution of pennies and dimes. The mechanism here is Schelling's rule: Pick a penny or dime at random. If at least one-third of its neighbors are like it, it stays. If not, it moves. The output of the model is the final arrangement of pennies and dimes.

- Although we can build a three-stage model, we are really only confident of two of the three stages. On the basis of the two pieces of information we have, we use the three-stage model to give us the third piece of information we don't have—the missing piece of the puzzle.

- **Prediction** uses information about input and mechanism to forecast new information about the future. **Retrodiction** uses information about mechanism and output to generate information about the past. **Explanation** uses information about input and output to generate possible mechanisms for linking the two. In each case, we build in what we know at two stages of the three-stage model and read off the information we need at the third.

Using Prediction

- As we said, prediction uses what we know at two stages to get new information about the future. Weather prediction uses models in precisely this way. We know what the weather readings are across the United States today. We know what they've been for the past week. Those are our input conditions.

- Over decades, we have developed certain principles that we think apply to the weather; for example, we know what happens when a dry cold front meets a wet warm front; we know something about wind and the dynamics of the atmosphere. All of these are part of the mechanism of our models. In prediction, we know the input conditions and are fairly confident of the mechanism. We read the new information we're after at the output stage: an estimate of what the weather is going to be like tomorrow.

- If we use an X for the information we have and an O for the information we're trying to get, the case of prediction looks like this:

	Input	Mechanism	Output
Prediction	X	X	O

Using Retrodiction

- Retrodiction, on the other hand, uses two different stages to give us information about the past. For example, there is still a great deal of debate about how the moon was formed. One theory is that the centrifugal force of the spinning earth ejected it as a chunk early in the history of the solar system. Another theory holds that the moon was a separate body captured in the earth's gravitational field. According to a third theory, the moon was formed when some other large body hit the early earth, blasting out material that became the moon.

- We can build a model to decide among those theories. We build into our mechanism a representation of physical laws: centrifugal force relative to rotation, the physics of velocity and impact, and gravity. The output is a given: a moon rotating around the earth in the way we know it does, with the composition we know it has.

- If any of those inputs produces our given output with the mechanism of natural law, we have a point in favor of that particular theory. If we can't get the model to work with a particular input, we have a point against it. In prediction, we enter input and mechanism and read off the information we're after from the output. In retrodiction, we assume mechanism and output and read off the information we're after from what input produces that output with that mechanism.

	Input	Mechanism	Output
Prediction	X	X	O
Retrodiction	O	X	X

Using Explanation

- Explanation uses the three stages in yet a different way. What mechanism might take us from just about any input—a fairly random one, for example—to the patterns of residential segregation we see around us?

- In Schelling's case we assume an input, observe an output, and ask what mechanism would take us from one to another. Put beside the others, the structure of explanation looks like this:

	Input	Mechanism	Output
Prediction	X	X	O
Retrodiction	O	X	X
Explanation	X	O	X

Limitations to Thinking with Models

- Thinking with models is a way of thinking through "if...then" connections in the real world. In order to be useful, models must be simpler than the reality that we're trying to understand. But they also have to match reality in relevant ways.

- If we want a reliable prediction of tomorrow's weather, our input must be correct, and our mechanisms must work the way the weather really does. That's a challenge that meteorologists have been dealing with in computer models for years.

- The case of retrodiction is perhaps even more challenging. We can probably rely on our observations of the moon, but here, too, we must be confident that our mechanism captures the relevant physical laws. Another problem with retrodiction is that there might be more than one set of inputs that are sufficient to produce the outcome.

- The same issue shows up in explanation. The fact that a particular mechanism leads from a particular input to an observed output doesn't tell us that it's the only mechanism that could produce that outcome. It also doesn't tell us that it's the mechanism that actually produced that result in reality.

Models and Scientific Experimentation

- Scientific experiments are also about constructing "if...then" scenarios. There is, in fact, a close connection between model building and experimentation. The models we've talked about so far are conceptual models or conceptual models instantiated in computer programs. But the history of physical models shows the same three-stage structure. There, too, the three stages can also be used for prediction, retrodiction, or explanation.

- In 1900, Orville and Wilbur Wright built a small wind tunnel to test models of miniature wings in preparation for building a flying machine. Over the course of several months, they tested more than 200 wing designs to enable them to predict which ones might actually work.

- But remember the limitations of models. At this point, the Wright brothers were working only with a model. They assumed that the forces in their miniature wind tunnel would scale up to the real world. But in physical models, as in conceptual models and computer models, that correspondence with reality is always a risk.

Hobbes's *Leviathan*

- Conceptual models have sometimes served merely as metaphors, as in Plato's image of the cave in *The Republic*, for example. But conceptual models have often served as something more in the history of philosophy.

- The philosopher Thomas Hobbes witnessed firsthand the ravages of the English civil war that raged from 1642 to 1651. What he wanted to understand were prospects for stability and peace. We could phrase Hobbes's question this way: Given what we know of human nature as input, what mechanism would produce stable peace as an outcome?

- Hobbes's view of human nature was that individuals were out for their own interests alone. That was his input, what he called a "state of nature." In a state of nature, life was "solitary, poor, nasty, brutish and short." What mechanism could lead us from that to a stable peace?

- The name of Hobbes's book describing his proposal for the necessary mechanism was *Leviathan*. In order to get from the state of nature to a state of stable peace, people would have to contract together. They'd have to give up their individual liberties to a central state—a leviathan (more specifically, to a monarchy).

- Hobbes's book exhibits thinking with a three-stage model, but like all models, it has risks. We need to know how real all three stages of the model are. Are people as egoistic as Hobbes treats them in his input condition? Would a social contract really give the desired output, or would it result in a centralized tyranny? Even if Hobbes's social contract mechanism would produce the outcome, is it the only way to do so? Is it the most desirable way to do so?

- Three-stage models are a way to put strategies of visualization and simplification to work in figuring out "if...then" scenarios. It often helps to think of things in terms of simplified input, mechanism, and output, whether that structure is instantiated physically, computationally, or conceptually.

- Thinking in models can be used for many purposes: to guide data collection, to explore the core dynamics of a process, to suggest analogies, to illuminate core uncertainties, to point up overlooked details, to offer options for change or intervention, to show the simple core of complex processes and the complexities behind simple ones, and perhaps most importantly, to suggest new questions.

Terms to Know

explanation: In a three-stage model, information about input(s) and output(s) is used to generate possible mechanisms for linking the two.

input: The first structural stage of a three-stage input-mechanism-output model: the initial setup, conditions, or changes on which the model operates.

mechanism: The second structural stage of a three-stage input-mechanism-output model: the operations that run on input in order to produce the output.

output: The third structural stage of a three-stage input-mechanism-output model: the result of a model run given a particular input.

prediction: In a three-stage model, information about input(s) and mechanism is used to forecast the future, that is, to generate information about probable output(s).

retrodiction: In a three-stage model, information about the mechanism and output(s) is used to generate information about the past (inputs).

Suggested Reading

Axtrell, Epstein, Dean, et al., "Population Growth and Collapse in a Multi-Agent Model of the Kayenta Anasazi in Long House Valley."

Cummings, Chakravarty, Singha, et al., *Toward a Containment Strategy for Smallpox Bioterror.*

Epstein, "Why Model?"

Questions to Consider

1. Models are now commonly used in weather forecasting (what the weather will be like tomorrow and next week), political forecasting (who is likely to win an election), and financial forecasting of economic and stock market trends. These models are often instantiated as computer programs. From your own experience, how well do you think those computer models currently do in making predictions in those three fields?

2. Schelling arranges dimes, pennies, and empty spaces randomly on a checkerboard grid. He then picks a coin at random. If the occupied spaces around it contain 30 percent or more of its own kind, it is "happy" and stays. If the occupied spaces contain less than 30 percent of its own kind, it is "unhappy" and moves to an empty spot close by. As the rules are progressively applied, segregated suburbs of pennies and dimes develop. What can legitimately be concluded from that model about de facto racial segregation? Just as important: What cannot be concluded from that model?

3. In response to people who ask why they should work with models, the computational modeler Josh Epstein likes to say: "You already model. You're working with a model in your head. The real question is just what assumptions you are making and how good that model is." Do you think Epstein is right about that, or not?

NetLogo is a simple programming environment designed and authored by Uri Wilensky, director of Northwestern University's Center for Connected Learning and Computer-Based Modeling. It comes with a wonderful Models Library of examples ready to run and easy to tweak. You can download it for free at http://ccl.northwestern.edu/netlogo/download.shtml and will find Schelling in the Models Library under "Social Science" and "Segregation."

Start by setting %-similar-wanted at different levels. What happens with segregation if neighbors are "happy" with just 25 percent of their own kind? What happens if they are "happy" only with 50 percent of their own kind?

Thinking with Models
Lecture 23—Transcript

Professor Grim: In this lecture I want to draw on the problem-solving strategy of visualization, the problem-solving strategy of simplification, the problem-solving strategy of thought experiments. We've worked with all of those before. I want to bring them together in order to apply them to the kind of if-then connections we talked about with regard to scientific experimentation. There is a conceptual tool that incorporates all of these. It has a long and distinguished history, but with the advent of computers I expect it to have an even more important future. This lecture is about thinking with models. All kinds of models.

I'll start with an example drawn from the work of the economist and political scientist Thomas C. Schelling. This is Schelling's segregation model, one of the contributions for which he received the Nobel Prize in 2005. Schelling was as familiar as the rest of us with the de facto residential segregation characteristic of American cities and suburbs. If you look around your neighborhood, it's unlikely that you'll find a statistical cross section of racial and ethnic groups. Are 15 percent of your neighbors Latino? Thirteen percent African American? Four percent Asian American? One percent Native American? Sixty-seven percent white non-Latino? Probably not. Ethnicity maps of American cities typically show White areas, African American areas, Latino and Asian areas.

Schelling's question was this: What does that say about racism? Do those ethnicity maps show that we're all more racist than we'd like to think? Schelling decided to construct a little model to help him think that question through. He drew a checkerboard grid on a piece of paper—just a lot of squares. He reached into his pocket for change and used dimes and pennies to represent two ethnic groups. He distributed those randomly across the checkerboard, making sure to leave some empty spaces. I encourage you to try this yourself. The dimes and pennies were the agents in his model. The checkerboard was an abstract model for residential space—where those agents lived. In an initial random distribution, pennies and dimes would have been pretty evenly distributed. What Shelling thought of as the neighbors of a penny or dime were those in the immediate squares around it. The neighborhood for any square are those eight squares touching it on each side and on the diagonals. Schelling then gave his pennies and dimes some low-level preferences. The dimes weren't racist. They did want to live in a neighborhood where a third of the people around them were of the same ethnicity, but they were perfectly happy to have two thirds of their neighbors be different. The same was true of the pennies. Two thirds unlike them? Fine, but they would like at least one third of their neighbors to share their background.

He picked a random penny or dime. It would have some other pennies or dimes in the immediate neighborhood around it. If at least a third of those were of its kind, then it could stay where it was. If it had less than a third of its own kind around them, it would move to a random empty space. So we have a random distribution. Nobody is particularly racist, but they do want at least a third of their neighbors to be like them, and they'll move if they don't. What happens as we play those preferences out? Even with those low preference levels, distributions on the board start to form little neighborhoods. There are little patches that are mostly all dimes. It starts to look a lot like the residential segregation characteristic of American cities and suburbs.

Much of contemporary computer modeling is just an extension of that kind of thinking. Here is a computer simulation, a computer instantiation, of Schelling's model, just larger and done by a computer rather than by hand. The same low-level preferences on the part of agents. The same resultant pattern of de facto residential segregation. Schelling's is a great example of what I mean by a model. In the original, there with dimes and pennies on a hand-drawn checkerboard, he was thinking with a model, in a sense, thinking through a model.

Let me emphasize some of the important characteristics of that case. The first is that Schelling didn't just think it through in his head. He manipulated pennies and dimes on a hand-drawn checkerboard. We've seen that lesson before. Thinking doesn't have to all be in your head. Thinking shouldn't be confined to the dark and lonely recesses of your skull. Thinking is something you can do with your eyes. That is the strategy of visualization. We have amazing abilities of pattern recognition, instant, detailed, and accurate. It would be a shame not to use those abilities in thinking. If we can convert an abstract problem into a picture—something we can see—we can often think it through much more effectively.

That is how Pythagoras may himself have thought through the Pythagorean theorem, thinking visually. Pythagoras may actually have manipulated tiles. Look at that. Sure enough. The area of the square on the hypotenuse is equal to the sum of the squares on the other two sides. Pythagoras may have thought the theorem through using a model. Pythagoras may have been thinking with his hands, just like Schelling with his pennies and dimes. In order to think, you don't have to strike a pose like Rodin's "Thinker." Indeed that's probably not be the most effective thinking pose. I'll bet Rodin, himself, didn't strike that pose when he thought about his sculpture. It's, in fact, very likely that he thought it through with a piece of clay; he thought it through with his hands. It's very likely that Rodin thought the sculpture through by using a model. Thinking is often most effective when you recruit your eyes to the task. Thinking is often most effective when you can manipulate things with your hands. Both of those show up in thinking with models.

A second, and obvious, characteristic of Schelling's model is that it uses the strategy of simplification. What he's out to understand is something very complicated, social patterns of housing by ethnicity. Those patterns are the result of innumerable careful and complex decisions by thousands of interacting individuals. We know those individuals are influenced in their decisions by employment patterns, demographics, the complexities of housing availability, mortgage practices, individual finances, and life histories. Schelling's model includes none of that detail. It doesn't even really have individuals who belong to ethnic groups. Just pennies and dimes. It includes no real representation that's anything like a map of any city, just a checkerboard drawn on a piece of paper.

Do you remember the Tower of Hanoi problem? You have three posts in a row. The middle and right posts are empty. On the left post are stacked a series of six disks ordered from largest at the bottom to smallest at the top. Your task is to end up with all six disks, in the same order, but on one of the other two poles. You can only move one disk at a time. You can only move a disk from one pole to another. And you can never put a larger disk on top of a smaller one. What is the least number of moves in which you can get all six disks in order from one post to another? An important strategy for that kind of problem was simplification. If you want to know how to do it with six disks, start with just two. That will give you a basic understanding of a pattern of moves required to solve the more complicated problem. Move the top small disk to one pole, then the one under it to another, and move the small disk on top. For three disks you can do the same thing for the top two, then move the third disk to another pole, and repeat the original dance with the two smaller disks—seven moves. By the time you get up to six disks, that compounds to 63 moves.

Schelling wants to understand something much more complicated than that and much more real—much more real than the Tower of Hanoi problem. He wants to understand why residential segregation might happen with real people in real cities. That makes the simplification of his model far more outrageous than the Tower of Hanoi with just two disks. But the strategy is the same. Try to understand something complicated by understanding something related but much simpler.

Now let's look at the structure of Schelling's thinking. The model works with an input and generates an output. It goes from input to output by way of what I'll call the mechanism. Three steps: input, mechanism, and output. The input is the original random distribution of pennies and dimes. The mechanism here is embodied in the Schelling's rules. Pick a penny or dime at random. If at least a third of its neighbors are like it, it stays. If not, it moves. The output of the model is the final arrangement of pennies and dimes, segregated patches. This is where the link to the if-then structure of experiments comes in. What the model explores is what happens if. What happens if we start with a random distribution and a one third preference rule? What Schelling's model shows is that residential segregation could occur even without any blatant racism. None of our dimes or pennies hates the other kind. There are no pennies that move out just because a dime moves into the neighborhood. All any of them want is to have some of their own kind around. Not even half, just one third.

Models often have that three-stage structure: input, mechanism, and output, but there are lots of different purposes to which models are put, and that three-stage structure is used in lots of different ways. That's important both for your own thinking with models and for evaluating the use of models by others. Here is the basic picture. We build a three-stage model but are really only confident of two of those stages. We use the model to tell us about the third stage. On the basis of two pieces of information we have, we use the three-stage model to give us the third piece of information we don't have—the missing piece of the puzzle. Prediction uses what we know at two stages to get new information about the future. Retrodiction uses two different stages to give us information about the past. Explanation uses the three stages in a still different way, but in each case we build in what we know at two stages of a three-stage model and read off the information we need at the third.

Let's start with prediction. Our three-stage model consists of input, mechanism, and output. In prediction we have some confidence in our input conditions. We also think we understand our mechanism. Our model uses input and mechanism to generate an output, and that output is what we take as the model prediction. Weather prediction uses models precisely this way. We know what the weather readings are across the United States today. We know what they've been for the past week. Those are our input conditions. Over decades we have developed certain principles that we think apply to the weather. We know what happens when a dry, cold front meets a wet, warm front. We know something about wind and the dynamics of the atmosphere. We know something about how the Rocky Mountains and the Great Plains enter the picture. All of those are part of the mechanism of our models.

These days the models for weather forecasting are instantiated as computer programs. The MM5 is a fifth-generation model at Penn State. The Weather Research Forecasting Model is billed as the next generation, backed by the National Center for Atmospheric Research, with the National Oceanic and Atmospheric Administration, the Air Force Weather Agency, and the Naval Research Laboratory. Those models are much more complicated than Schelling's, but the basic idea is the same, a set of principles that take us from input to output. In prediction, we know our input conditions and are fairly confident of our mechanism. We read the new information we're after at the output stage—an estimate of what the weather is going to be like tomorrow.

If we use an X for information we have and an O for the information we're trying to get, the case of prediction looks like this.

Prediction is one thing the three-stage structure of models is good for. But sometimes what we want to know is not what is going to happen in the future, but what happened in the past. We want, not prediction, but postdiction or retrodiction. Here's an example. There is still a great deal of debate as to how the moon was formed. One theory is that the centrifugal force of the spinning earth ejected it as a chunk early in the history of the solar system. There is another theory that the moon was a separate body captured in the earth's gravitational field. According to a third theory, the moon was formed when some other large body hit the early earth, blasting out material that became the moon.

How do we decide between those theories? How do we figure out what happened? Here's one way, build a model. We build into our mechanism a representation of physical laws: centrifugal force relative to rotation, the physics of velocity and impact, and gravity. We take that mechanism as given. We also take the output as given: a moon rotating the earth in the way we know it does, with the composition we know it has. The question for the model is now, take what the theories say happened as input. Will that hypothesized input produce that output with that mechanism? Could the early earth have rotated in such a way as to eject the moon as a chunk? Could the earth have captured another body in its gravitational field? Could the moon have been formed by a collision with another body? If any of those inputs produces our given output with the mechanism of natural law, we have a point in favor of that particular theory. If we can't get the model to work with a particular input, we have a point against it.

In prediction we put in input and mechanism and read off the information we're after as the output. In retrodiction we assume mechanism and output and read off the information we're after from what inputs produce that output with that mechanism. So we have two different ways of using the three-part structure of models. Put next to each other, they look like this:

In this case, by the way, contemporary computer models favor the third theory. Given the mechanisms of physical law, a giant impact in the early

Solar System is that input that does best at producing an output that matches the size, orbit, and composition of the moon.

Schelling's model uses the three-part structure in a third way. He isn't after a prediction. He isn't after a retrodiction. What he wants is an explanation. What mechanism might take us from just about any input—a fairly random one, for example—to the patterns of residential segregation we see around us? In Schelling's case we assume an input, observe an output, and ask what mechanism would take us from one to another. Put beside the others, the structure of explanation looks like this.

Models quite standardly come with those three stages: input, mechanism, and output. But we can get three different cognitive payoffs by using that three-part structure in different ways. We may be looking for new information in model output, that's prediction. We may be considering what happens with different inputs; that's what we want for retrodiction. We may be looking for mechanisms that take us from input to output; that's what explanation is all about.

Thinking with models is a great tool. But it's also a risky one with important limitations. It's important to keep those in mind when evaluating the information that models give us. Thinking with models is a way of thinking through if-then connections in the real world. We're trying to understand something real: what the weather is going to be like, how the moon was formed, or what it is that accounts for persistent residential segregation. In order to be useful, models have to be simpler than the reality that we're going to understand, but they also have to match reality in the relevant ways. If we want a reliable prediction of tomorrow's weather, our inputs have to be right. Our mechanisms have to work the way the weather really does. That's a challenge that meteorologists have been dealing with in computer models for years.

The case of retrodiction is perhaps even more challenging. We can probably rely on our observations of the moon, but here, too, we have to be confident that our mechanism captures the relevant physical laws. Another problem with retrodiction, there might be more than one set of inputs that are sufficient to produce the outcome. Put it this way. Models are made to mirror an if-then connection. If a Mars-size body collided with the early earth, it could have thrown out a moon that matches ours in composition and orbit. But even if our model tells us that, it doesn't tell us that is the only chain of events that could have produced that outcome. It doesn't tell us this is the only input that would give us that output.

The same issue shows up in explanation. The fact that a particular mechanism leads from a particular input to an observed output doesn't tell us that it's the only mechanism that could produce that outcome. It also doesn't tell us that it's the mechanism that actually produced that result in reality. Schelling was pretty careful about all of that. He only claimed that low levels of quite ordinary preference could have produced residential segregation. That is, in fact, all that a model like his can show. It can't show that low-level preference is the only thing that could have produced that result. It can't show that that low-level preference is what actually produced the result.

Models are all about constructing an if, then. As we saw in an earlier lecture, that's what scientific experiments are about too. There is, in fact, a close connection between model building and experimentation. The models we've talked about so far are conceptual models, or conceptual models instantiated in computer programs. But the history of physical models shows the same three-stage structure. There, too, the three stages can also be used for prediction, or retrodiction, or explanation. In 1900, Orville and Wilbur were trying to figure out how to build a flying machine. It was going to have wings, but what kind of wings would work? What kind of wings would allow controlled flight? The Wright brothers could have tried working entirely inside their skulls. Other people had, but they didn't. They were bicycle mechanics by trade. They were used to thinking with a wrench in their hands. Their thinking about wings was hands on too. They built a wind tunnel, 16 inches wide, 16 inches tall, and six-feet long. It was a finicky apparatus. Wilbur said, "Occasionally I had to yell at my brother to keep him from moving even just a little in the room because it would disturb the air flow." The Wright brothers moved with models of miniature wings in that wind tunnel, none over nine inches long. As Wilbur said, "it is doubtful if anyone would have ever developed a flyable wing without it".

They were after an if, then; if a wing has this structure, then it will fly like this. Over the course of several months they tested over 200 wing designs. They wanted to be able to predict which wing designs might actually work. But remember the limitations of models. At this point, they were only working with a model. They assumed that the forces in their miniature wind tunnel would scale up to the real world, but in physical models, as in conceptual models, as in computer models, that correspondence with reality is always a risk. Wilbur says, "We finally stopped our wind tunnel experiments just before Christmas, 1901. We really concluded them rather reluctantly because we had a bicycle business to run."

There have been some wonderful physical models. One of my favorites is the Mississippi Basin Model in Clinton, Mississippi. The 1920s and 1930s had seen devastating flooding from the Mississippi River. In the 1940s, in order to figure out what to do about it, the Army Corps of Engineers built a miniature model of the entire river basin. Begun with labor by German prisoners of war, the model covered some 200 acres. Trees were modeled with folded screen wire at a height taken from aerial photographs, and the roughness of river bottoms was modeled by scoring and brushing concrete channels in the model. Three days on the river could be modeled in five minutes in the model. That model was used for a whole range of if, thens. Can we prevent flooding in New Orleans by redirecting floodwaters into the Atchafalaya Basin? Which are the levees most at risk? The model was used in real time to guide policy in 1973 in response to 77 days of flooding from St. Louis on down. A similar physical was constructed for the San Francisco Bay in the 1950s in order to evaluate a proposal to drain portions of the bay. What would happen if we did that? The Army Corps of Engineers built physical models in searching for an explanation of why the 17th Street Canal levee gave way under the impact of hurricane Katrina.

Conceptual models have sometimes served merely as metaphors. I think this is true of Plato's image of the Cave in the Republic, for example. Plato constructs a model to illustrate his theory of the relations between appearance and reality. We are like prisoners chained in a cave. We think we see reality, but all we really see are shadows cast by light coming in from outside. But conceptual models have often served as something more in the history of philosophy, something more like Schelling's.

The philosopher Thomas Hobbes witnessed first hand the ravages of the English civil war that raged from 1642 to 1651. What he wanted to understand were prospects for stability and peace. You could put his question this way: Given what we know of human nature as input, what mechanism would produce stable peace as an outcome? His view of human nature was of individuals out for their own interests alone. That was his input, what he called a state of nature. In a state of nature, life was "solitary, poor, nasty, brutish and short." What mechanism could lead us from that to a stable peace? Hobbes called his book *Leviathan*, and that was his proposal for the necessary mechanism. In order to get from the state of nature to a state of stable peace, people would have to contract together. They'd have to give up their individual liberties to a central state, to a Leviathan; more specifically, to something like a monarchy. That's thinking with a three-stage model. Like all models, it has risks. We want to know how real all three stages of the model are. Are people as egoistic as Hobbes treats them in his input condition? Would a social contract really give the desired output, or would it just result in a centralized tyranny? Even if Hobbes's social contract mechanism would produce the outcome, is it the only way to do so; the most desirable way to do so?

Game theory is another instance of model thinking. It assumes players' rationality and self-interest as input. It assumes a matrix of payoffs as a mechanism. It looks to see what the outcome can be expected to be with certain structured penalties and incentives. We talked about some of the risks of that kind of model thinking in the lecture on game theory. Do people behave in the way game theory assumes? Are we really Homo Economicus? That's a challenge to input assumptions. What, exactly, are the values of our interactions? Why think they are captured in the matrix for the Prisoner's Dilemma? That's a challenge to the mechanism assumed in game-theoretic models.

The take-home message is this: Three-stage models are a way to put strategies of visualization and simplification to work in figuring out if, thens. It often helps to think of things in terms of simplified input, mechanism, and output, whether that structure is instantiated physically, computationally, or conceptually. If you want to predict what's going to happen, work from input and mechanism. If you want to figure out what happened in the past, figure out what inputs would give you observed outputs with known mechanisms. Explanation? Figure out what mechanism would lead from known inputs to known outputs.

Learning to think in models can also take you farther. My friend Josh Epstein is a contemporary researcher who has built models intended to cast light on how diseases transfer across populations, how revolutions happen, and what may have happened to the ancient Anasazi in the American southwest. In a piece called "Why Model?" he offers a list of other things thinking in models is good for. Thinking with models offers tools for prediction, retrodiction, and explanation. But thinking in models can also be used to guide data collection; to explore the core dynamics of a process; to suggest analogies for understanding, what are other things that work this way; to illuminate core uncertainties, where is our knowledge most vulnerable; to point up overlooked details that might make a difference; to offer options for change or intervention; to show the simple core of complex processes and the complexities behind some simple ones; last but not least, to suggest new questions.

The next lecture is our last. I'll use that for a quick review of some of what we've done in terms of lessons from the great thinkers.

Lessons from the Great Thinkers
Lecture 24

Examining the philosopher's toolkit has been the aim of these lectures, that is, learning a set of conceptual techniques and strategies, tips, methodologies, and rules of thumb for thinking more clearly, with greater focus, more effectively, and more accurately and for building stronger and more rational patterns of inference and argument. In this lecture, we'll talk about the philosopher's toolkit in terms of some of the great thinkers who have used it and added to it. What was it that made these thinkers great?

Plato

- The influence of Plato has been enormous. Alfred North Whitehead said that all of Western philosophy was "a series of footnotes to Plato." But what made Plato so great?

- First and foremost, Plato saw the power of abstraction. Abstraction gives us the power of generality. In abstraction, we simplify away the differences of different cases—the sand of triangles drawn in sand, the chalk of triangles in chalk—and think simply in terms of the shape things have in common. In order to think abstractly, you don't add on; you subtract—you simplify.

- But Plato didn't talk about abstractions as removed from reality. He thought that the realm of abstraction was the true reality. For Plato, the world we experience around us is just an illusory, imperfect, and distant echo of the true reality of abstraction.

- Another element that made Plato great is that his works are dialogues, not monologues. Crucial to the dialogues is the interaction between multiple points of view, multiple opinions, and multiple people. The emphasis is on social rationality—ideas as something we work through together.

Aristotle

- Aristotle founded logic by seeing things with a very particular vision. Deep in the complicated and messy process that is thinking, deliberation, discussion, and argument, Aristotle saw something amazingly simple: structure. The ability to see that structure gave him the idea of systematizing thought.

- If we could systematize thought, we might be able to avoid some of our biases. We could make our thinking faster, more accurate, and more effective. That vision we owe to Aristotle, and it's a vision that extends even into contemporary computers.

- Aristotle is also a great thinker because, in contrast to Plato, he insisted on the lessons of experience. He was a real-world observer. Although there aren't any true experiments in Aristotle—that came much later—he believed that real knowledge of the world demands looking at what the real world is really like.

- This contrast between Plato and Aristotle is captured in Raphael's painting *The School of Athens*. At the center, Raphael shows Plato and Aristotle in deep discussion. Plato, the great teacher, is pointing upward—toward the abstract. Aristotle, his rebellious student, is gesturing with a level hand toward the concrete reality before him.

Galileo

- Early experimentation was often used to question the authority of Aristotle. That was certainly true of Galileo, who has to rank as one of the great thinkers on anyone's list. The major lesson from the great experimenters is: Don't trust authority—not even Aristotle's. Don't rely on what most people think. Design an experiment and put it to the test.

- Galileo was an aggressive and indefatigable observer. We've mentioned Galileo's legendary experiment, dropping a musket ball and a cannonball from the top of the Leaning Tower of Pisa. Whatever Aristotle thought, whatever anybody thought, Galileo's experiment showed that things fall at the same rate regardless of weight.

- But precisely how do things fall? In order to investigate, Galileo built a model. What he really wanted to know was how things accelerate in free fall. But his tools weren't precise enough to measure that. Instead, he rolled balls down an inclined plane and measured the speed with which objects passed successive marks on the plane. He found principles that applied whatever the angle of the ramp and generalized from those to answer the question about acceleration in free fall.

Descartes
- If Galileo is the founding figure in modern science, René Descartes may be the founding figure in both modern philosophy and modern psychology.

- One of his major works is the *Discourse on the Method of Rightly Conducting One's Reason and of Seeking Truth in the Sciences.* His messages:
  - "Divide each of the difficulties under examination into smaller parts."

  - "Starting with what is simplest and easiest to know, ascend little by little to the more complex."

  - "Never… accept anything as true which [you do] not clearly know to be such… avoid precipitancy and prejudice."

  - "To be possessed of a vigorous mind is not enough; the prime requisite is rightly to apply it."

Newton
- Sir Isaac Newton is known as a great formalizer. But he was a great experimentalist, as well. When the plague hit England in 1664, Newton went home to his mother's house in Lincolnshire and began experimenting with prisms.

- The prevailing view was that a prism stained light into different colors. But Newton was able to show that what the prism did was bend different colors of light differently—what we call different wavelengths.

- We know from his notebooks that Newton didn't follow any rigid scientific method. He simply tried many things, inventively and on the fly, to see what happened. His example carries an important lesson about the two things effective thinking requires.
  - One requirement is creative exploration. Good thinking—whether experimental or not—requires that we take time to play with ideas. Genuine creative thinking cannot be regimented. Children know this intuitively.

  - But creative conceptual play isn't enough. Good thinking also requires a second phase in which one squints critically at the creative results. That is the stage at which Newton asks what precisely has been shown. Can we formulate those results more clearly? How exactly does the experiment refute established doctrine?

Thinking Better: Lessons from the Great Thinkers

We can sum up the lessons from the great thinkers in seven maxims.

- First, simplify.
  - We've seen this lesson in Newton and Descartes: Break a complex problem into simple parts; start with the simple, build up to the complex. We emphasized that strategy early on in the Tower of Hanoi problem, in going to extremes, and in the strategy of thought experiments.

  - If you can't solve a hard problem, use a simpler model. That is what Galileo did. Simplification is what model building is all about, from the Wright brothers' wind tunnel to the Mississippi River basin model to Schelling's segregation model.

  - Plato's abstraction represents the power of generality you can get not by adding detail but by subtracting it.

New thinking, in science as well as anywhere else, must be based in creative play—we need creative thinking to give us material worth working with.

- Second, look for patterns, but watch out for them, too.
  - Visualization allows us to harness visual pattern recognition. It lets us think faster, more intuitively, and more powerfully. Aristotle knew the power of visualization. He used it in his square of opposition. Einstein was one of the great visualizers. The heuristics that make us smart are often forms of pattern recognition.

  - But the biases that lead us astray are often patterns, as well. Our vivid images can mislead us as to true probabilities. We miss the gorilla in the room because we're paying attention to some other pattern. Patterns are extremely powerful. That's why it's important to put them to work and equally important to handle them with care.

- Third, think systematically.

- To systematize thought is the Aristotelian vision. Thinking systematically is what gave Newton his edge. We have developed a number of tools for systematic thought: tools in statistics, probability, and decision theory.

- One simple lesson in thinking systematically goes all the way back to the bat and ball example in the first lecture. Want to think better? Check your work.

- Fourth, put it to the test.
  - Respect for reality runs through the whole history of experiment. Charles Sanders Peirce thought of science as the one truly self-correcting technique for fixation of belief. Karl Popper emphasized that genuine science demands risk—the risk of falsifiability. Galileo is the patron saint of experimentation. Simple and beautiful experiments go all the way from Newton's prism to Richard Feynman's O-rings.

  - Do you really want to know what reality entails? Ask. Put your thinking about reality to the test.

- Fifth, think socially.
  - Many of the complexities we face are social complexities. They require thinking in terms of multiple perspectives and multiple players. Thomas Hobbes knew the importance of thinking socially, as did Jean Jacques Rousseau. That kind of thinking is the whole goal of contemporary game theory.

  - Inevitably, there are social threats to good thinking, as well. We have to watch out for mere rhetoric, standard fallacies, and the tricks of advertisers. Plato's dialogues are the first great works in philosophy: They emphasize this social side to thinking.

- Sixth, think with both sides of your brain.
  - Newton combined the power of creative play with systematic critique. Both parts need to be present: the discipline and the imagination.

○ We emphasized early on that emotion need not be the enemy of rationality if used in the right domain. We couldn't do without fast-and-frugal heuristics, but for long-range decisions, we need careful calculation, as well. Remember Daniel Kahneman's system 1 and system 2 thinking.

- Finally, give yourself time to think.
  ○ Descartes must have been a phenomenally quick thinker, yet he said, "Those who travel slowly may yet make far greater progress than those who run."

  ○ Google's engineers are given one day a week to work on something entirely of their own. It's unfortunate, but these days, our last bastions for free-range thinking seem to be the shower and the commute to work. Real thinking demands conceptual exploration, and that just takes time.

More Great Thinkers to Come
- Each of these great thinkers worked with a conceptual toolkit, but none of them built it alone. All inherited a set of conceptual tools from their predecessors. Newton said, "If I have seen farther, it is by standing on the shoulders of giants." All added to the toolkit and then passed it on.

- That cultural heritage has now been passed to you: the philosopher's toolkit—to use, to apply, to add to, to improve, to pass on. It's wonderful to peer into the future and realize that there are more great thinkers to come.

Suggested Reading

Baggini and Stangroom, *Great Thinkers A–Z.*

Boorstein, *The Discoverers.*

Bronowski, *The Ascent of Man.*

1. One of the lessons drawn from the great thinkers in this lecture—with Newton as an example—was that good thinking requires both (a) creative conceptual play and (b) systematic and critical review. Which of those do you think you are better at? Do you have a friend who is better at the other?

2. In Raphael's *School of Athens* (http://mv.vatican.va/4_ES/pages/z-Patrons/MV_Patrons_04_01.html), Plato points heavenward, emphasizing abstract conceptualization. Aristotle spreads his hand ahead, emphasizing attention to the real world. Which of those do you think is most important in good thinking? Would it be possible to have good thinking with just one and not the other?

3. What do you think we will learn from the next great thinker in our history? What do you think the next conceptual breakthrough will be?

Sit down with a piece of paper and make your own list of great thinkers—the 10 greatest, perhaps. Who do you think have been the greatest in the history of Western thought?

The second question is harder, but more important: What exactly made those figures great?

Lessons from the Great Thinkers
Lecture 24—Transcript

Professor Grim: What we've been assembling in the course of these lectures is the philosopher's toolkit. We've been pulling together a set of conceptual techniques and strategies, tips, methodologies, and rules of thumb for thinking more clearly, thinking with greater focus, more effectively, more accurately, and for building stronger and more rational patterns of inference and argument. In this lecture I want to take a different perspective on some of those techniques and on a few more. This time around I want to talk about the philosopher's toolkit in terms of some of the great thinkers who have used it and added to it. This time around I want to look at some important thinking tools by looking at some of the great thinkers.

Everybody will have their own selection of the greats, of course, and there are plenty to choose from. I'll pick just a few, treating them chronologically and with one central question in mind: What was it that made these thinkers great?

Let's start with Plato. Though Plato, you may be surprised to learn, wasn't even his real name. He was Aristocles, son of Ariston. Plato was a nickname. It means stocky or broad shouldered. So let's start with that great thinker, Stocky. Plato's influence has been enormous. All of western philosophy has been characterized as "a series of footnotes to Plato." But what made Plato so great? I think there are two things that stand out. First and foremost, he saw the power of abstraction. It's amazingly hard to say what abstraction is, though we all know how to do it. When you think, not in terms of this specific triangle scratched in the sand, or that one drawn in chalk on the board, or this other one cast as a shadow, but about that shape in general, you're abstracting. When you think, not of a Valentine's box of chocolates, or a birthday or graduation present, but rather of giving in general, you're abstracting. Abstraction gives us the power of generality. It does so by simplifying. In abstraction we simplify away the differences of different cases—the sand of triangles drawn in sand, the chalk of triangles in chalk—and think simply in terms of the shape they have in common. In order to think abstractly, you don't add on, you don't complexify, you subtract—you simplify. The realm of abstractions is what people talk of as Plato's Heaven.

But Plato didn't think he was talking about abstractions removed from reality. He thought the realm of abstraction was the true reality. For Plato, the world we experience around us is just an illusory, imperfect, and distant echo of the true reality of abstraction.

Abstraction is a crucial tool in all individual thought But I think there is another thing that makes Plato great, and it's entirely different. Plato's works are dialogues, not monologues. Crucial to the Dialogues is interaction between multiple points of view, multiple opinions, and multiple people. The emphasis is on social rationality—ideas as something we work through together.

What about Aristotle? What makes him great? Aristotle founded logic by seeing things with a very particular vision. Deep in the complicated and messy process that is thinking, deliberation, discussion, and argument about all the complicated and messy things that are humanly important, Aristotle saw something amazingly simple. He saw structure. The ability to see that structure gave him an idea. Maybe we can systemize thought. Thinking is often difficult. Because of built-in psychological biases, our thinking often goes wrong. We connect dots that aren't really there. But if we could systematize thought, we might be able to make it easier. If we could systematize it, we might also be avoid some of those biases. We could make our thinking faster, and more accurate, and more effective. That vision we owe to Aristotle. It's a vision that extends even into contemporary computers. Systematizing, formalizing, mechanizing thought is what computers are all about. They're all built from the basic elements of logic. It's not quite Aristotle's logic, but we wouldn't have it without his precedent.

But there's something else that makes Aristotle great. I think it's in direct contrast to Plato's abstraction; it's a necessary counter-balance. Plato thought it was the abstract that was truly real. Aristotle insisted on the lessons of experience. He was a real-world observer. Many of his works are cited as the first in biology: *The Parts of Animals*, *The Generation of Animals*, *The Movement of Animals*. There aren't any true experiments in Aristotle, that came much later, but he thinks that real knowledge of the world demands more than a glimpse into Plato's heaven. Real knowledge demands looking at what the real world is really like. This contrast between Plato and Aristotle is captured in Raphael's painting, The School of Athens. At the center, Raphael has painted Plato and Aristotle deep in discussion. Plato, the great teacher, is pointing upward toward the abstract. Aristotle, his rebellious student, is gesturing with a level hand toward the concrete reality before him.

Any attempt to track down the first scientific experiment is bound to create controversy—both historical controversy and controversy as to what constitutes an experiment. But there is little controversy that experimentation first comes into its own during the Renaissance. Early experimentation was often used to question the authority of Aristotle. That was certainly true of Galileo, who has to rank as one of the great thinkers on anyone's list. The major lesson from the Great Experimenters is, don't trust authority, not even Aristotle's. Don't rely on what most people think. Don't be satisfied with what seems reasonable in an armchair. Design an experiment and put it to the test. Galileo was an aggressive and indefatigable observer. He didn't invent the telescope, for which there are patent applications in the Netherlands in 1608. But by 1609 he had built himself increasingly powerful versions, and he had turned his telescope to the heavens. Night after night he carefully recorded craters on the moon, the phases of Venus. He not only saw the moons of Jupiter but figured out what they really were. We've mentioned Galileo's legendary experiment in dropping a musket ball and a cannonball from the top of the Leaning Tower of Pisa. Whatever Aristotle thought, whatever anybody thought, Galileo's experiment showed that things fall at the same rate regardless of weight. But, precisely, how do things fall? Do they reach a designated speed and stay there? That was Aristotle's answer, by the way, or do they keep moving faster? Precisely how much faster?

In order to investigate those questions, Galileo built a model. What he really wanted to know was how things accelerate in free fall, but his tools weren't precise enough to measure that. Instead he rolled balls down an inclined plane—a wooden ramp with a groove down the middle. With as much accuracy as his tools would allow, he measured the speed with which objects passed successive marks down the plane. He found principles that applied whatever the angle of the ramp and generalized from those to answer the question about acceleration in free fall. So one measure of Galileo's greatness? Putting the question to the test. Another measure? Using a question he could address in order to get a handle on a question he couldn't.

If Galileo is the founding figure in modern science, Descartes may be the founding figure in both modern philosophy and modern psychology. He is known for the phrase, "I think therefore I am." He was also the inventor of Cartesian coordinates, a triumph both of visualization and conceptual linkage. Algebraic equations translate into visual shapes in the Cartesian plane. One of Descartes's major works is the *Discourse on Method* of 1637. Its full title shows how it ties in with these lectures: *Discourse on the Method of Rightly Conducting One's Reason and of Seeking Truth in the Sciences.* And what is that method? Descartes says, "Divide each of the difficulties under examination into smaller parts." We called that divide and conquer. He says, "Starting with what is simplest and easiest to know, ascend little by little to the more complex." He says, "...never...accept anything as true which [you do] not clearly know to be such...avoid precipitancy and prejudice..." The *Discourse on Method* is full of good advice. "...to be possessed of a vigorous mind is not enough; the prime requisite is rightly to apply it."

I'd certainly put Sir Isaac Newton on the list of great thinkers. But not, perhaps, for the reasons you'd expect. Newton is known as a great formalizer, but he was a great experimentalist as well. When Bubonic plague hit England in 1664, Cambridge University closed its doors for two years. Newton went home to his mother's house in Lincolnshire. There, in a darkened room, he started playing with prisms. Everyone knew that light through a prism casts the spectrum of a rainbow. If you let light from a circular hole through a prism—as Newton did—the result is an oval or oblong shape that distributes the colors top to bottom. But what did that mean? The prevailing view was that the prism stained the light into different colors, but that wouldn't explain why the shape was an oblong. By ingeniously playing with prisms, Newton was able to show that what the prism did was bend different colors of light differently, what we call different wavelengths. White light isn't one color. It's a composition of many. We know from his notebooks that Newton didn't follow any rigid scientific method. He just tried lots of things, inventively and on the fly, just to see what happened. He played with a prism just to see what happened. Only against that background did he set out systematically to try to explain what he had seen. Newton's example carries a very important lesson about the two things effective thinking requires. Each has its place. Each is needed. Each is worthy of respect.

One requirement is creative exploration. Good thinking, whether experimental or not, requires that we take time to play with ideas. Creative conceptual play can't be forced. It can't be scheduled. It has to be play. Genuinely creative thinking can't be regimented. Children know this intuitively. But creative conceptual play isn't enough. Good thinking also requires a second phase in which one squints critically at the creative results. One tries to tack them down and figure out what they really mean. That's the stage at which Newton asks, what precisely has been shown? Can we formulate those results more clearly? How exactly does the experiment refute established doctrine? Both of those sides of thinking, the creative and the systematic, appear in Newton's notebooks. When you read a finished scientific article, it often looks as if the work was systematic through and through, but don't be deceived. New thinking, in science as well as anywhere else, has to be based in creative play. Creative play alone isn't enough— we need to tack it down. But tacking it down isn't enough, either. We need creative thinking in order to give us material worth working with.

I said Newton was also a great formalizer. In the 1680s several members of the Royal Society had floated an idea as to how gravity works. The idea was that the farther two bodies were apart, the less the gravitational attraction between them. How much less? The suggestion was that their attraction fell off with the square of the distance between them. It was suggested that such a principle might explain the elliptical path of the planets around the sun, but that's as far as the idea went. Nobody could tack the idea down mathematically. Edmond Halley was a participant in that discussion. In August of 1684 he went to visit Isaac Newton in Cambridge. "What kind of curve," Halley asked him, "would be described by the planets supposing the force of attraction toward the sun to be reciprocal to the square of their distance?" Newton said, "an ellipse." "How do you know?" Halley asked. "I have calculated it," Newton said. Halley asked to see the calculations, but Newton couldn't find them in his messy stacks of papers. They arrived several months later. Newton had been perfecting some details. Those calculations became the core of Newton's famous *Principia*.

In Newton, the great formalizer, we again we see the power of simplification. He simplifies the problem of gravitational forces by first considering the attraction between just two bodies. Calculating gravitational attraction for three or more turns out to be astoundingly difficult. Newton also simplifies the problem by treating them as points rather than planets. Halley used the principles of Newton's *Principia* to predict that a comet that had been recorded in 1531, 1607, and 1682 would reappear in 1758. We know it as Halley's comet. It appears every 75 to 76 years. If you missed it in 1986— that was a terrible year for visibility—tell your grandkids to check it out in 2061.

Newton's life overlaps with another great thinker, Gottfried Wilhelm Leibniz. Ironically, the thing they have in common is the thing that made them bitter enemies: the calculus. Who invented calculus? They both did, independently. Newton developed his form first, but he kept the technique secret, using it to find results which he then presented in more standard mathematical form. Leibniz developed his form a little later but published first. These two great formalizers could have been the best of friends, but they weren't. They fell victim to all the traps of polarization with charges and counter charges of priority and plagiarism. Indeed, their defenders continued that polarization long after both Newton and Leibniz were dead. That's a negative lesson from the great thinkers. Avoid unnecessary polarization. It was Descartes who said, "The greatest minds, as they are capable of the highest excellences, are open likewise to the greatest aberrations." Newton thought space and time were absolutes. That all changes with Einstein, who we've used as an example of great thinking throughout. Einstein was the great visualizer, amazingly flexible in looking at things an entirely different way.

Now wait a minute. I've been trying to track some of the great thinkers, but I notice that I keep coming up with either philosophers or physicists. What about the great thinkers in biology, psychology, or the social sciences? Remember that history of philosophy that I sketched in the very first lecture. In the beginning, everything was philosophy. It was only over centuries that different disciplines split off on their own. Aristotle was the first great biologist. Descartes clearly plays a major role in the history of psychology. Darwin was one of the greats—perhaps the great—in the history of biology. How is one to explain speciation, the development of species? Darwin uses a strategy of analogy. The analogy is laid out clearly in the opening passages of *The Origin of Species*. He says that people who have bred and worked with animals have long known that you can cultivate particular traits by selective breeding. Perhaps that same kind of selection happens in nature. That analogy is what the theory of natural selection is all about.

Freud is clearly one of the greats in psychology. His major insight? That so much of our mental processing goes on below the level of consciousness, and that's a lesson we've returned to again and again. As disciplines peeled off from philosophy in the 20th century, as philosophy of mind became psychology, as philosophy of man became anthropology, as parts of social philosophy became sociology, something else happened in our intellectual history. Something that carries an important lesson of its own. Single great thinkers, working alone were slowly replaced by research communities working together. We have used important results from social psychology throughout the course. But that work doesn't come from a single great thinker. It's a product of social rationality. Sometimes it's not a single great thinker you need but a Great Intellectual Community.

Even in the collaborative science of the 20th century, there are individuals who stand out. Let me end by returning to my candidate for the smartest man of the 20th century: John von Neumann. What makes von Neumann great is the fact that he worked so flexibly across so many disciplines. He is known for contributions in mathematics, in economics, in physics, in computer science. We have said that he deliberately memorized formulae at the back of textbooks. Why? So that he could recognize a similar formalism applicable across fields, so that he could see the links between very different disciplines. The lesson? Stay open to unexpected connections; von Neumann clearly prepared for unexpected connections. As Louis Pasteur said, "Chance favors the prepared mind."

I want to apply some of the tools we've talked about to a von Neumann problem. If you do this right, you should be able to outsmart von Neumann. We've emphasized visualization. We've emphasized simplification. We've emphasized flexibility, the flexibility required to stand back and see a problem differently. I want you to use all of those but pay particular attention to simplification.

Von Neumann loved parties. At Princeton, he would throw a party for his colleagues about once a week, heavy-drinking 1950s style. Because of his reputation, people would come to the party with problems for von Neumann to solve. They'd pose him a problem. Standing there with a drink in his hand, he would rock back and forth and try to come up with the answer, which he almost always did. Here's a problem that they gave von Neumann one night. The important thing is that there's a trick to it. If you can visualize the problem flexibly and with an eye to simplification, you should be able to see the trick.

Two bicycles are approaching each other from 20 miles apart. Each is going 10 miles an hour. A fly leaves the front tire of one of the bicycles at 15 miles per hour, flies to the front tire of the other, and turns around. It goes back to the first bicycle, still at 15 miles per hour, hits the front tire, turns around. The problem is idealized, of course. The bicycles go straight towards each other at a constant speed of 10 miles per hour, and the fly loses no time in turning around. The question, how far will the fly have flown by the time it is crushed between the front tires of the two bicycles?

As I say, there's a trick to the question. If you tried to work it out straight, without flexibility, you'd be doing this: You'd figure out the distance from one bicycle to where the other is by the time the fly gets there, plus the distance back to where the first bicycle is then, plus the distance back to the other bicycle, all calculated in terms of that 15 miles per hour. Figuring out just that first distance is hard. Figuring out the fly's total distance would mean adding up all those progressively smaller distances, taking the infinite sum. Whew! But there's a much simpler way to get the answer. Can you see the trick? If you want to press pause for one last time, go ahead. Visualize the problem by sketching it out, and remember, simplification is the key.

Here's the trick. The bicycles are 20 miles apart, approaching each other at 10 miles per hour. How long until the bicycles hit? One hour. The fly's speed is 15 miles per hour. So how far will it have flown by the time it is crushed between the two front tires? 15 miles. If you saw the trick, you did it the easy way. You outsmarted von Neumann. When his colleagues posed the problem, von Neumann thought about it for two minutes, three minutes, and then said, "15 miles." They said, "Oh, so you knew the trick?" "What trick?" he said. "I just took the infinite sum."

Here, then, are some tips from the great thinkers. We've really been talking about them throughout these lectures. First tip, simplify, simplify. That's how to outsmart von Neumann. We've seen that lesson in Newton and Descartes, break a complex problem into simple parts. Start with the simple, build up to the complex. We emphasized that strategy early on in the Tower of Hanoi problem, in going to extremes, and in the strategy of thought experiments. If you can't solve a hard problem, use a simpler model. That's what Galileo did. Simplification is what model building is all about, from the Wright Brother's wind tunnel, to the Mississippi River Basin model, to Schelling's segregation model. Plato's abstraction represents the power of generality you can get, not by adding, but by subtracting detail—by simplifying.

Second tip: Look for patterns, but watch out for them too. Visualization allows us to harness something we're great at, visual pattern recognition. It lets us think faster, more intuitively, and more powerfully. Aristotle knew the power of visualization. He used it in his Square of Opposition. Einstein was one of the great visualizers. The heuristics that make us smart are often forms of pattern-recognition. But the biases that lead us astray are often patterns as well. Our vivid images, like images of shark attacks, can mislead us as to true probabilities. We miss the gorilla in the room because we're paying attention to some other pattern. Patterns are extremely powerful. That's why it's important to put them to work and why it's important to handle them with care.

Third tip: Think systematically. Maybe we can systematize thought. That is the Aristotelian vision. Thinking systematically is what gave Newton his edge. We have developed a number of tools for systematic thought, tools in statistics, in probability, in decision theory. One simple lesson in thinking systematically goes all the way to the bat and ball example in the first lecture. Want to think better? Check your work.

Fourth Tip: Put it to the test. A respect for the reality behind the abstractions is the difference between Plato and Aristotle. Respect for reality runs through the whole history of experiment. Charles Sanders Peirce thought of science as the one really self-correcting technique for fixation of belief. Popper emphasizes that genuine science demands risk, the risk of falsifiability. Galileo is the patron saint of experimentation. Simple and beautiful experiments go all the way from Newton's prism to Richard Feynman's O-rings. Do you really want to know what reality entails? Ask her. Put it to the test.

Fifth tip: Think socially. Many of the complexities we face are social complexities. They require thinking in terms of multiple perspectives and multiple players. Thomas Hobbes knew the importance of thinking socially, so did Jean Jacques Rousseau. That kind of thinking is the whole goal of contemporary game theory. Inevitably, there are social threats to good thinking as well. We have to watch out for mere rhetoric, for standard fallacies, for the tricks of advertisers. Newton and Leibniz are an unfortunate example of the pitfalls of polarization even for great minds. We've tried to emphasize ways of escaping those traps. Plato's dialogues are the first great works in philosophy. They emphasize this social side to thinking. Plato takes an optimistic view. He thinks we often think most powerfully when we think together. Despite the pitfalls, I want to think he's right.

Sixth tip: Think with both sides of your brain. Newton combined the power of creative play with systematic critique. Great thinking requires disciplined imagination. Both parts need to be there, both the discipline and the imagination. We emphasized early on that emotion need not be the enemy of rationality if used in the right domain. We couldn't do without fast and frugal heuristics, but for long-range decisions we need careful calculation as well. Remember Kahneman's System one and System two. We need both.

Let me add a Seventh Tip that appears in all the great thinkers. Give yourself time to think. Descartes must have been a phenomenally quick thinker, and yet he says, "those who travel slowly may yet make far greater progress than those who run." Google's engineers are given one day a week to work on something entirely of their own. It's unfortunate, but these days our last bastions for free-range thinking seem to be the shower and the commute to work. Real thinking demands conceptual exploration, and that just takes time. Do like Google. Do like all the great thinkers did. Give yourself time to think.

Of course the great thinkers have more lessons to teach. Don't trust authority. Don't trust the majority. Don't even trust your own prejudices as to what is the obvious answer. Think for yourself. That's a lesson clear in Galileo, Descartes, and Einstein. Stay flexible. The right way to think about something may be one no one has used before. Be prepared to try a new view. Stay open to the unexpected. Each of the great thinkers worked with a conceptual toolkit. But none of them built it alone. All inherited a set of conceptual tools from their predecessors. Newton said, "If I have seen farther, it is by standing on the shoulders of giants." All added to the toolkit and then passed it on.

That cultural heritage has now been passed to you, the philosopher's toolkit, to use, to apply, to add to, to improve, to pass on. It's wonderful to peer into the future and realize there are more great thinkers to come.

Glossary

abstraction: The thought process that allows us to derive general concepts, qualities, or characteristics from specific instances or examples.

ad hominen: A fallacy that depends on an attack against the person making a claim instead of the claim that is being made.

amygdala: A small region deep within the brain that is associated with emotion.

anchor-and-adjustment heuristic: A common strategy used in calculating probabilities, but one that depends on how a question is phrased. Information given in the question is taken as a starting point, or anchor; individuals tend to adjust their responses upward if the anchor seems too low or downward if the anchor seems too high, arriving at an answer that is less extreme than the information given in the question but that may have little connection to the real answer.

appeal-to-authority fallacy: A fallacy in which the opinion of some prominent person is substituted for rational or evidential support; often used in advertising by linking a product to a celebrity "expert" rather than providing rational or evidential support for a claim about the product.

appeal-to-emotion fallacy: A fallacy in which positive or negative emotional tone is substituted for rational or evidential support; an argument strategy intended to cross-circuit the ability of the listener to assess whether a rational link exists between premise and conclusion by "pushing emotional buttons."

appeal-to-honesty advertising strategy: Use of a spokesperson who appears credible and trustworthy.

appeal-to-ignorance fallacy: A fallacy in which absence of information supporting a conclusion is taken as evidence of an alternative conclusion. This fallacy acts as if ignorance alone represents some kind of positive evidence.

appeal-to-majority fallacy: An argument that treats majority opinion as if that alone constituted evidence supporting a conclusion or gave a reason for belief. This fallacy ignores the fact that people, even large numbers of people, are fallible.

appeal-to-prestige advertising strategy: Linking a product to status symbols in order to enhance the product's desirability. Foreign branding is a specific form of the appeal-to-prestige strategy.

Aristotelian logic: Aristotle's attempt to systematize thought by outlining a set of formal relations between concepts and propositions. These relations can be visualized by his square of opposition and his treatment of arguments as syllogisms. See **square of opposition, proposition**.

attention bias: Overlooking the unexpected because we are attending to the expected.

"attractiveness attracts" advertising strategy: Enhancing the desirability of a product by using spokespeople who are more attractive than average, by showing the product in beautiful settings, or by doctoring the product's image or the product itself so that it is more photogenic. The flip side, "unattractiveness detracts," is common in political campaign literature, which often seeks unflattering images of the opponent.

availability heuristic: The tendency for individuals to assume that things that are easier to bring to mind must be more common or occur more frequently; the tendency to generalize from simple and vivid images generated by single or infrequent cases and to act as if these are representative.

axiom: A claim accepted as a premise without proof and from which other claims are derived as theorems. Euclid's geometry is a familiar example, in which theorems are derived from a small number of initial axioms.

axiomatic: Organized in the form of axioms and derivations from them. Euclidean geometry is an example of an axiomatic system. See **axiom**.

Brownian motion: The random motion of particles in a gas or liquid, such as motes in sunlight or grains of pollen in water.

Cartesian coordinates: The position of a point on a plane as indicated in units of distance from two fixed perpendicular lines, the x (horizontal) axis and the y (vertical) axis.

categorical proposition: In Aristotelian logic, a simple proposition that combines two categories using "all are," "none are," "some are," or "some are not." Categorical propositions can be visualized using two circles in a Venn diagram.

category: Any group of related things; the grouping is based on what are perceived as important similarities between those things.

causation: The act of producing an effect. In a cause-and-effect relationship, the presence of one variable, the effect, can be established as a direct result of the other variable, the cause. Opposed to **correlation**.

change blindness: The propensity for individuals not to perceive unexpected changes, particularly when attention is focused on something else.

Chicken: In game theory, a two-person matrix game in which the value of mutual cooperation is greater than the value of joint defection, the value of defecting against a cooperator is greater than the value of mutual cooperation, but the value of cooperating against a defector is higher than the value of mutual defection, as illustrated in the following matrix:

	B	
	Cooperate	**Defect**
Cooperate	*3*, 3	*1*, 5
Defect	*5*, 1	*0*, 0

A appears to the left of the Cooperate/Defect row labels.

Chicken Matrix

In Chicken, the real loser is mutual defection. For alternative payoff scenarios, see **Prisoner's Dilemma** and **Stag Hunt**.

combined probability: The rules for calculating the odds that two or more events will happen. The probability of either one or another of two mutually exclusive events happening can be calculated by adding their individual probabilities. The probability of two independent events both happening can be calculated by multiplying their individual probabilities. Other rules apply for dependent events; see **independent events**, **mutually exclusive events**.

complement: In logic, given any category, the complement comprises all those things that do not fall in that category. For example, "senators" and "non-senators" are complements.

complex-question fallacy: A "trick question" presenting a false dilemma, or forced-choice alternative, presented in such a way that any answer is incriminating. For example: "Answer yes or no: Have you stopped beating your wife?" If you say yes, you have essentially admitted that at one time, you did beat your wife; if you say no, you have admitted that you are still beating her.

concepts: Ideas, the basic elements or "atoms" of thought, as distinct from the words that represent those ideas.

concept tree: The hierarchical structure that visualizes the relationships within a set of related concepts.

conclusion: The endpoint of an argument; in a logical argument, the claim to which the reasoning flows is the conclusion. See also **premise**.

connotation: The emotional tone or "flavor" associated with the ideas or things that words label.

context of discovery: Used by philosophers of science to label the process by which a hypothesis first occurs to scientists, as opposed to the context of justification, the process by which a hypothesis is tested. The consensus is that context of justification follows clear, rational rules, but context of discovery need not do so.

context of justification: Used by philosophers of science to label the process by which a hypothesis is tested or established, as opposed to how it is first invented or imagined. The consensus is that context of justification follows clear, rational rules, but context of discovery need not do so.

contradiction: A statement that both asserts and denies some proposition, P, often represented in the form "P and not-P." If either part of a contradiction is true, the other cannot be true, and thus, a contradiction P and not-P is treated as universally false.

contradictories: The relationship between propositions on the diagonals of Aristotle's square of opposition. It is a contradiction for both propositions on a diagonal to be true; if one proposition of the diagonal is true, the other must be false. See **square of opposition**.

contrapositive: A way of transforming categorical propositions by switching subject and predicate and replacing each with its complement. For some categorical propositions, the result is an immediate inference: the truth or falsity of the proposition is not altered. The contrapositive transformation preserves equivalence only for propositions in the upper left and lower right on the square of opposition: the universal positive ("All S are P") and particular negative ("Some S are not P"). See **square of opposition, universal proposition, particular proposition, complement**.

contraries: The relationship between propositions at the top left ("All S are P") and right ("No S are P") of Aristotle's square of opposition. If two propositions are contraries, it is not possible for both propositions to be true, but it is possible for both propositions to be false. See **subcontraries**, **square of opposition**.

controlled experiment: A test that demonstrates a causal connection between a variable (X) and an outcome (Y) by establishing both "if X, then Y" and "if not X, then not Y." In essence, a controlled experiment is a double test of two cases side by side, one with and one without X.

converse: A way of transforming categorical propositions by switching subject and predicate. For some categorical propositions, the result is an immediate inference: the truth or falsity of the proposition is not altered. The converse preserves equivalence only for propositions in the upper right and lower left on the square of opposition: the universal negative ("No S are P") and the particular positive ("Some S are P"). See **square of opposition**, **universal proposition**, **particular proposition**.

correlation: A direct or inverse relationship between two variables. In a direct correlation, the strength or frequency of both variables increases proportionately to each other; in an inverse correlation, the strength or frequency of one variable increases as the strength or frequency of the other decreases. However, correlation does not establish that one variable causes the other; one variable is not necessarily the cause and the other the effect. See also **causation**.

decision theory: A theory of how to make rational decisions by maximizing expected utility.

deductive validity: A deductively valid argument is one in which it is logically impossible for all premises to be true and the conclusion to be false.

deferred gratification: The ability to restrain oneself from taking an immediate payoff in order to obtain a larger payoff later.

demarcation: The philosophical problem of precisely defining the differentiation between two concepts. Specifically, the problem of precisely differentiating science from pseudoscience.

denotation: The things that a concept or word applies to.

dependent reasons: Premises that support the conclusion only when they are both present; propositions or claims that function together but are insufficient alone as support for the conclusion.

descriptive: Used to designate a claim that merely reports a factual state of affairs rather than evaluating or recommending a course of action. Opposed to **normative**.

diminishing marginal utility: The concept that units of value may not amass in equal increments; additions of one unit may not always be the same as additions of one unit of value. For example, one dollar may not be worth as much to a billionaire as to an individual who has no savings at all.

diversion fallacy: Also known as a "red herring"; the diversion fallacy arrives at a conclusion after diverting the listener's attention from relevant considerations to the contrary.

double-blind experiment: In experimentation, a way of controlling the possibility that unconscious bias of subjects and investigators may influence the results by assigning subjects so that the experimental condition is hidden from them. In a double-blind experiment, neither the subjects nor the investigators who interact with them know which experimental condition is being tested until after the experiment is complete.

epidemiology: The study of the distribution, patterns, dynamics, and causes of health-related events in a population.

epithumia: Appetite or passion; according to Plato, the second element of the soul.

equiprobable: Having an equal mathematical or logical probability of occurrence.

ethics: The field of philosophy that focuses on moral issues: ethically good actions, ethically right actions, rights, and obligations.

ethos: Character of the speaker; according to Aristotle, the first quality of a persuasive presentation is that the speaker must appear knowledgeable and wise.

eugenics: A social movement that advocates practices and direct interventions aimed at changing, or ostensibly "improving," the genetic characteristics of a population.

expected utility: In economics, calculated by multiplying the potential benefit of an outcome by its probability.

explanation: In a three-stage model, information about input(s) and output(s) is used to generate possible mechanisms for linking the two. See **input, mechanism, output**.

extension: In philosophy, the set or totality of things that a concept applies to.

fallacy: A form of argument in which the premises appear to support a conclusion but, in fact, do not; the term is often used to refer to familiar types of logical mistakes that may be used to trick or mislead.

false alternative: A fallacy in which a problem is presented as an either/or choice between two alternatives when, in fact, those are not the only options. Also called a "false dilemma."

falsifiability: Karl Popper's criterion for distinguishing real science from pseudoscience was that real scientific theories can be shown to be false. Real science takes risks in that theories are formulated so that they are open to disconfirmation by empirical testing or empirical data. See also **verifiability**.

flow diagram: A systematic sketch of a train of thought illustrating the lines of support between premises and conclusions in a rational argument; when one claim is intended as support for a second claim, an arrow is drawn from the first to the second. See **premise, conclusion**.

foreign branding: In advertising, associating a product with a foreign country in order to increase its desirability; a particular type of prestige advertising.

formalization (systematization): The process of characterizing abstract relations, physical processes, or chains of thought in terms of explicit axioms, principles, or rules. Euclidean geometry systematizes spatial relations in a certain way; logic formalizes patterns of rational inference and valid argument.

gambler's fallacy: Treating independent events as if they were dependent events. Someone who thinks black is "bound to come up" in roulette because of the appearance of a string of blacks (or reds) has committed the gambler's fallacy.

game theory: The mathematically idealized study of rational interaction in competitive and cooperative situations.

grammar: The study of the proper structure of language in speech or writing; along with logic and rhetoric, an element of the classical medieval curriculum known as the *trivium*.

hasty generalization fallacy: Also known as jumping to conclusions. This fallacy occurs when one jumps to a conclusion about "all" things from what is known in a small number of individual cases. Racism and sexism often take the form of hasty generalizations.

heuristics: Simple guides to action or rules of thumb that allow us to act or make a decision without calculation or deliberation.

Homo economicus: The name given to a "species" of agents who would act rationally in the way economic game theorists say that individuals should act in order to maximize their own gain.

implication: In Aristotelian logic, the relationship moving from the top to the bottom left corners or the top to the bottom right corners of the square of opposition. If the proposition on the top left corner is true, then the proposition on the bottom left corner is also true; if the proposition on the top right corner is true, then the proposition on the bottom right corner is also true. Expressed as "if all S are P, then some S are P" for the left side of the square of opposition and "if no S are P, then some S are not P" for the right side of the square of opposition. See **square of opposition**.

independent events: Events that are isolated probabilistically; the fact that one event happens is not linked to and will not affect the fact that the other event happens.

independent reasons: A group of premises, or reasons, that are given as support for a conclusion, each of which could support the conclusion on its own.

independent standard: In negotiation or the attempt to reduce polarization, a deciding touchstone or court of appeal that is not open to manipulation and that can be agreed on by both parties in advance. In establishing a fair price for a house, for example, both parties might agree in advance to use the price that similar houses have recently sold for in the neighborhood.

induction (inductive reasoning): Thinking that proceeds from individual experiences to theoretical generalizations.

inference: In logic, the derivation of a conclusion from information contained in the premises.

input: The first structural stage of a three-stage input-mechanism-output model: the initial setup, conditions, or changes on which the model operates.

intuitive approach: An approach to problem solving that focuses on the general outline of a problem, attending to a few critical issues and generating a quick decision after consideration of relatively few alternative solutions. Opposed to an analytic approach, which examines a wide range of specifics and generates an exhaustive list of alternative solutions.

Kripkean dogmatist: An individual who believes that his or her position is right and, on that ground alone, is prepared to reject any and all evidence to the contrary.

law of large numbers: In probability theory, the fact that as the number of trials increases, the outcome will approach the mathematically expected value. For example, in a situation where there are two equiprobable outcomes, such as heads or tails when flipping a coin, the longer the run, the more likely the outcome will be a 50/50 split.

logic: The study of patterns of rational inference and valid argument.

logical fallacy: See **fallacy**.

logos: Logic; according to Aristotle, the third quality of a persuasive presentation is that it will lay out the argument clearly and rationally, step by step.

matrix: A rectangular array; a visualization technique using a checkerboard to represent how two variables align. One set of values is represented by rows in the matrix; a second set of values is represented by columns.

mean: The average of a set of numbers.

means-end reasoning: Describes the practical problem-solving process that links actions to expected ends or results; means-end reasoning does not help in evaluating which ends or results are desirable.

mechanism: The second structural stage of a three stage input-mechanism-output model: the operations that run on input in order to produce the output.

median: The midpoint of a set of numbers that has been ordered from lowest to highest; that point at which there are as many below as above.

mental set: In perception, the background expectation that may influence what is perceived; that is, when one is expecting a normal deck of cards, mental set may lead one to ignore altered cards, such as a red ace of spades.

Mill's joint method of agreement and difference: In John Stuart Mill's *System of Logic*, evidence for causal linkage between X and Y is stronger when conditions for both Mill's method of agreement and Mill's method of difference are present; that is, Y appears whenever X is present and disappears when it isn't.

Mill's method of agreement: In John Stuart Mill's *System of Logic*, if two cases are alike only in X, and Y occurs in both, we have evidence that X and Y are causally linked.

Mill's method of concomitant variation: In John Stuart Mill's *System of Logic*, applies to cases in which there is a range of values for X and Y. If the amount or strength of Y varies as the amount or strength of X varies, there is evidence for a causal link between X and Y.

Mill's method of difference: In John Stuart Mill's *System of Logic*, if the presence or absence of X is the only difference between two cases, with Y a result in the first case and not in the second, we have an indication that X causes Y.

mode: The most frequently occurring value in a set of numbers.

mutually exclusive events: Events are mutually exclusive if the presence of one categorically excludes the presence of the other; both cannot occur.

mysticism: A claim to an immediate, intuitive, and nonexperiential knowledge of reality.

necessary condition: X is a necessary condition for Y if one must have X in order to have Y; one cannot have Y without X. See also **sufficient condition**.

negotiation strategy: An approach to conflict resolution that may attempt to remove the "contest of wills" characteristic of positional negotiation. Among other techniques, negotiation strategies may include employing ego-distancing; talking about issues without identifying with a particular position; "going to the balcony," that is, trying to put emotions aside and view the problem from a distance; appealing to independent standards;

coming to some agreement about what kinds of objective criteria could help to clarify or settle the issue; and replacing debate with collaborative research on the topic.

normative: Used to designate a claim that is evaluative in nature or recommends a course of action, as opposed to descriptive.

nous: Reason; according to Plato, the first element of the soul and that which should rule.

null hypothesis: In experimental design, the prediction that there will be no detectable differences between the experimental conditions.

open file: In chess, a file is a vertical row with no pawns of either color on it, allowing a clear route into enemy territory for the queen or a rook.

outlier: An extreme value; something that is out of the ordinary. In statistics, a number that is extremely divergent from all the others in a set of numbers; outliers have misleading impact on the mean, or average, of a set of numbers.

output: The third structural stage of a three stage input-mechanism-output model: the result of a model run given a particular input.

overconfidence heuristic: The psychological tendency for people to overestimate their own abilities, also known as the Lake Wobegon effect.

particular proposition: In logic, a proposition about "some" rather than "all": "Some S are P" (a particular positive, e.g., some cleaning products are poisons) or "Some S are not P" (a particular negative, e.g., some cleaning products are not poisons). The particular positive occupies the lower-left corner of Aristotle's square of opposition; the particular negative occupies the lower-right corner.

Pascal's wager: A decision-theoretic argument offered by Blaise Pascal in the 17th century with the conclusion that it is rational to believe in God. If you believe in God and he exists, your payoff is eternal bliss; if you believe in God and he does not exist, your payoff is being wrong, a small loss; if you do not believe in God and he does not exist, you have the pleasure of being right, a small gain; if you do not believe in God and he does exist, your payoff is eternal damnation. Given the potential gains and losses, expected utility dictates that the most rational thing to do is believe in God.

pathos: Emotion; according to Aristotle, the second quality of a persuasive presentation is that it resonates emotionally with the listener.

pattern recognition: The ability to recognize a set of stimuli arranged in specific configurations or arrays, for example, to recognize faces as faces rather than patches of color or melodies as melodies rather than merely sequences of notes.

perceptual bias: A "hard-wired" tendency in our perceptual processing that forces us to perceive things in particular ways. Our color perception does not track pure wavelengths of light or actual lengths in a stimulus, for example, because our visual processing has evolved to interpret input immediately in terms of contextual cues regarding shadow and perspective.

poisoning-the-well fallacy: A fallacy that depends on an attack against a person's motives for saying something rather than a refutation of the claims being made; a subtype of *ad hominem*.

polarization: Radical or extreme disagreement between groups with no apparent willingness to compromise and/or with few individuals representing a middle group between the extreme positions. Polarization normally implies a wide gap between positions and increased uniformity within positions. Political polarization refers to extreme positions taken by political organizations, either major political parties or smaller interest groups; cultural polarization refers to extreme differences in attitudes of the general public that may or may not be expressed politically.

positional negotiation: In conflict resolution scenarios, an approach in which people are ego-involved or identify with their specific positions. Those involved in positional negotiation often end up in one of two roles: as the "soft" negotiator, who tries to avoid conflict by giving in and winds up feeling exploited, or as the "hard" negotiator, who is out to win at all costs and, thus, starts with an absurd extreme, allowing room to make some concessions and still hit his or her initial target.

post hoc ergo propter hoc: "After it, therefore because of it"; a fallacy based on the claim that because something followed another thing, it must have been because of that other thing. This fallacy overlooks the possibility of coincidental occurrence. Abbreviated as *post hoc*.

prediction: In a three-stage model, information about input(s) and mechanism is used to forecast the future, that is, to generate information about probable output(s). See **input, mechanism, output**.

prefrontal cortex: The anterior portion of the brain, which lies just behind the forehead, heavily involved in complex planning and decision making.

premise(s): The proposition(s) or claims that are given as support for a conclusion; in a rational argument, the reasoning flows from the premises to the conclusions. See **conclusion**.

Prisoner's Dilemma: In game theory, a two-person game in which the value of mutual cooperation is greater than the value of joint defection, the value of defecting against a cooperator is greater than the value of mutual cooperation, but the value of cooperating against a defector is lower than the value of mutual defection, as illustrated in the following matrix:

B

		Cooperate	Defect
A	Cooperate	3, 3	0, 5
	Defect	5, 0	1, 1

Prisoner's Dilemma Matrix

In the Prisoner's Dilemma, the real loser is cooperation in the face of defection; for alternative payoff scenarios, see **Chicken** and **Stag Hunt**.

probability: The ratio calculated by dividing the number of possible outcomes in a particular category by the total number of possible outcomes.

proposition: A claim, statement, or assertion; the message or meaning behind the words in a written or spoken sentence; the information a sentence expresses or conveys.

prospect theory: In the work of Daniel Kahneman and Amos Tversky, a descriptive theory concerning how people make decisions in terms of prospective loss and gain rather than final outcome alone.

pseudoscience: A set of practices or tenets that is made to resemble science and claims to be scientific but violates scientific standards of evidence, rigor, openness, refutation, or the like.

Ptolemaic astronomy: The view of astronomy put forward by Ptolemy (A.D. 90–168); based on a geocentric (earth-centered) model of the universe.

Pythagorean theorem: In Euclidean geometry, the relation of the three sides of a right triangle, represented by the formula $a^2 + b^2 = c^2$. The area of the square whose side is the hypotenuse of the right triangle (the side opposite the right angle) is equal to the sum of the area of the squares of the other two sides.

quantum mechanics: Developed early in the 20th century, quantum mechanics is a sophisticated theory of physics at the subatomic scale; a mathematically elegant, empirically established but philosophically puzzling theory of the very small.

questionnaire bias: A way of phrasing a question that influences the way people answer the question. Depending upon how the question is phrased, respondents may be more likely to respond in a negative or a positive direction than they would if the question were phrased in an unbiased, or neutral, way.

randomization: In experimentation, a procedure for assigning groups so that the differences between individuals within groups distribute with equiprobability across experimental conditions.

rational argument: A way of presenting and supporting claims that relies on logical transition from premises to conclusion.

rationality: Exercising reason, that is, analytical or logical thought, as opposed to emotionality.

recognition heuristic: A fast-and-frugal rule of thumb for decision making in which one picks the alternative or orders alternatives in terms of recognition; often most effective when one is working on very limited information about alternatives.

relativity theory: Proposed as an alternative to Newtonian physics, Einstein's theory of relativity (general and special) holds that matter and energy are interchangeable, that time moves at a rate relative to one's rate of speed, and that space itself can be curved by gravity. In large part, relativity theory is a series of deductions from the assumption that some things are not relative. For example, assuming that the speed of light is a constant, Einstein showed that observable simultaneity will be relative to the comparative motion of two observers.

retrodiction: In a three-stage model, information about the mechanism and output(s) is used to generate information about the past (inputs). See **input, mechanism, output**.

retrograde motion: In astronomy, the apparent movement of a body opposite to that of other comparison bodies. Plotted against constellations night by night, Mars moves in one direction for a while but then appears to move backwards before it again seems to reverse directions. Retrograde motion does not fit with the predictions of a simple geocentric theory, which holds that the sun and planets circle around the earth.

rhetoric: The skills of effective speaking and presentation of ideas; also, the techniques of persuasion, fair or foul, for either good ends or bad.

risk aversion: In economics, an investor is considered risk averse if he or she avoids risk as opposed to being risk taking. People's choices involving possible losses are often more risk averse than similar choices regarding possible gains.

risk taking: In economics, an investor is considered risk taking if he or she is not opposed to taking risks. Choices involving possible gains are often more risk taking than similar choices regarding possible losses.

sampling issues: Problems with the group from which statistics are derived that may lead to misrepresentations about the general characteristics of the group. Some sampling issues are size (the sample group is too small relative to the population to which it is being generalized) and bias (the sample group has been selected in a way that overrepresents certain subgroups and underrepresents other subgroups of the population).

satisficing: A heuristic, or rule of thumb, for decision making in which one picks a realistic goal or set of goals, then selects the first course of action that meets the goal or goals. This heuristic emphasizes that one is not seeking perfection, just something that is "good enough."

Scholastics: Philosophers practicing the dominant Western Christian philosophy of the European Middle Ages.

selective memory: The phenomenon in which recall of past events is uneven, with particular classes of events overemphasized because of emotional tone, individual interest, and personal history. Gamblers have been shown to have a selective memory that favors past winnings over past losses.

sentence: A sentence is a series of words, spoken or written, that expresses a claim, statement, or assertion. See also **proposition**.

set theory: The branch of mathematics that studies the properties of sets and their interrelations—abstract properties of collections, regardless of what they contain.

simplification: Conceptual replacement of a complex concept, event, or phenomenon with a stripped-down version that is easier to understand or to manipulate mentally.

sound argument: An argument that is both valid—that is, the premises lead by strong logic to the conclusion—and in which the premises are true.

square of opposition: Aristotle's visualization of the logical relations between categorical propositions.

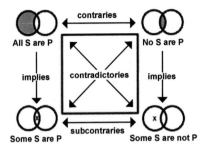

Stag Hunt: In game theory, a two-person game in which the value of mutual cooperation is greater than the value of either defecting against a cooperator or mutual defection, as illustrated in the following matrix:

		B	
		Cooperate	Defect
A	Cooperate	5, 5	0, 1
	Defect	1, 0	1, 1

Stag Hunt Matrix

In the Stag Hunt, the real winner is mutual cooperation; for alternative payoff scenarios, see **Chicken** and **Prisoner's Dilemma**.

statistical significance: A mathematical calculation that is used to determine whether the results of an observation or experiment might have occurred by chance alone.

Stoics: The Stoics held that the rational road to tranquility is to control one's emotional reactions to such an extent that one becomes impervious to the slings and arrows of outrageous fortune.

straw man fallacy: The rhetorical strategy of exaggerating or misrepresenting the claims of the opposition in order to more easily appear to refute those claims.

stress test: A technique used to examine the strength or stability of an entity under operational conditions that are more extreme than what is expected normally. In analyzing argument flow, a technique for detecting dependency between reasons by eliminating them individually in order to see whether the argument still goes through.

subcontraries: The relationship between propositions at the bottom left ("Some S are P") and right ("Some S are not P") of Aristotle's square of opposition. If two propositions are subcontraries, it is possible for both propositions to be true, but it is not possible for both propositions to be false. See also **contraries**, **square of opposition**.

sufficient condition: X is a sufficient condition for Y if X is enough; once one has X, Y is guaranteed. See also **necessary condition**.

syllogism: An argument using three categories (A, B, and C) that are linked in pairs to make categorical propositions (e.g., "All A's are B's," or "Some B's are not C's"), which are then combined into a three-step argument. Some syllogisms are valid and some are not. See **categorical proposition**, **validity**.

thermodynamics: A branch of physics that is concerned with the relationships of heat, pressure, and energy.

thumos: Spirit or fortitude shown in battle; according to Plato, the third element of the soul.

tit for tat: In game theory, a strategy for playing an iterated game in which a player mirrors the play made by the opponent in the previous round; the player cooperates if the opponent cooperated in the previous round but defects if the opponent defected in the previous round.

tripartite theory of the soul: Plato's theory that the soul includes three parts: *nous*, *epithumia*, and *thumos*.

trivium: The basis of a classical education throughout the Middle Ages; a three-part curriculum composed of grammar, logic, and rhetoric.

tu quoque: A fallacy in reasoning that tries to defuse an argument by claiming, "You did it, too."

Ultimatum Game: In game theory, a game in which two participants are given the task of splitting a payoff, for example, a certain amount of money. One participant is randomly selected to be the Proposer of how to make the split, and the other, the Responder, must decide whether to accept the offer. If the Responder accepts, the payoff is split between them as proposed; if the Responder rejects the offer, no one gets a payoff.

universal generalization: A statement about "all" of a particular class: "All X's are Y's."

universal proposition: In logic, a universal proposition refers to a claim either in the form "All S are P" (universal affirmative, e.g., all snakes are reptiles) or in the form "No S are P" (universal negative, e.g., no snakes are reptiles). The universal affirmative occupies the upper-left corner of Aristotle's square of opposition; the universal negative occupies the upper right.

valence: A psychological term describing the positive or negative value (often emotional) that we ascribe to things.

validity: An argument is valid if the conclusion follows from the premises, if the premises offer sufficient logical or evidential support for the conclusion. An argument is deductively valid if it is impossible for all premises to be true and the conclusion to be false.

value: Desirability; psychological research indicates that relative valuations tend to be inconsistent over time and sensitive to context.

Venn diagram: A way of visualizing the relations between concepts and their extensions through the use of overlapping circles.

verifiability: The criterion for what distinguishes science from non-science, proposed by a group of 20th-century scientists and philosophers called the Vienna Circle. Science, they said, is that which can be confirmed by empirical or experiential data. See also **falsifiability**.

visualization: The process of using diagrams or imagery as an aid to thought in representing a problem, calculation, or set of possibilities.

warrant: A general underlying principle that licenses an inference from data to a conclusion. In a probability warrant, the strength of the link between premise and conclusion is expressed in terms of probabilistic connection (e.g., 90 percent of the time, premise A is linked to conclusion B). In a definitional warrant, the premises are linked to conclusion as a matter of definition (e.g., whales are mammals by definition because they breathe air and give live birth). A legal warrant relies on a point of law as the link between the premise and conclusion (e.g., a contract requires a signature; thus, this unsigned document is unenforceable). An ethical warrant relies on an underlying ethical belief (e.g., if there is a shared belief that one should not deceive, then the conclusion that a deliberately deceitful act was wrong is warranted).

Wason selection task: A psychological task used to examine the psychology of reasoning. If shown four cards, one displaying the letter E, one the letter K, one the number 4, and one the number 7, using a minimum number of cards, which cards do you turn over to test the theory "If a card has a vowel on one side, it has an even number on the other"? It has been demonstrated that people tend to seek confirmation when working with unfamiliar or abstract concepts, such as "vowels" and "even numbers," but will seek falsifiability in similar tasks using familiar situations, such as underage drinking.

wisdom-of-crowds heuristic: A rule of thumb in which one bases one's decision on the most popular answer selected by a random group of people. In order for this heuristic to be effective, certain conditions of randomness must be met. The reason it works can be explained statistically by the fact that in an appropriately random group of people, most will be ignorant about the topic, but a few will actually know the answer. The answers of the "ignorant" will distribute randomly and cancel each other out, allowing the answers of those who really know to tip the scale.

word: A linguistic representation of an idea; as distinct from **concept**.

Bibliography

Abbott, Edwin A. *Flatland.* New York: Dover Publications, 1952. The 1884 story of A Square, whose two-dimensional world is invaded by three-dimensional forces. An inspiration for trying to think from entirely new perspectives.

Aczel, Amir D. *Chance: A Guide to Gambling, Love, the Stock Market, and Just About Everything Else.* New York: Thunder's Mouth Press, 2004. Entertaining but thorough.

Aristotle. *Rhetoric.* Translated by W. Rhys Roberts. Mineola, NY: Dover Publications, 2004. The classic source. A valuable read even if skimmed.

————. *The Organon, or Logical Treatises, of Aristotle.* Berkeley: University of California Library, 2012. The *Organon* includes the *Categories, Topics, On Interpretation, Prior Analytics,* and *Posterior Analytics.* This is Aristotle straight, often difficult but always rewarding to read. The sources are also available in many other forms. See Aristotle, *Categories and De Interpretatione,* translated by J. L. Ackrill, Oxford: Clarendon Press, 1963; Aristotle, *The Organon,* Shenandoah Bible Ministries, 2009; and Richard McKeon, ed., *The Basic Writings of Aristotle,* New York: Random House, 1941.

Axelrod, Robert. *The Evolution of Cooperation.* New York: Basic Books, 1984. A very readable but fully sophisticated treatment of the triumphs of the strategy tit for tat in the iterated Prisoner's Dilemma.

————. *The Complexity of Cooperation: Agent-Based Models of Competition and Collaboration.* Princeton, NJ: Princeton University Press, 1997. A later contribution from Robert Axelrod, emphasizing the complexity of richer models over the simplicity of tit for tat.

Axtell, R. L., J. M. Epstein, J. S. Dean, G. J. Gumerman, A. C. Swedlund, J. Harberger, S. Chakravarty, R. Hammond, J. Parker, and M. Parker. "Population Growth and Collapse in a Multi-Agent Model of the Kayenta Anasazi in Long House Valley." *Proceedings of the National Academy of Sciences*, Colloquium 99 (2002): 7275–7279. A fascinating attempt to construct a model indicating why (and why not) a population of ancient Pueblo people may have disappeared from the Long House Valley in 1350. Was it just climate change or something else?

Baggini, Julian, and Jeremy Stangroom. *Great Thinkers A–Z*. New York: Continuum, 2004. One hundred biographical snapshots, with recommendations for further reading in each.

Baggini, Julian, and Peter S. Fosl. *The Philosopher's Toolkit: A Compendium of Philosophical Concepts and Methods*. Oxford: Blackwell Publishing, 2003. Despite the title, a volume unaffiliated with this course. A valuable and very readable source, more closely tied to specifics in the history of philosophy but less devoted to cultivating thinking skills in general.

Baggini, Julian. *The Pig That Wants to Be Eaten: 100 Experiments for the Armchair Philosopher*. New York: Penguin, 2005. A delightful compendium of thought experiments lifted from the history of philosophy.

Bargh, John A., Mark Chen, and Lara Burrows. "Automaticity of Social Behavior: Direct Effects of Trait Construction and Stereotype Activation on Action." *Journal of Personality and Social Psychology* 71 (1996): 230–244. The source of the study on "elderly" stereotype words and their influence on walking speed to an elevator.

Berkun, Mitchell M. "Performance Decrement under Psychological Stress." *Human Factors: The Journal of the Human Factors and Ergonomics Society* 6 (1964): 21–30. The classic studies on how rationality fails under stress.

Best, Joel. *Damned Lies and Statistics: Untangling Numbers from the Media, Politicians, and Activists*. Berkeley: University of California Press, 2001. The source of "the worst statistic ever": the claim that the number of children killed with guns has doubled each year since 1950.

————. *More Damned Lies and Statistics: How Numbers Confuse Public Issues*. Berkeley: University of California Press, 2004. A treatment of further examples of statistics gone wrong, both entertaining and enlightening.

Bishop, Bill. *The Big Sort: Why the Clustering of Like-Minded America Is Tearing Us Apart*. Boston: Houghton Mifflin Harcourt, 2009. Bishop makes the case that polarization is the result of our increased tendency to associate only with others like ourselves.

Boorstein, Daniel. *The Discoverers: A History of Man's Search to Know His World and Himself*. New York: Vintage, 1985. A beautiful exploration of our intellectual history, portrayed in broad and compelling strokes.

Bronowski, Jacob. *The Ascent of Man*. London: Ebury Printing, BBC Books, 2011. This wonderful book is very close to a word-for-word transcript of a television series originally produced in 1973 and still available on video. Both are well worth a second visit.

Brownstein, Ronald. *The Second Civil War: How Extreme Partisanship Has Paralyzed Washington and Polarized America*. New York: Penguin Press, 2007. An analysis of polarization in the American political scene.

Burger, Dionys. *Sphereland: A Fantasy about Curved Spaces and an Expanding Universe*. Translated by Cornelie J. Reinboldt. New York: Barnes & Noble Books, 1965. A further exploration in the spirit of Abbott's *Flatland*.

Carroll, Lewis. *Lewis Carroll's Symbolic Logic*. Edited by William Warren Bartley III. Hassocs, Sussex, England: Harvester Press, 1977. A look into the logical (but still lighthearted) side of Charles Dodgson, the author of *Alice in Wonderland*.

Clark, Ronald W. *Einstein: The Life and Times*. New York: Avon Books, 1972. Perhaps the best biography of the master.

Crease, Robert P. *The Prism and the Pendulum: The Ten Most Beautiful Experiments in Science*. New York: Random House, 2004. What are the most beautiful experiments? Crease is a columnist for *Physics World* and asked that question of his readers. Here, he offers a philosophical discussion of some of the foremost candidates.

Cummings, D. A. T., S. Chakravarty, R. M. Singha, D. S. Burke, and J. M. Epstein. *Toward a Containment Strategy for Smallpox Bioterror: An Individual-Based Computational Approach.* Washington, DC: Brookings Institute Press, 2004. A fascinating case study in computational modeling, applying known facts about smallpox to simple "what if?" scenarios.

Damasio, Antonio. *Descartes' Error: Emotion, Reason, and the Human Brain.* New York: Penguin Books, 1994. Contemporary neuroscience applied to questions in philosophy and psychology; as engaging as it is informative.

Darwin, Charles. *The Expression of Emotions in Man and Animals.* London: John Murray, 1872. Available from Digireads.com Publishing, Stilwell, Kansas, 2005. Published later than *The Origin of Species*, this aspect of Darwin's work is only now being given its due.

David, F. N. *Games, Gods and Gambling: A History of Probability and Statistical Ideas.* London: Charles Griffin & Co. Ltd., 1962; reprinted, Mineola, NY: Dover Publications, 1998. An intriguing introduction to the history of probability theory.

Davidson, D., and H. Aldersmith. *The Great Pyramid: Its Divine Message: An Original Co-ordination of Historical Documents and Archaeological Evidences.* London: Williams and Norgate, 1948. A clear example of pseudoscience, but one that may challenge attempts at demarcation.

Einstein, Albert. *Relativity: The Special and General Theory.* New York: Crown Publishers, 1961. The theories in the master's own words, geared to a popular audience.

———. *The World as I See It.* Translated by Alan Harris. New York: Quality Paperback Books, 1990. Einstein in his own words.

Epstein, Joshua M. *Generative Social Science: Studies in Agent-Based Computational Modeling*. Princeton: Princeton University Press, 2006. An accessible discussion of important new forms of computational modeling.

———. "Why Model?" *Journal of Artificial Societies and Social Simulation* 11 (2008). Available online at http://jasss.soc.surrey.ac.uk/11/4/12.html. A short and easy commentary on the wide variety of purposes to which models can be put.

Feynman, Richard P. *"Surely You're Joking, Mr. Feynman!" Adventures of a Curious Character*. As told to Ralph Leighton, edited by Edward Hutchings. New York: Bantam Books, 1985. A wonderful read, revealing some of the content but even more of the character of a marvelous mind.

Feynman, Richard P., Robert B. Leighton, and Matthew Sands. *The Feynman Lectures on Physics*. Reading, MA: Addison-Wesley, 1963. Physics that can be read for the entertainment value alone. The best and most accessible introduction to aspects of physics that anyone has produced or is likely to.

Feynman, Richard P. *Six Easy Pieces: Essentials of Physics Explained by Its Most Brilliant Teacher*. Reading, MA: Addison-Wesley 1995. Excerpted from *The Feynman Lectures*, with some of the same beautiful insights in physics. Also available in audio recordings.

Fineman, Howard. *The Thirteen American Arguments: Enduring Debates That Define and Inspire Our Country*. New York: Random House, 2009. Not an examination of debates per se but a wonderful discussion of issues that have been controversial through our history—the stuff of debates.

Fiorina, Morris P., with Samuel J. Abrams and Jeremy C. Pope. *Culture War? The Myth of a Polarized America*. Upper Saddle River, NJ: Pearson Longmans, 2010. Despite polarization within the political elite, Fiorina argues that wider popular polarization is a myth.

Fisher, Roger, and William L. Lury. *Getting to Yes: Negotiating Agreement without Giving In*. New York: Penguin Group, 1981. Fisher and Lury phrase their book in terms of practical hints on polarized negotiation, but their suggestions apply to polarized discussion generally.

Frederick, Shane. "Cognitive Reflection and Decision Making," *Journal of Economic Perspectives* 19 (2005): 25–42. The source of the "bat and ball" example of easy miscalculation.

Galileo. *Dialogue Concerning the Two Chief World Systems.* Translated with revised notes by Stillman Drake, with a foreword by Albert Einstein. Berkeley: University of California Press, 1967. A short classic everyone should read.

Gardner, Martin. *Fads and Fallacies in the Name of Science.* Mineola, NY: Dover, 1957. A dated but classic examination of hollow-earth theories, scientology, medical fads, and the like. The traps of pseudoscience become even clearer with temporal distance.

Gazzaniga, Michael S. *The Ethical Brain.* New York: Dana Press, 2005. Gazzaniga includes important empirical studies on the runaway trolley case.

Gigerenzer, Gerd, Peter M. Todd, and the ABC Research Group. *Simple Heuristics That Make Us Smart.* Oxford: Oxford University Press, 1999. An anthology of papers emphasizing the positive side of cognitive rules of thumb.

Gladwell, Malcolm. *Blink: The Power of Thinking without Thinking.* New York: Little, Brown and Company, 2005. Gladwell is interested in how gut reaction is sometimes superior to analysis.

Grim, Patrick, ed. *Philosophy of Science and the Occult.* 1st and 2nd eds. Albany, NY: SUNY Press, 1982, 1991. The core of the book is devoted to the problem of demarcation, with other sections that focus on particular topics, such as astrology and parapsychology. UFOlogy has a section in the first edition, replaced with one on quantum mysticism in the second.

Grim, Patrick, Gary Mar, and Paul St. Denis. *The Philosophical Computer: Exploratory Essays in Philosophical Computer Modeling.* Cambridge, MA:

MIT Press, 1998. Logic and game theory as seen with the techniques of computer modeling.

Groarke, Louis F. *"Aristotle's Logic." Internet Encyclopedia of Philosophy.* Available online at www.iep.utm.edu/aris-log/. A good introduction to important elements of Aristotelian logic.

Hacking, Ian. *An Introduction to Probability and Inductive Logic.* Cambridge: Cambridge University Press, 2001. Solid, thorough, and philosophically rich.

Hallinan, Joseph T. *Why We Make Mistakes: How We Look without Seeing, Forget Things in Seconds, and Are All Pretty Sure We Are Way above Average.* New York: Broadway Books, 2009. A rich discussion of a wealth of conceptual biases, very accessibly presented.

Henrich, Joseph, Robert Boyd, Samauel Bowles, Colin Camerer, Ernst Fehr, Herberg Gintis, and Richard McElreath. "In Search of Homo Economicus: Behavioral Experiments in 15 Small-Scale Societies." *American Economic Review* 91 (2001): 73–78. A contemporary classic on the limitations of game theory.

Hicks, Stephen R. C., and David Kelley. *Readings for Logical Analysis.* 2nd ed. New York: W. W. Norton, 1998. A collection of editorials and short essays designed for practice in argument analysis.

Hoffman, Donald D. *Visual Intelligence: How We Create What We See.* New York: W. W. Norton, 1998. A wonderful volume on optical illusions, better than many sources in that it contains a great deal of suggestive analysis.

Hoffman, Paul. *Archimedes' Revenge: The Joys and Perils of Mathematics.* New York: Fawcett Crest, 1988. A collection of entertaining mathematical bits, with history attached.

Huff, Darrell. *How to Lie with Statistics.* New York: W. W. Norton, 1954. A classic on the topic but incredibly racist and sexist to contemporary eyes.

Hume, David. "Of the Passions." In *An Enquiry Concerning Human Understanding*, 1748. A contemporary version is edited by Tom L. Beauchamp in the Oxford Philosophical Texts series (New York: Oxford University Press, 1999). This is the classic source for "emotion, not reason, ought to rule."

Kahneman, Daniel, and Amos Tversky. "Prospect Theory: An Analysis of Decision under Risk." *Econometrica* 47 (1979): 263–292. The original paper on the psychological unreality of classical decision theory. Available online at http://www.princeton.edu/~kahneman/docs/Publications/prospect_theory.pdf.

Kahneman, Daniel. *Thinking Fast and Slow*. New York: Farrar, Straus and Giroux, 2011. Kahneman includes his work with Tversky and much more, outlining system 1 and system 2 of cognitive processing, their strengths and weaknesses.

Kasner, Edward, and James Newman. *Mathematics and the Imagination*. New York: Simon and Schuster, 1940. An examination of the creative core of mathematical thought.

Kelley, David. *The Art of Reasoning*. 3rd ed. W. W. Norton, 1998. There is no current informal logic book that I can recommend without reservation, but Kelley's has two good chapters on Venn diagrams for categorical propositions and syllogisms.

Kitcher, Philip. *Science, Truth, and Democracy*. Oxford: Oxford University Press, 2001. A serious antidote to anti-science tendencies in public debate.

Larkin, Jill H., and Herbert Simon. "Why a Diagram Is (Sometimes) Worth Ten Thousand Words." *Cognitive Science* 11 (1987): 65–100. A great piece on the power of visualization, coauthored by one of the masters of both computer science and cognitive science.

Levine, Marvin. *Effective Problem Solving*. Prentice Hall, 1993. Levine's useful book includes a good section on visualization.

Levitt, Steven D., and Stephen J. Dubner. *Freakonomics: A Rogue Economist Explores the Hidden Side of Everything*. New York: HarperCollins, 2005. An entertaining and eye-opening introduction to the real statistics on guns, swimming pools, the economics of drug dealing, and what they mean.

The Lincoln-Douglas Debates. New York: BBC Audiobooks America, 2009. Available in audio form, with David Strathairn as Lincoln and Richard Dreyfus as Douglas.

Loftus, Elizabeth. *Eyewitness Testimony*. Cambridge, MA: Harvard University Press, 1979, 1996. A classic in applied psychology of perception, with lessons for the legal system from experimental research in the limitations of attention, memory, and cross-race identification.

Lord, C. G., L. Ross, and M. R. Lepper. "Biased Assimilation and Attitude Polarization: The Effects of Prior Theories on Subsequently Considered Evidence." *Journal of Personality and Social Psychology* 37 (1979): 2098–2109. The classic study on belief-based rejection of contrary evidence.

Ludlow, Peter, Yujin Nagasawa, and Daniel Stoljar, eds. *There's Something about Mary: Essays on Phenomenal Consciousness and Frank Jackson's Knowledge Argument*. Cambridge, MA: MIT Press, 2004. An anthology of articles on the black-and-white Mary example.

Miller, George A. "The Magical Number Seven, Plus or Minus Two: Some Limits on Our Capacity for Processing Information." *Psychological Review* 63 (1956): 81–97. The standard reference on cognitive limitations.

Mussweiler, Thomas. "Doing Is for Thinking! Stereotype Activation by Stereotypic Movements." *Psychological Science* 17 (2006): 17–21. When forced to walk like an elderly person, one recognizes stereotypical "elderly" words more readily.

Packard, Vance. *The Hidden Persuaders*. Brooklyn, NY: Ig Publishing, 2007. A 1957 classic on postwar advertising, rereleased for its 50th anniversary. Includes a critical introduction by Mark Crispin Miller.

Peirce, Charles Sanders. "The Fixation of Belief." Originally published in *Popular Science Monthly* in 1877, this piece has established itself as a touchstone in epistemology, philosophy, and pragmatism. Among other places, it appears in the wide-ranging collection *Philosophical Writings of Peirce*, edited by Justus Buchler (Mineola, NY: Dover, 2011).

Plato. *The Republic*. Plato's three-part theory of the soul, with a theory of society to match. Available in numerous editions.

Polya, G. *How to Solve It*. Princeton, NJ: Princeton University Press, 1945, 2004. The mathematician's often-recommended handbook on problem strategies, useful well beyond the bounds of mathematics.

Popper, Karl. *Conjectures and Refutations*. New York: Harper and Row, 1968. Popper's central statement on the problem of demarcation.

Poundstone, William. *Prisoner's Dilemma: John von Neumann, Game Theory, and the Puzzle of the Bomb*. New York: Doubleday, 1992. An entertaining book that combines a biography of von Neumann with the essentials of game theory.

Priest, Graham. *Logic: A Very Short Introduction*. Oxford: Oxford University Press, 2000. A wonderful short course in standard logic, alternative logic, and related areas. Chapter 13 includes a nice discussion of Pascal's wager and the "many gods" objection to it.

Purtill, Richard. *A Logical Introduction to Philosophy*. Englewood Cliffs, NJ: Prentice-Hall. Chapter 2 of the book analyzes simple arguments in Plato's dialogues. Chapter 4 concentrates on syllogistic argument and assumptions.

Purves, Dale, and R. Beau Lotto. *Why We See What We Do: An Empirical Theory of Vision*. Sunderland, MA: Sinauer Associates, Inc., 2003. Includes a particularly nice examination of our built-in color processing in context.

Resnik, Michael. *Choices: An Introduction to Decision Theory*. Minneapolis, MN: University of Minnesota Press, 1987. A good, solid introduction to the field.

Bibliography

Salmon, Merrillee. "Index of Fallacies." Appendix 2 in *Logic and Critical Thinking*, 4th ed. Belmont, CA: Wadsworth, 2002. Salmon's catalog of fallacies is short and selective but broad-ranging.

Salsburg, David. *The Lady Drinking Tea: How Statistics Revolutionized Science in the 20th Century*. New York: Henry Holt & Co., 2001. An enjoyable introduction to the core concepts with the lightest of mathematical touches.

Schopenhauer, Arthur. *The Art of Controversy*. New York: Cosimo Classics, 2007. Also freely available online. Schopenhauer is often infuriating; a valuable read for precisely that reason.

Searle, John. "Minds, Brains, and Programs." *Behavioral and Brain Sciences* 3 (1980): 417–424. One of the earliest of Searle's presentations of the Chinese room and one of the best.

Shermer, Michael. *Why People Believe Weird Things: Pseudoscience, Superstition, and Other Confusions of Our Time*. London: Souvenir Press, 2007. Focused more on the "weird things" than on why people believe them, Shermer's entertaining book reflects his status as editor in chief of *Skeptic* magazine.

Shoda, Yuichi, Walter Mischel, and Philip K. Peake. "Predicting Adolescent Cognitive and Self-Regulatory Competencies from Preschool Delay of Gratification: Identifying Diagnostic Conditions." *Developmental Psychology* 26 (1990): 978–986. The Stanford marshmallow test.

Simons, Daniel J. *Surprising Studies of Visual Awareness*. Champaign, IL: VisCog Productions, 2008. This DVD includes full video of the "invisible gorilla" and a number of astounding change-blindness experiments.

Skyrms, Brian. *Choice and Chance: An Introduction to Inductive Logic*. 3rd ed. Belmont, CA: Wadsworth, 1986. A deeper and more difficult text on probability and scientific inference.

Sorensen, Roy A. *Thought Experiments*. New York: Oxford University Press, 1992. Sorensen argues that thought experiments are literally that: experiments. He includes a range of wonderful examples in the discussion.

Sterrett, Susan G. *Wittgenstein Flies a Kite: A Story of Models of Wings and Models of the World*. New York: Pi Press, 2006. This is a profound book, both theoretical and historical, on the obstacles conceptual modeling faces in scaling up to the real world. Well worth reading, despite some technically difficult later chapters.

Surowiecki, James. *The Wisdom of Crowds: Why the Many Are Smarter Than the Few and How Collective Wisdom Shapes Business, Economies, Societies and Nations*. New York: Little, Brown, 2004. An entertaining read; the examples of Galton's ox and *Who Wants to Be a Millionaire?* are drawn from this source.

Taleb, Nassim Nicholas. *The Black Swan: The Impact of the Highly Improbable*. New York: Random House, 2010. A reminder that the unexpected is exactly that and how important the unexpected can be.

Tetlock, Philip E. *Expert Political Judgment: How Good Is It? How Can We Know?* Princeton, NJ: Princeton University Press, 2005. A systematic attempt to answer the questions in the title, often with surprisingly negative results.

Thagard, Paul. *Hot Thought: Mechanisms and Applications of Emotional Cognition*. Cambridge, MA: MIT Press, 2006. A serious attempt to give emotional cognition its due within the loop of rationality.

Toulmin, Stephen. *The Uses of Argument*. Cambridge: Cambridge University Press, 1958, 2003. Pages 87–100 of section III, "Uses of Argument," offer a good outline of Toulmin's approach to the role of data and warrant.

Turner, Paul. E. "Cheating Viruses and Game Theory." *American Scientist* 93 (2005): 428–435. An intriguing introduction to game theory from a biological perspective.

University of Illinois Advertising Exhibit. http://www.library.illinois.edu/adexhibit/technology.htm. Sometimes it's easier to see through advertising ploys in older ads. Here, you will find a great selection of old ads by topic (celebrities, tobacco, diet).

von Neumann, John, and Oscar Morgenstern. *Theory of Games of Economic Behavior.* Princeton, NJ: Princeton University Press, 1944; commemorative edition, 2007. The 20th-century source from which all game theory springs.

Wason, P. C. "Realism and Rationality in the Selection Task." In *Thinking and Reasoning: Psychological Approaches*, edited by J. S. B. T. Evans, pp. 44–75. London: Routledge and Kegan Paul, 1983. The source of the "4 and 7" selection task that so many people seem to get wrong.

Weston, Anthony. *Creativity for Critical Thinkers.* New York: Oxford University Press, 2007. One of the few attempts to encourage creativity as an explicit element of critical thinking.

White, Jamie. *Crimes against Logic.* New York: McGraw-Hill, 2004. An entertaining survey of fallacies (broadly construed) from "politicians, priests, journalists, and other serial offenders."

Zeisel, Hans. "The Deterrent Effect of the Death Penalty: Facts v. Faith." In *The Supreme Court Review, 1976*, edited by Philip B. Kurland, pp. 317–343. Chicago: University of Chicago Press Journals, 1979. Reprinted in *The Death Penalty in America*, 3rd ed., edited by Hugo Adam Bedau, pp. 116–138. Oxford: Oxford University Press, 1982. A nice critical analysis of an important statistical study.

Answers

Lecture 1, Exercise

Lines 1, 7, and 12 come from John Keats, "Ode to a Nightingale."

Lines 2, 5, and 10 come from Robert Burns, "A Red, Red Rose."

Lines 3, 8, and 11 come from Walt Whitman, "Out of the Cradle Endlessly Rocking."

Lines 4, 6, and 9 come from T. S. Eliot, "The Love Song of J. Alfred Prufrock."

Lecture 3, Exercise

In a sketch, the initial information lays out like this:

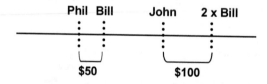

If John makes $900 a week, twice Bill's salary is $1,000, and half of that is $500. Phil makes $50 less; thus, he must make $450 a week.

Lecture 6, Exercise

Here, we've rearranged Aristotle's square of opposition. Your sketch should look something like this:

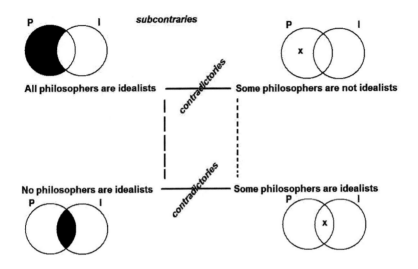

Lecture 7, Exercises

Here are progressive diagrams for premises and conclusions in the exercises:

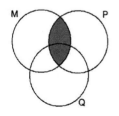 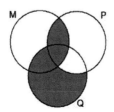 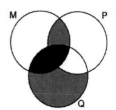

Step 1: "No M are P" means that nothing is in the intersection of M and P—we blank that out.

Step 2: "All Q are P" means that nothing in Q is outside of the P area. We add that information in the second diagram.

Step 3: The conclusion, "No M are Q," would mean that nothing is in the overlap of M and Q. But that information is already in the second diagram. Given the premises, the conclusion must be true already; the argument is valid.

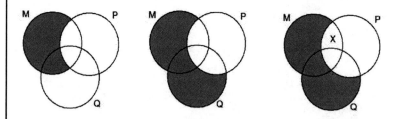

Step 1: "All M are P" means that nothing is in M outside of the P area.

Step 2: We add "All Q are P": There are no Q's outside of the P area.

Step 3: The conclusion would tell us that there is definitely something in the M area outside of Q. That would be new information—something that doesn't follow from the premises alone. The argument is invalid.

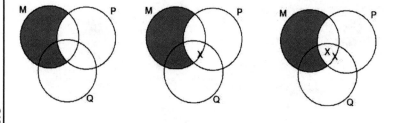

Step 1: We sketch in "All M are P."

Step 2: "Some P are Q" tells us that there is at least one thing in the overlap between P and Q. But it doesn't tell us where in that overlap, so we leave it "indefinite" on the line. It might be on either side.

Step 3: The conclusion tells us that some M are Q: that there is an X on that particular side of the line. That goes beyond the information we already have and, thus, doesn't follow from the premises. The argument is invalid.

Lecture 8, Questions to Consider

(a) Collect 1985 pennies and you will have $19.85.

(b) No, because he would be dead, and it's not legal for a dead person to marry.

(c) Meat.

(d) Those letters can be rearranged to form: one word.

(e) They are two of a set of triplets.

Lecture 10, Exercise

(a) Rio de Janeiro

(b) Bill Gates, Larry Ellison, Christy Walton

(c) Sudan

(d) Ferrell, Drysdale, Johnson

On which questions did the recognition heuristic work for you? On which questions did it fail? Can you tell why?

Lecture 14, Questions to Consider

Post hoc ergo propter hoc. Because the magician "disappears" one coin at point A and then produces a second coin at point B, we are led to believe that the second event happened because the first one did: that the disappearance "caused" the appearance and that a single coin has "traveled" from point A to B.

Lecture 17, Exercise

There are lots of ways to construct the quiz scores, but here's an example. Let the children's scores be:

1 1 1 1 2 3 8 9 10 10 10

(a) The mode is 1. Create a "most people" category first and aim for something different than the "middle," which you can expect to be more like the mean or the median.

(b) The median is 3, with as many above as below.

(c) The mean is 5.09. Having kept the median low, you can make the mean significantly higher by adding high numbers at the end.

Lecture 19, Exercise

Here's how to start thinking about the exercise:

Option 1 has a 20 percent chance of gaining $1 million over investment and an 80 percent chance of gaining nothing. Probability multiplied by desirability is 0.2 x $1 million = $200,000 expected utility.

Option 2 has a 40 percent chance of gaining $500,000 over investment. Probability multiplied by desirability is 0.4 x $500,000 = $200,000 expected utility.

Option 3 has a 90 percent chance of gaining $150,000 over investment. Probability multiplied by desirability is 0.9 x $150,000 = $135,000 expected utility.

But that may be only the start of your deliberations. Are there other considerations you would take into account?

Notes

Notes

Notes

Notes

Notes